Transgenesis and the Management of Vector-Borne Disease

ADVANCES IN EXPERIMENTAL MEDICINE AND BIOLOGY

Recent Volumes in this Series

Volume 619
CYANOBACTERIAL HARMFUL ALGAL BLOOMS: STATE OF THE SCIENCE AND RESEARCH NEEDS
Edited by H. Kenneth Hudnell

Volume 620
BIO-APPLICATIONS OF NANOPARTICLES
Edited by Warren C.W. Chan

Volume 621
AXON GROWTH AND GUIDANCE
Edited by Dominique Bagnard

Volume 622
OVARIAN CANCER
Edited by George Coukos, Andrew Berchuck, and Robert Ozols

Volume 623
ALTERNATIVE SPLICING IN THE POSTGENOMIC ERA
Edited by Benjamin J. Blencowe and Brenton R. Graveley

Volume 624
SUNLIGHT, VITAMIN D AND SKIN CANCER
Edited by Jörg Reichrath

Volume 625
DRUG TARGETS IN KINETOPLASTID PARASITES
Edited by Hemanta K. Majumder

Volume 626
GENOMIC IMPRINTING
Edited by Jon F. Wilkins

Volume 627
TRANSGENESIS AND THE MANAGEMENT OF VECTOR-BORNE DISEASE
Edited by Serap Aksoy

Transgenesis and the Management of Vector-Borne Disease

Edited by
Serap Aksoy, PhD
Professor and Head of the Division of Epidemiology and Microbial Diseases, Yale University School of Public Health, New Haven, Connecticut, USA

Springer Science+Business Media, LLC
Landes Bioscience

Springer Science+Business Media, LLC
Landes Bioscience

Printed in the USA

Springer Science+Business Media, LLC, 233 Spring Street, New York, New York 10013, USA
http://www.springer.com

Please address all inquiries to the Publishers:
Landes Bioscience, 1002 West Avenue, 2nd Floor, Austin, Texas, USA 78701
Phone: 512/ 637 6050; FAX: 512/ 637 6079
http://www.landesbioscience.com

Transgenesis and the Management of Vector-Borne Disease, edited by Serap Aksoy, Landes Bioscience / Springer Science+Business Media, LLC dual imprint / Springer series: Advances in Experimental Medicine and Biology, vol. 627.

ISBN: 978-0-387-78224-9

Library of Congress Cataloging-in-Publication Data

Transgenesis and the management of vector-borne disease / edited by Serap Aksoy.
p. ; cm. -- (Advances in experimental medicine and biology ; v. 627)
Includes bibliographical references.
ISBN 978-0-387-78224-9
1. Insects as carriers of disease. 2. Vector control--Biological control. 3. Insect pests--Biological control. 4. Transgenic animals. I. Aksoy, Serap. II. Series.
[DNLM: 1. Communicable Disease Control--methods. 2. Communicable Diseases--transmission. 3. Gene Transfer Techniques. 4. Insect Vectors--genetics. W1 AD559 v.627 2008 / WA 110 T772 2008]
RA639.5.T73 2008
614.4'32--dc22

2008001892

ABOUT THE EDITOR...

SERAP AKSOY, PhD received her BS at Vassar College and her PhD in Biology from Colombia University. She conducted post-doctoral training at Yale School of Medicine where she became an Assistant Professor in 1990. Dr. Aksoy is the current Division Head of Epidemiology of Microbial Diseases at Yale School of Public Health. Her main research interests center on control of microbial infectious diseases in the developing world. Her laboratory efforts focus on African trypanosomiasis and aim to develop novel approaches for control of tsetse flies, the insect vector for this disease. She has pioneered a paratransgenic microbial technology where parasite inhibitory products can be synthesized in the gut symbiotic flora of tsetse in order to block parasite transmission. She is the director of International *Glossina* Genomics Initiative, funded through WHO/TDR to promote the genome sequencing efforts and functional genomics activities in tsetse vector. Professor Aksoy is a Deputy Editor of *PloS Neglected Diseases* and serves on the Editorial Advisory Boards of *Insect Biochemistry and Molecular Biology* and *Parasites and Vectors*. Dr. Aksoy serves on many national and international panels and grant study sections in the area of infectious diseases and vector biology. She is the current Chair of Innovative Vector Control program at WHO/TDR. Dr. Aksoy has initiated and participated in many International Health Projects funded through NIH and WHO as well as foundations including the Ambrose Monell Foundation, Robert Leet and Clara Gutherie Patterson Trust, Culpeper Foundation Biomedical Initiative, Li Foundation, McKnight Foundation Biotechnology Grant, MacArthur Foundation and Sigma Xi.

PREFACE

Parasitic, bacterial and viral agents continue to challenge the welfare of humans, livestock, wild life and plants worldwide. The public health impact and financial consequences of these diseases are particularly hard on the already overburdened economies of developing countries—especially in the tropics. Many of these disease agents utilize insect hosts (vectors) to achieve their transmission to mammals. In the past, these diseases were largely controlled by insecticide-based vector reduction strategies. Now, many of these diseases have reemerged in the tropics, recolonizing their previous range, and expanding into new territories previously not considered to be endemic. Habitat change, irrigation practices, atmospheric and climate change, insecticide and drug resistance as well as increases in global tourism, human traffic and commercial activities, have driven the reemergence and spread of vector-borne diseases. While these diseases can be controlled through interventions aimed at both their vertebrate and invertebrate hosts, no effective vaccines exist, and only limited therapeutic prospects are available for their control in mammalian hosts. Molecular technologies such as transgenesis, which is the subject of this book, stand to increase the toolbox and benefit disease management strategies.

The debate on the feasibility and acceptability of the genetically modified insect technology continues. There is no doubt that the technology has revolutionized the study of insect sciences and promises to improve the reproduction and viability of beneficial insects and the use of insects as bioreactors—but will it meet the expectations as a disease control strategy? In the contributed chapters in this book, we begin by providing a historical perspective to transgenic technology applied with the goal of modification of vector competence (Chapter 1). Recent advances utilizing viral transducing systems (Chapter 2) as well as symbionts (Chapters 3 and 4) are described in more detail for the introduction and expression of anti-pathogen products in several vector systems. New applications for use of transposons are presented in Chapter 5. With expression systems well underway, effector molecules must now be chosen that will be toxic to the pathogen of interest when expressed in insects. In this regard, advances made in basic insect physiology and immunology, coupled with genomics information and gene mapping strategies (Chapter 6), are most relevant. Of importance for the success of genetically modified insects is the identification of gene drive systems to deliver desirable phenotypes into natural populations such that the modified disease resistant insects can establish and lead to a reduction in disease transmission. For this purpose, *Wolbachia*-mediated drive systems continue to be promising candidates, and Chapters 9 and 10 address the recent developments in these fields. In addition, certain strains of *Wolbachia* are currently being used to modify insect age structures, a highly promising and innovative approach to reducing disease transmission (Chapter 11). Transgenic applications can also advance biological control methods such as sterile insect technique (Chapter 7) and genetic sexing and sterilization approaches (Chapter 8).

It is likely that fine-tuning the remaining technological challenges will make these systems safer for eventual field applications. Building on laboratory successes, the challenge now is to plan field-based investigations to test the ability of genetically modified insects to interfere with disease transmission cycle. While the integration of transgenic approaches for human disease management may face significant cultural and ethical challenges, a strong economic engine may drive the application of this technology against agricultural diseases as discussed in Chapter 12. Experience from the agricultural systems may provide important data that can guide the future applications in the context of human disease systems. Interdisciplinary field-based investigations are now necessary to describe the interactions between insects, pathogens and symbionts in natural populations to predict the feasibility of using modified insects to reduce the spread of harmful zoonotic diseases. We must gain a better understanding of the ecological factors, which influence disease epidemiology and insect population dynamics (Chapter 13). This information is essential to better predict how and where transgenic technologies, possibly in combination with other control approaches, may best be implemented.

A most important area not addressed in this book concerns the social and political implications of transgenic application. In this regard, it would be important for the scientific evidence to inform and steer the development of public health policy guidelines. The scientific discoveries and their potential benefits as well as limitations need to be shared with the general public through continued education and advocacy, especially in disease endemic countries where these technologies can have the highest impact on public health. Ultimately, acceptability and applicability will depend on the choices made by the populations where these diseases take their greatest toll.

Serap Aksoy, PhD

ACKNOWLEDGEMENTS

I am very grateful to the many colleagues who agreed to contribute chapters in this book. Thank you for sharing your ideas, results and perspectives.

PARTICIPANTS

Serap Aksoy
Department of Epidemiology
and Microbial Diseases
Yale University School of Public Health
New Haven, Connecticut
USA

Alberto Alma
Dipartimento di Valorizzazione
e Protezione delle Risorse
Agroforestali
Università degli Studi di Torino
Grugliasco
Italy

Luke Alphey
Department of Zoology
Oxford University
Oxford
and
Oxitec Limited
Abingdon
UK

Nina Alphey
Department of Zoology
Oxford University
Oxford
and
Oxitec Limited
Abingdon
UK

Peter W. Atkinson
Department of Entomology
Institute for Integrative Genome Biology
University of California
Riverside, California
USA

Geoffrey Attardo
Yale University School of Medicine
Department of Epidemiology
and Public Health
New Haven, Connecticut
USA

Claudio Bandi
Dipartimento di Patologia Animale,
Igiene e Sanità Pubblica Veterinaril
Università degli Studi di Milano
Milan
Italy

Mark Q. Benedict
NCZVED/CCID Entomology Branch
Atlanta, Georgia
USA

William C. Black IV
Department of Microbiology,
Immunology and Pathology
Colorado State University
Fort Collins, Colorado
USA

Kostas Bourtzis
Department of Environmental
and Natural Resources Management
University of Ioannina
Agrinio
Greece

Peter E. Cook
School of Integrative Biology
The University of Queensland
Brisbane, Queensland
Australia

Daniele Daffonchio
Dipartimento di Scienze e Tecnologie, Alimentari e Microbiologiche
Università degli Studi di Milan
Milan
Italy

Martin J. Donnelly
Vector Group
Liverpool School of Tropical Medicine
Pembroke Place, Liverpool
UK

Norma Gorrochetegui-Escalante
Department of Microbiology, Immunology and Pathology
Colorado State University
Fort Collins, Colorado
USA

Guido Favia
Dipartimento di Medicina Sperimentale e Sanità Pubblica
Università degli Studi di Camerino
Camerino
Italy

Brian D. Foy
Arthropod-borne and Infectious Diseases Laboratory
Department of Microbiology, Immunology and Pathology
Colorado State University
Fort Collins, Colorado
USA

Norma Gorrochetegui-Escalante
Department of Microbiology, Immunology and Pathology
Colorado State University
Fort Collins, Colorado
USA

Alfred M. Handler
United States Department of Agriculture
Agricultural Research Service
Center for Medical, Agricultural and Veterinary Entomology
Gainesville, Florida
USA

Laura C. Harrington
Department of Entomology
Cornell University
Ithaca, New York
USA

Bart G.J. Knols
Laboratory of Entomology
Wageningen University and Research Centre
Wageningen
The Netherlands

David J. Lampe
Biological Sciences Department
Duquesne University
Pittsburgh, Pennsylvania
USA

Carol R. Lauzon
Biological Sciences Department
California State University East Bay
Hayward, California
USA

Massimo Marzorati
Dipartimento di Scienze e Tecnologie, Alimentari e Microbiologiche
Università degli Studi di Milan
Milan
Italy

Conor J. McMeniman
School of Integrative Biology
The University of Queensland
Brisbane, Queensland
Australia

Thomas A. Miller
Entomology Department
University of California
Riverside, California
USA

Ilaria Negri
Dipartimento di Valorizzazione e Protezione delle Risorse Agroforestali
Università degli Studi di Torino
Grugliasco
Italy

Derric Nimmo
Oxitec Limited
Abingdon
UK

David A. O'Brochta
Center for Biosystems Research
University of Maryland Biotechnology Institute
Rockville, Maryland
USA

Sinead O'Connell
Oxitec Limited
Abingdon
UK

Ken E. Olson
Arthropod-borne and Infectious Diseases Laboratory
Department of Microbiology, Immunology and Pathology
Colorado State University
Fort Collins, Colorado
USA

Scott L. O'Neill
School of Integrative Biology
The University of Queensland
Brisbane, Queensland
Australia

Nadine P. Randle
Vector Group
Liverpool School of Tropical Medicine
Pembroke Place, Liverpool
UK

Jason L. Rasgon
The W. Harry Feinstone Department of Molecular Microbiology and Immunology
Bloomberg School of Public Health
Johns Hopkins University
and
The Johns Hopkins Malaria Research Institute
Baltimore, Maryland
USA

Irene Ricci
Dipartimento di Medicina Sperimentale e Sanità Pubblica
Università degli Studi di Camerino
Camerino
Italy

Alan S. Robinson
Joint FAO/IAEA Programme of Nuclear Techniques in Food and Agriculture
Insect Pest Control Sub-Programme
International Atomic Energy Agency
Agency's Laboratories
Seibersdorf
Austria

Luciano Sacchi
Dipartimento di Biologia Animale
Università degli Studi di Pavia
Pavia
Italy

Thomas W. Scott
Department of Entomology
University of California
Davis, California
USA

Willem Takken
Laboratory of Entomology
Wageningen University and Research Centre
Wageningen
The Netherlands

Brian Weiss
Department of Epidemiology and Public Health
Yale University School of Medicine
New Haven, Connecticut
USA

CONTENTS

1. PERSPECTIVES ON THE STATE OF INSECT TRANSGENICS 1

David A. O'Brochta and Alfred M. Handler

Introduction 1
Transposon Vectors and Insect Transformation 2
Transformation Markers 4
Targeting and Stabilization 5
Random Transgene Insertion 5
Transgenic Strains for Biocontrol 8
Fluorescent Protein Genetic Markers 8
Conditional Regulation for Sterility and Lethality 9
Paratransgenesis 10
Viruses 10
Perspectives on TE Spread and Genetic Manipulation for Disease Vector Control ... 11
Broadening Our Perspective 12

2. ALPHAVIRUS TRANSDUCING SYSTEMS 19

Brian D. Foy and Ken E. Olson

Introduction 19
Alphaviruses 20
ATS Construction 22
Currently Used ATSs 22
Alphavirus/Mosquito Interactions 24
Gene Expression with ATSs 27
ATS's for Induction of RNA Interference 27
Biosafety Considerations for Using ATSs 30
Conclusion 30

3. PARATRANSGENESIS APPLIED FOR CONTROL OF TSETSE TRANSMITTED SLEEPING SICKNESS 35

Serap Aksoy, Brian Weiss and Geoffrey Attardo

Introduction 35
Symbiosis in Tsetse 36
Symbiont-Host Interactions During Development 37
Tsetse's Reproductive Biology and Symbiont Transmission 38

Biology of African Trypanosome Transmission in Tsetse 38
Paratransgenic Gene Expression in Tsetse 39
Reconstitution of Tsetse with Modified Symbiont Flora 41
Effector Genes with Trypanocidal Activity 42
Gene Driver System 43
Applications with Parasite Resistant Tsetse 43
Application of Paratransgenesis for Control of Other Insect Transmitted Diseases 44
Future Directions 45

4. BACTERIA OF THE GENUS ASAIA: A POTENTIAL PARATRANSGENIC WEAPON AGAINST MALARIA 49

Guido Favia, Irene Ricci, Massimo Marzorati, Ilaria Negri, Alberto Alma, Luciano Sacchi, Claudio Bandi, and Daniele Daffonchio

Introduction 49
Malaria and Symbiotic Control Strategies 50
a-Proteobacteria of the Genus *Asaia* Dominate the Microflora of *An. stephensi* 51
Asaia* is Localized in Different Organs of *An. stephensi 53
***Asaia*: A Self-Spreading Carrier of Potential Antagonistic Factors in Mosquitoes** 55
Conclusions and Perspectives 56

5. PROPOSED USES OF TRANSPOSONS IN INSECT AND MEDICAL BIOTECHNOLOGY 60

Peter W. Atkinson

Introduction 60
Genetic Control Strategies and Transposon Immobility 63
Transformation Efficiency 63
Target Site Preference 65
Post Integration Stability 65
Regulation of Transposition by piRNA 66
Genetic Control Strategies and Transposon Drive 67
Conclusions 68

6. THE YIN AND YANG OF LINKAGE DISEQUILIBRIUM: MAPPING OF GENES AND NUCLEOTIDES CONFERRING INSECTICIDE RESISTANCE IN INSECT DISEASE VECTORS 71

William C. Black IV, Norma Gorrochetegui-Escalante, Nadine P. Randle and Martin J. Donnelly

Abstract 71
Mapping of Genome Regions and SNPs Conferring Insecticide Resistance 72
Terminology in Linkage Disequilibrium and QTN Mapping 73
Measuring Linkage Disequilibrium 73

Patterns of Linkage Disequilibrium in Vectors 74
The Yin and Yang of Linkage Disequilibrium 77
QTL Mapping 78
Microarray Technology 79

7. IMPACT OF TECHNOLOGICAL IMPROVEMENTS ON TRADITIONAL CONTROL STRATEGIES 84

Mark Q. Benedict and Alan S. Robinson

Introduction 84
Applications 85
Are Transgenic Insects Compatible Partners in the IPM Mix? 90
Conclusions 90

8. INSECT POPULATION SUPPRESSION USING ENGINEERED INSECTS 93

Luke Alphey, Derric Nimmo, Sinead O'Connell and Nina Alphey

Introduction 93
Genetic Sexing and Genetic Sterilization 95
Molecular Biology of Repressible Lethal Systems 100
Regulatory Issues and Concluding Remarks 100

9. *WOLBACHIA*-BASED TECHNOLOGIES FOR INSECT PEST POPULATION CONTROL 104

Kostas Bourtzis

Introduction 104
***Wolbachia* Induced Cytoplasmic Incompatibility** 105
***Wolbachia*-Based Applications** 106
Conclusions and Future Challenges 109

10. USING PREDICTIVE MODELS TO OPTIMIZE WOLBACHIA-BASED STRATEGIES FOR VECTOR-BORNE DISEASE CONTROL 114

Jason L. Rasgon

Introduction 114
***Wolbachia* Endosymbionts** 115
More Realistic Population Dynamics of *Wolbachia* Infections 118
Can Modeling Highlight a Better Way to Control Disease Using *Wolbachia* Infections? 119
Using *Wolbachia* to Drive Nuclear Traits? 122
Conclusions 124

11. MODIFYING INSECT POPULATION AGE STRUCTURE TO CONTROL VECTOR-BORNE DISEASE 126

Peter E. Cook, Conor J. McMeniman and Scott L. O'Neill

Introduction 126
Entomological Components of Pathogen Transmission 127
Wolbachia pipientis 128
Life-Shortening *Wolbachia* 128
Experimental Transfer of *Wolbachia* into Disease Vectors 129
Temperature and the Impact of Life-Shortening *Wolbachia* 130
Molecular Basis of Life-Shortening in *w*MelPop 131
Entomopathogenic Fungi 132
Densonucleosis Viruses 132
Evaluating of the Efficacy of Strategies Targeting Vector Longevity 133
Evolutionary Consequences of Strategies That Reduce Vector Longevity 134
Conclusion 135

12. TECHNOLOGICAL ADVANCES TO ENHANCE AGRICULTURAL PEST MANAGEMENT 141

Thomas A. Miller, Carol R. Lauzon and David J. Lampe

Introduction 141
Sterile Insect Technique 142
Symbiosis and Pierce's Disease 143
Bacterial Transgenesis and the Suppression of Horizontal Gene Transfer 144
Anti-*Xylella* Factors 144
Ecological Microbiology 145
Regulatory Issues 146
Conclusions 148

13. **APPLICATIONS OF MOSQUITO ECOLOGY FOR SUCCESSFUL INSECT TRANSGENESIS-BASED DISEASE PREVENTION PROGRAMS 151**

Thomas W. Scott, Laura C. Harrington, Bart G. J. Knols and Willem Takken

Introduction 151
Mating Behavior and Male Biology 152
Assessing Fitness 158
Population Biology 159
Regulatory Issues 161
Conclusions 162

INDEX 169

Chapter 1

Perspectives on the State of Insect Transgenics

David A. O'Brochta* and Alfred M. Handler

Abstract

Genetic transformation is a critical component to the fundamental genetic analysis of insect species and holds great promise for establishing strains that improve population control and behavior for practical application. This is especially so for insects that are disease vectors, many of which are currently subject to genomic sequence analysis, and intensive population control measures that must be improved for better efficacy and cost-effectiveness. Transposon-mediated germ-line transformation has been the ultimate goal for most fundamental and practical studies, and impressive strides have been made in recent development of transgene vector and marker systems for several mosquito species. This has resulted in rapid advances in functional genomic sequence analysis and new strategies for biological control based on conditional lethality. Importantly, advances have also been made in our ability to use these systems more effectively in terms of enhanced stability and targeting to specific genomic loci. Nevertheless, not all insects are currently amenable to germ-line transformation techniques, and thus advances in transient somatic expression and paratransgenesis have also been critical, if not preferable for some applications. Of particular importance is how this technology will be used for practical application. Early ideas for population replacement of indigenous pests with innocuous transgenic siblings by transposon-vector spread, may require reevaluation in terms of our current knowledge of the behavior of transposons currently available for transformation. The effective implementation of any control program using released transgenics, will also benefit from broadening the perspective of these control measures as being more mainstream than exotic.

Introduction

The goal of this introductory chapter is to give an overview of the state of the art of insect transgenesis, with emphasis on recent advancements in methodology and applications and especially as they relate to vectors of disease. Many of these topics are reviewed in much greater detail in the forthcoming chapters.

The ability to genetically transform the germ-line of non-drosophilid insects was first achieved, as a routine process, only within the last decade and impressive strides have been made in the methodology, practical applications and the types of insects amenable to the process. Some of these advances include defining vector host ranges, vector-stabilization methods, "marker" gene development and paratransgenesis. Significant progress has been made as well, in the somatic or non-heritable forms of transgenesis that preceded development of germ-line transformation. Somatic transformation is particularly useful in the laboratory and has allowed rapid determinations of promoter function and sequence-function relationships. Paratransgenesis, achieved by expressing genes that alter insect phenotypes in microbial symbionts, holds great promise as an insect

*Corresponding Author: David A. O'Brochta—University of Maryland Biotechnology Institute, Center for Biosystems Research, Rockville, MD, USA. Email: obrochta@umbi.umd.edu

Transgenesis and the Management of Vector-Borne Disease, edited by Serap Aksoy.

control tool. In addition, certain microbial symbionts provide unique opportunities to preferentially spread transgenes through insect populations. These advances come at a particularly propitious time, coinciding with the availability of genomic sequence data from some of the most important species in terms of human health, agriculture, and fundamental research. Transgenesis fulfills the need of insect molecular biologists to understand sequence function as part of ongoing functional genomic analyses, and to develop new approaches for the control of pest populations, their behavior, and/or their competence to transmit pathogens. Less explored, but of enormous significance is the potential of transgenic insect technology to improve the reproduction and viability of beneficial insects, and to use insects as bioreactors.

The first great challenge for germ-line transformations was the development of functional vector systems for non-drosophilid species. Initial hope that the *P* element system widely used in Drosophila could be similarly applied to other insects met with great frustration, with the inability of many researchers to find true *P*-mediated transformants.[1-4] We know now that *P* has a highly restricted host range although we do not understand the basis of this limitation.[5] Interestingly initial and more recent successes in insect transgenesis have resulted equally from advancements in marker gene development as well as vector system development. Indeed, the first insect transformations utilized both existing and newly discovered transposable elements as vectors, and markers created from newly isolated genes that could complement eye color mutations, as was done routinely in Drosophila. The yellow fever mosquito *Aedes aegypti* was transformed with *Hermes*[6] and *mariner*[7] transformation vectors containing the *kynurenine hydroxylase* gene that complemented the *kh*^{w} allele.[8,9] At about the same time, the tephritid fruit fly pest, *Ceratitis capitata*, was transformed with *Minos*[10] and *piggyBac*[11] vectors using the *white* (or *white eye*) gene cDNA (under *hsp70* promoter control) isolated from the same species.[12] In these experiments, successful transformants were identified by a mutant-rescue effect that restored the wild type eye color phenotype in the mutant host. The utility of these markers was limited to species with appropriate mutant strains and clones of the complementing wild-type allele. In the past decade these four transposon vectors remain the primary vehicle for all insect transformations, as most have a broad range of transpositional activity. Yet the ability to use these vectors in hosts not amenable to mutant-rescue selection depended on more broadly applicable marker genes and promoters. This has been achieved by use of a variety of dominant-acting fluorescent protein genes,[13] first tested using a viral-based somatic transformation system,[14] under the regulation of functionally conserved transcriptional promoters and regulatory systems.

Despite the significant progress in transformation of disease vectors and agricultural pests, some species have yet to be transformed or have been so only with great difficulty. Suboptimal performance of insect transformation may be a function of technical expertise or effort, but certainly for Anopheles, species such as *An. stephensi* are transformed routinely,[15-20] while the highly important species *An. gambiae* requires much more effort.[21,22] Similarly, some lepidopteran species, such as *Bombyx mori*, are transformed routinely,[23-25] while others such as *Spodoptera frugiperda* and *Heliothis virescens* have yet to be transformed. To expand the number of species amenable to transformation, significant challenges remain in improving gene vector delivery to germ cells, increasing integration rates, extending host ranges, understanding post-integration behavior of gene vectors and developing transposable element-based gene spreading technologies.

Transposon Vectors and Insect Transformation

All transformation systems of insect germ-lines utilize Class II transposable elements as vectors that transpose by a DNA-mediated "cut-and-paste" process.[26] These mobile genetic elements are typically 1 to 5 kb in length and have terminal sequences that are inverted repeats of one another. The terminal inverted repeats (TIR) are usually 30 base pairs or less, but some are several hundred base pairs and some terminal regions also have requisite sub-terminal inverted repeat sequences. The TIRs and the DNA that lies between them are excised and reinserted into new chromosomal positions as part of the transposition process. In between the TIR regions are DNA sequences that encode for a transposase enzyme that catalyzes both the cut and paste processes. Elements with TIRs and a transposase coding region are said to be self-mobilizing or autonomous, although they may require other host-encoded nuclear proteins. Importantly, while the TIRs and transposase gene are normally linked, the TIRs and any intervening DNA can be mobilized by an unlinked source of transposase. This feature has allowed the development of non-autonomous vectors that include the TIRs, marker genes, and genes-of-interest but with the transposase gene either mutated or deleted.[27,28] Non-autonomous vectors are mobilized when transposase is provided in trans - either from a plasmid-encoded transposase gene, transposase RNA or transposase protein. When vectors and transposase sources are coinjected into preblastoderm embryos, transposase catalyzes vector transposition and integration, but because the source of transposase is not contained on a vector it does not become integrated and is diluted and degraded over time. In the absence of transposase, vectors that have chromosomally integrated cannot undergo further transposition and are expected to be stable.

The first transposable element developed for use as a gene vector in *Drosophila melanogaster*, *P*,[29] was not effective in other species.[30] Interestingly most of the functional transposable elements that have been used as gene vectors in non-drosophilids were discovered fortuitously. Only recently have directed searches for functional transposable elements been fruitful as a source of new gene vectors. For example, the *Minos* element was discovered in *D. hydei* during the analysis of ribosomal RNA genes, and it was found to be closely related to *Tc* elements from nematodes.[31] The *Mos1 mariner* element from *D. mauritiana* was discovered during the analysis of an unstable allele of the *white* gene.[32] *Hermes*, a member of the *hobo, Ac, Tam3* (*hAT*) family was discovered in the housefly, *Musca domestica* during the evaluation of the *D. melanogaster hobo* element as a gene vector in that species.[33] *Hermes* is widely functional, but quite importantly, it has been shown to functionally interact with *hobo*,[34] providing some of the strongest experimental evidence to support the need for methods to stabilize transgene integrations. Other *hAT* elements have been found in tephritids that also functionally interact with *hobo*[35] and the *hopper* element from *Bactrocera dorsalis* has the potential for vector development.[36] Currently, the most widely used insect transposable element gene vector is the *piggyBac* element. This element was discovered during the analysis of mutations in baculoviruses and was subsequently found to have originated in the genome of the cabbage looper moth, *Trichoplusia ni*, which had served as a host for the virus.[37,38] Recently, bioinformatic analysis of public genomic DNA sequence databases aimed at finding functional transposable elements have resulted in the development of a number of new insect gene vectors e.g., *Herves*[39] and *Buster* (P.W. Atkinson and D.A. O'Brochta, unpublished). Non-insect transposable elements have also been used as gene vectors in insects. The bacterial transposable element *Tn5* has been used as a gene vector in *Ae. aegypti* although with very limited effectiveness.[40]

Enormous progress has been made in the study of the mechanism of action of at least two insect transposable elements that are used as gene vectors. Successful efforts to express and purify the transposases of the *Mos 1* and *Hermes* elements have enabled a wealth of biochemical information to be obtained regarding how these proteins perform the reactions involved in element excision and transposition. The successful determination of transposase crystal structures has provided unique insights into the mechanism of transposition and the relationship of these elements to other 'integrase' proteins.[41] In vitro assays have aided in assessing the biochemical requirements for transposable element mobility, and embryonic- and cell-line-mobility assays continue to be highly informative for assessing transposon mobility in a specific cellular environment. The development of powerful assay systems has allowed the testing of numerous mutations on both transposase enyzymatic activity and their binding specificities.[42] These advancements in functional analysis should not only result in enhancement of vector function in the near future but also the improved ability to modulate transposition for particular applications.

Early during the study of transposable element systems it was realized that most were members of larger families of elements that shared structural and functional similarities. The study of the evolutionary and natural history of transposable elements has been extended to the elements currently used for insect transgenesis. The extensive analysis of the *mariner/Tc* family has led to a greater understanding of the role of horizontal transfer in the life history of transposable elements and the role of this phenomenon on the preservation of functional elements despite their negative influence on the survival of their hosts.[43,44] The *Minos* element, which is widely distributed among drosophilids,[45,46] is strongly related to *Tc* elements originally discovered in nematodes,[31] and has exhibited function in invertebrate and vertebrate systems.[47,48] The *hAT* elements *Hermes* and *Herves* were discovered by virtue of their relationship to other *hAT* elements, and so their structural and functional similarities to *hATs* is understood. However, *Hermes* was discovered using a degenerate-PCR approach using discrete amino acid homologies,[49] while *Herves* used a sophisticated algorithm to discover existing database sequences fitting specific criteria.[39] While the *piggyBac* element has been used as a vector in the widest range of insect species, analysis of its distribution or the existence of a related family of elements is most recent relative to the other systems. Unlike the other vector transposons, elements nearly identical to the original *piggyBac* discovered in *T. ni*, have been found in other noctuid moths[50] in addition to distantly related tephritid fruit flies in the Bactrocerid family.[51] Yet similar to the other vectors, somewhat distantly related *piggyBac*-like elements have been found in *Bombyx*[52] and *Heliothis*,[53] with more distantly related sequences found in a large range of organisms.[54]

The continued discovery of transposable elements related to current insect gene vectors in many other organisms has important implications for the use of these vectors. First, the broad commonalities of structure and function undoubtedly led to our ability to effectively harness these elements as vectors for use in a wide range of host species. These commonalities, however, now become a challenge for practical application of transgenic insects, since this also increases the risk of vector remobilization by the same or related systems that might preexist in a host species. Such remobilizations would undoubtedly negatively impact desired strain characteristics or the ability to safely release such strains into the environment.

Transformation Markers

Reliable phenotypic markers for identifying transformed insects have been critical to the development of insect transformation technology and will also be essential for monitoring transgenic insects intentionally released into the field as part of future programmatic operations. The first genes used as transformation markers complemented eye-color mutations in mutant hosts, and while these were effective, their use was limited to species for which appropriate mutant strains were available.[55] Furthermore, these eye-color complementation markers were typically sensitive to chromosomal position effects.

Chemical-based selections have not been adopted for routine production of transgenic insects although two systems have been tested.[56] The first attempts at non-drosophilid transformations used the *neomycin phosphotransferase II* gene that confers resistance to the neomycin analog, Geneticin® (or G418). This system was first developed as an effective screen in Drosophila; yet in non-drosophilids false positives were common making it highly unreliable. It is possible that for some insects, such as tephritid flies, bacterial populations in the insect gut that were already neomycin resistant and may have allowed some non-transgenic insects to survive Geneticin® treatment (H. Genc and A.M. Handler, unpublished). A similar system based on *organophosphorus dehydrogenase* (*opd*) was developed using the insecticide paroxon for selection, but it also proved unreliable and further complicated by chemical exposure and waste disposal issues.[56] Nevertheless, a visible marker in conjunction with a reliable chemical selection system that allows mass selection could greatly simplify the creation of transgenic insects, particularly in insects where existing gene vectors are highly inefficient and require many progeny to be produced before a transgenic is found.

The insect transformation markers that have had the greatest impact on efforts to create transgenic insects have been based on fluorescent protein genes that result in a dominant neomorphic phenotype, do not depend on preexisting mutations and can be unambiguously identified. Currently, the four major fluorescent color variants (green, yellow, cyan and red) are optically distinguishable from one another and are highly stable.[57,58] When tested in Drosophila in parallel with the popular eye color marker gene *white*$^+$, enhanced green fluorescent protein (EGFP) was much less affected by position effect suppression of marker gene expression. In some transformants GFP was easily detected in flies in which the eye color marker gene produced no visible phenotype.[59] Autofluorescent proteins, however, do have limitations. For example, the time for protein cyclization necessary for maximal fluorescence can be long, interfering with real-time evaluation of gene expression. The benefits of fluorescent protein markers can only be realized by their appropriate expression, and a number of constitutive and tissue-specific promoters have been tested. Constitutive promoters have come from *actin*[60] and *ubiquitin*[61] genes, with the *polyubiquitin* promoter being especially useful for marker detection of tephritid flies in the field due to strong thoracic flight muscle expression.[62] The artificial 3xP3 promoter that binds the Pax-6 regulatory protein is active in embryonic, larval and adult central nervous tissue.[57] In many adult transgenic insects, 3xP3-regulated fluorescent protein gene expression can be detected in pigmented eyes, though it is difficult for some insects such as *Ae. aegypti*.[63,64]

Targeting and Stabilization

The significant advances in basic and applied studies that genetically transformed insects may provide must, however, be viewed in light of limitations that are inherent to the gene-transfer vector systems used to integrate transgenes into the host genome. New vectors that surmount many of these limitations likely represent the most tangible advances in transgenesis technology in recent years. For example, the transposon-based vectors for heritable germ-line transformation provide advantages over other types of vectors and transformation strategies, especially for functional analysis based on insertional mutagenesis, but this potential for remobilization can also compromise the stability of the transgenic host strains in practical applications, adversely affecting program effectiveness.[65] While the transposase enzyme required for transposon movement is typically eliminated after integration, the unintended or undetected presence of the transposase or related enzymes within the host can result in vector remobilization. While generally not considered to be a problem for small-scale laboratory studies, the rearing and release of many millions of insects for bio-control programs increases the probability that such rare events may occur.[66]

Another major caveat relates to the generally random nature of transposon integration into host genomes. Localized genomic effects on gene expression can result in variable transgene expression depending upon the integration site.[67] Vector integrations into coding regions or important regulatory regions can result in mutations having deleterious effects on the transformed host's fitness and viability. Thus random integrations make true comparative gene expression studies impossible, and greatly reduce the efficiency of creating optimal strains for applied use.

Random Transgene Insertion

A major difficulty in creating optimal transgenic strains for bio-control is decreased fitness and viability due to the random nature of transposon vector integrations that can disrupt vital gene functions and diminish or alter transgene expression due to genomic position effects. Most transposons integrate randomly except for short nucleotide sequence specificity (e.g., TTAA for *piggyBac* and TA for *Minos* and *mariners*), or general biases for genomic regions or gene structures that are not well understood (e.g., *P* integrations in 5' regulatory sequences).[68] These insertions cause mutations due to disruption of coding or regulatory sequences, and indeed, this is the basis for transposon-based insertional mutagenesis such as transposon-tagging and enhancer- or exon-trapping. Such mutations, however, are a significant drawback to the development of transgenic strains for applied use, and fitness costs have been found in some transgenic strains of Mediterranean fruit flies (G. Franz, A.S. Robinson and A.M.Handler, unpublished) and mosquitoes.[69,70] However, some transgenic strains, such as in screwworm[71] exhibit strong fitness parameters when transformed with the same or similar vectors.

Random transgene integrations also result in variable transgene expression resulting from chromosomal position or enhancer effects. Chromatin structure in eukaryotes has a regional effect on gene expression, most clearly evidenced by transcriptional suppression of genes inserted proximal to heterochromatin.[67] Transgene insertions are also affected by nearby promoters and enhancers or other epigenetic influences resulting in altered expression with respect to developmental and tissue specificities. The variability in transgene expression is reflected in several studies in Drosophila where conditional lethal constructs have been tested[72,73] and in our own studies testing temperature-dependent lethality by DTS-5 mutant alleles in Medfly and Caribfly (A.M. Hander, unpublished). In all of these studies, only some of the reported transgenic strains yielded lethality greater than 95% under non-permissive conditions. While all of these strategies can be further optimized, for some purposes only 100% lethality from heterozygous transgenes will be acceptable.[74] This creates a great challenge for transgenic strain development, and minimizing the variabilities introduced by genomic position effects should greatly enhance the ability to produce effective strains. Position effect suppression can be alleviated by surrounding the transgene with insulator elements,[75] but expression from different insertion sites is still expected to vary, and the use of insulators in one lethality study in Drosophila also yielded variability among strains.[73]

Vector Targeting

Both detrimental mutations and genomic position effects can be minimized by having transgene integrations limited to defined genomic target sites known to be devoid of coding or regulatory sequences or unusual effects on gene expression. Targeting systems have been a goal of most genetic model systems, with four recently developed for Drosophila based on either serine or tyrosine recombinases.[76-79] The serine integrase ϕC31 used genomically inserted *attB* sites as a target for *attP* sites within an introduced plasmid, and this system was subsequently integrated into *Aedes aegypti*.[80] While this provides the first targeting system for a non-drosophilid, it has not been reported if any of the target sites have been characterized for insertional mutations or negative effects on transgene expression. Undoubtedly this will follow in the near future. Limitations of integrase systems such as ϕC31 include insertion of the entire donor vector and the potential for mis-directed insertion at pseudo-*attB* sites within the genome.[76]

A more defined targeting system allows only the uni-directional integration of genes of interest, and two systems for recombinase-mediated cassette exchange (RMCE) have been described for Drosophila. RMCE is based upon double recombination between small heterospecific recombination sites within a genomic target site, and a plasmid donor sequence. Two examples of these systems are the *FRT*/FLP recombinase system from the 2μm plasmid of yeast[81] and the bacteriophage Cre-*loxP* system.[82] Functional *FRT* and *loxP* recombination sites consist of two 13 bp inverted repeats separated by an 8 bp spacer that specifically recombine with one another in the presence of FLP or Cre recombinase, respectively. Variations in the 8 bp spacer create heterospecific sites, and only identical sites can recombine with one another. One RMCE system inserted a target site vector with *loxP* recombination sites in Drosophila, which recombined with a donor vector in the presence of Cre recombinase at a relatively high frequency.[78] The target vector was integrated with the *P* transformation system and recombinants were marked eye color mutations, making this system specific for Drosophila.

Another RMCE system tested in Drosophila used a *piggyBac* target site acceptor vector to introduce heterospecific *FRT* recombination sites, that recombined with a donor vector in the presence of FLP recombinase at a frequency of ~23%.[77] A Drosophila *linotte* homing sequence was added in an effort to facilitate donor/target recombinations, but the lack of a homing sequence in the Cre/*loxP* system suggests that this is not critical. To ensure that only donor/target recombination events were selected, a promoter replacement system ensured that the promoterless-marker within the donor vector would only be detected after recombination by RMCE. Finally, the addition of a 3' *piggyBac* terminal sequence to the donor vector allowed subsequent stabilization of the target site (see below).

It is obvious that vectors that allow integrase or RMCE-based genomic targeting will greatly enhance the effectiveness of transgenic strains for biocontrol. They will also greatly reduce the time and effort needed to create these strains, and allow direct "allelic" comparisons of different recombinant DNA strategies. The addition of a post-integration stabilization component will serve to minimize ecological risks, and allow the direct use of optimized strains for transgene expression in field release studies and programs. The RMCE strategy will be equally important to comparative gene expression for functional genomics analysis, both for model systems and insects important to agriculture and human health. It will expand the tools available for reverse genetics, complementing current transposon-based germ-line transformation technology. Repetitive anchoring of different transgenes successively at identical loci will be possible, allowing true allelic comparisons of gene expression without restriction on gene size or number. This advancement should also greatly facilitate the study of regulatory elements such as enhancers, silencers and insulators. Transposon-based vectors are inherently limited by the size insert they can carry but recombinase systems are not, and indeed, they facilitate chromosomal rearrangements.[83] It is thus expected that RMCE will allow rapid genomic integration of large vectors such as BACs, greatly facilitating their functional analysis.

Vector Immobilization

Immobilization or stabilization of a transposon vector is most straightforwardly achieved by deleting or rearranging DNA within the vector that is required for transposition.[65] This includes the TIR sequences and possibly additional adjacent DNA. For most transposons this includes up to 100 bp of terminal sequence (but longer for *Minos* having TIRs of 255 bp). Deletion or rearrangement of these sequences is most simply achieved by introducing short recombination site sequences into the vector. Depending upon their location and orientation, *FRT* recombination can result in chromosomal rearrangements or the targeting of a plasmid carrying an *FRT* to a genomic *FRT* site.[83,84] Recombination of *FRT*s in direct orientation results in the deletion of the intervening DNA, while *FRTs* in the opposite orientation results in inversions. It should thus be possible to position *FRTs* in vectors to create rearrangements within a vector, or between two independent vectors after their genomic insertion by injection of plasmid-encoded FLP recombinase. While theoretically attractive, use of recombination systems for vector stabilization is not straightforward since recombination sites placed within subterminal sequences may negatively affect transposase-mediated integration. Rearrangement between independent vectors is more plausible, but requires two vector integrations on the same chromosome in a specific orientation. While this approach can be achieved in Drosophila (E. Wimmer pers. comm.), it is currently too formidable for routine use in most other insects.

Terminal Sequence Deletion

Another approach to terminal sequence deletion avoids the use of recombinase systems, and more simply uses the mobilizeable sequences that exist within the transposon.[85] The first of these systems provided a tandem duplication of the 5' or left terminus of the *piggyBac* element within the vector, though either terminus could have been duplicated. This creates a vector with a single 3' terminus that can be mobilized with either the 5' terminus or its internal 5' duplication, and these events could be distinguished by using independent markers between each set of terminal sequences. When the complete vector integrates using the external termini, the internal 5' terminus and the 3' external terminus can be subsequently remobilized by providing transposase in the form of transposase helper plasmid injection or by mating to a jumpstarter strain having a genomic source of the transposase. After remobilization, only the duplicated 5' external terminus and DNA between the duplicated termini, including marker genes and genes of interest, remains integrated. In the absence of the other external terminus, this remaining vector sequence should be stable with respect to transposase activity. This was tested by mating the stabilized line to the jumpstarter strain that showed that no remobilization occurred (by loss of phenotype) in several thousand progeny assayed.[85] This compared to a ~5% remobilization rate of the original internal vector, demonstrating that the transgene was stabilized owing to the loss of the 3' *piggyBac* terminus.

For most practical purposes, deletion of a single vector terminus should stabilize the remaining vector sequence, though a formal possibility exists that the stabilized sequence could exist proximal to a resident transposon, that might provide the requisite sequences for mobility that were deleted from the vector. This possibility was addressed by creating a vector that consists of two vectors with a spacer sequence between them, with the spacer including the transgene(s) of interest.[86] In this system, once the complete dual vector and spacer are integrated, the external vectors could then be remobilized leaving only the spacer including the transgenes, without any transposon terminal sequences. While this dual terminus deletion system worked reasonably efficiently, its practical advantage versus the single terminus deletion system remains to be evaluated.

Transgenic Strains for Biocontrol

A motivating force towards development of transgenic strains for agriculturally important insects was the desire to introduce attributes that would improve biological control systems, and the sterile insect technique (SIT) in particular. SIT involves the field release of many millions of sterile males, that render mated females non-reproductive.[87] While likely the most effective biological control strategy for insects, SIT still suffers from inefficiencies in implementation and cost, and transgenic strains could improve both.[66] First, females are generally superfluous and genetic sexing by female-specific lethality or male selection has been a high priority to decrease rearing costs and minimize mating competition with targeted females in the field. Male sterilization by irradiation is effective, but results in somatic cellular damage and fitness costs, and lastly, marking of released insects is typically achieved by dusting with fluorescent powders which can be unreliable and a safety concern for workers. Genetic-marking could be achieved most simply by using the same fluorescent markers currently used to identify transformants. Genetic-sexing and male sterility can both be achieved by the directed expression of lethal genes, including those that encode toxins, lethal gene products, or cell death genes.

Fluorescent Protein Genetic Markers

Use of fluorescent proteins to detect newly created transgenic insects can also be used for genetic-marking to detect released insects in the field, especially using constitutive promoters from genes such as ubiquitin that are expressed in most tissues throughout development, and especially well in thoracic flight muscles.[62] These markers may be ideal since fluorescent proteins are highly stable allowing their detection for a month or more after death in dry traps (R.A. Harrell and A.M. Handler, unpublished). Fluorescent proteins may also be used to mark specific tissues such as sperm within testes, which is relevant to assessing program effectiveness involving released males. Marked sperm could allow a determination of whether female insects trapped in the field had mated with released males, and in lab studies to assess sperm precedence. One approach to sperm marking is to link a fluorescent protein gene to a testes-specific promoter, such as *β2-tubulin*.[88] This was tested in several insects resulting in red or green fluorescing sperm bundles in the testes with unambiguous detection of sperm in the seminal receptacle or spemathecae of mated females[89,90] (Handler, A.M. and R.A. Harrell, unpublished). In *An. stephensi*, EGFP expression in larval testes could also be used for sexing in a COPAS™ fluorescent sorter.[89,91] This is particularly interesting as it represents an efficient means of male selection early in development. The use of testes-specific promoters may also find use in strains for male sterility, which would result from the testes-specific expression of gene products that cause cellular lethality, as discussed further on.

Conditional Regulation for Sterility and Lethality

A particularly interesting opportunity for the use of transgenics to control insect populations is the specific expression of genes that cause cellular lethality, resulting in organismal death when expressed in vital tissues or sterility when expressed solely in reproductive tissue. Lethal effector genes include those for toxin subunits, programmed cell death or apoptosis genes, and dominant, lethal mutations. To maintain breeding populations of transgenic strains, these genes must be under conditional regulation so that lethality or sterility only occurs under non-permissive conditions.

Tet-Off Regulation

Conditional expression can be achieved by use of suppressible or specifically inducible gene expression systems such as the tetracycline-resistance operon of the *E. coli* Tn10 transposon. This system can either activate or inhibit gene expression in the presence of dietary tetracycline or derivatives.[92,93] Several model systems for tet-repressed lethality have been tested in Drosophila using the programmed cell death gene, *hid*,[94] as the lethal effector. Two studies used the female-specific *yolk protein* gene promoter linked to the tTA regulatory protein to drive *hid* expression (linked to the tetracycline response element or TRE) in the female fat body, a vital organ, to specifically cause female lethality in late pupae and early adults.[72,95] Another approach used the promoters *nullo* and *serendipity* to cause death during embryonic development.[73] For both systems, most insects failed to survive only when removed from a tetracycline containing diet. Successful testing of these systems has yet to be reported for non-drosophilids species, possibly due to the need for species-specific promoters, if not cognates for cell death genes. An interesting permutation of the tet-repressible lethality system that has been tested in the medfly, takes advantage of a lethal effect of the tTA regulatory protein at high concentrations. When placed in a positive regulatory loop where tTA is linked to the TRE, tTA drives its own expression in the absence of tetracycline, that resulted in >98% death in medfly larvae and pupae.[96] The same or similar systems have been proposed for conditional lethal control of mosquito disease vectors.[97]

Gal4-UAS

The yeast Gal4 operon system allows conditional regulation when lethal genes are linked to Upstream Activating Sequences (UAS) that only promote transcription when bound to the GAL4 regulatory protein.[98] Thus lethal gene expression would depend upon the regulation of the *Gal4* gene. Typically the UAS and Gal4 constructs exist in different strains, with the desired expression achieved only in their progeny, which would be cumbersome if not impracticable for population control. However, the system has been made more versatile by use of a temperature-sensitive *Gal4* mutation, *Gal4*^{ts80} that provides conditional temperature-dependent regulation as well.[99]

Temperature-Dependent Lethality

More direct temperature-dependent regulation is provided by toxin subunit gene mutations such as the DTA diphtheria heat sensitive mutations,[100] and the ricin cold-sensitive mutations.[101] Similarly dominant gain-of-function mutations exist in Drosophila that result in temperature-dependent lethality, such as the *Notch*^{60g11} cold-sensitive mutation that is used in a system known as autocidal biological control (ABC).[102] The DTS-5 and DTS-7 heat sensitive lethal mutations exist for 20S proteasome subunit genes,[103,104] and DTS-5 has been introduced into medfly, with two independent lines homozygous for the transgene exhibiting 90-95% larval-pupal lethality at 30°C (R. Krasteva, and A.M. Handler, unpublished). The strategy for these genes would be the rearing of transgenic insects at permissive temperatures, with their offspring dying in the field under non-permissive conditions of elevated temperatures.

Paratransgenesis

For some species not amenable to direct genetic transformation, it is possible to create insects with a desired phenotype due to expression from a transformed symbiont. This is known as paratransgenesis, and was first demonstrated by the transformation of the bacterial symbiont, *Rhodococcus rhodnii*, of the kissing bug, *Rhodnius prolixius* which is a triatomine vector of Chagas disease.[105] This was first achieved with an episomal shuttle plasmid containing an anti-microbial cecropin A peptide gene that reduced trypanosome number in the bacterium.[106] Plasmid stability was later improved by use of plasmids capable of integrating into the bacterial genome using an *L1* mycobacteriophage integrase system.[107] While paratransgenesis can still be applicable to the control of many medically and agriculturally important insects, reports of new systems have been lacking in recent years.

Wolbachia bacterial symbionts that infect many species also have the potential to act as a vehicle for paratransgenesis, with the added advantage of conferring cytoplasmic incompatibility (CI) providing its hosts with a selective reproductive advantage.[108] Females uninfected with the bacterium are rendered sterile when mated to an infected male, while infected females are reproductive with infected or uninfected males. It has been proposed that this advantage conferred by Wolbachia could be used to drive beneficial or desired genes into a population.[109] Thus far however, Wolbachia has not been amenable to transformation, which is also hindered by an inability to culture the bacterium in vitro. It has been possible to infect Wolbachia into receptive insect strains by injection of a bacterium from another species. *Aedes albopictus* was newly infected with a Wolbachia from *Drosophila simulans*[110] and *Ceratitis capitata* was infected with a strain from another tephritid, *Rhagoletis cerasi*.[111] For both species, these newly introduced infections have been proposed as a method to control or suppress natural populations by the release of infected laboratory reared males.

Viruses

The efficiency with which viruses can deliver DNA to particular cells has made them very appealing to those interested in developing gene vectors. A variety of viral systems have been developed for vertebrates including retroviruses, adenoviruses, adeno-associated viruses and herpes simplex virus. Unfortunately a similar arsenal of viral vectors is not available for insect transformation and this is unfortunate because it limits insect germ-line transformation to those insects whose young embryos can be directly microinjected with vector DNA, and dependence on this DNA delivery system severely limits the use of insect transformation technology. Nonetheless viral vectors are available for introducing genes into somatic tissues and their use is expanding. Sindbis viruses continue to be used not only in mosquitoes but have been shown to function in a variety of phylogenetically diverse species of insect including Lepidoptera, Hemiptera and in non-insect arthropods such as brine shrimp.[112] Baculoviruses continue to be used in a limited way as transient expression systems in non-host species[113] and densoviruses are also commanding attention as an insect gene delivery systems for somatic gene expression.[114] They too have host range limitations but seem to be useful in a wide variety of non-host systems, especially those refractory, thus far, to germ-line transformation methods. Insects tend to be lacking retroviruses with few exceptions and this appears to be generally true based on available genome sequence data. The few insect retroviruses that do exist, e.g., *gypsy*, have not been developed into gene vectors and there have been no reported experiments looking at the host range of these viruses. Testing of pantropic retroviruses developed from vertebrate retroviruses has also not gone beyond use for somatic expression.[115]

Perspectives on TE Spread and Genetic Manipulation for Disease Vector Control

Beyond the P element Paradigm

Insect transformation technology has developed and evolved in the shadow of *Drosophila melanogaster* transformation technology, which has been based largely on the Class II transposable element, *P*. The *P* element has been considered a model Class II element and consequently the P system, within the context of non-drosophilid insect transformation research and development, has come to be seen as the paradigm for Class II elements and transformation systems. The paradigmatic stature of the *P* element system has shaped thinking and has had an enormous influence over research programs and resource allocation decisions. Indeed, research programs at institutions such as NIH, WHO and the MacArthur Foundation were initiated and supported based on a view of transposable element biology that rested heavily on the *P* element paradigm.[116-118] A notable idea in the area of insect-transmitted diseases over the last twenty years has been the idea that diseases like malaria might be controlled by genetically engineered insects. Pathogen resistant insects, it was claimed, could be created in the laboratory and introduced into natural populations where they would sweep to fixation and convert native insects such as mosquitoes into harmless biting nuisances as opposed to transmitters of disease-causing pathogens. The idea of altering the genotype/phenotype of human disease vector populations emerged in the early 1960's,[119] with the genetic engineering version of the idea appearing in the early 1980's,[120] gaining considerable support because proponents pointed to the natural history of the *P* element.[116] The *P* element provided a contemporary example of a transposable element entering a genome and in a matter of decades sweeping through essentially every population of *D. melanogaster* in the world.[121,122] In the decade or so since the first transformation of a non-drosophilid insect it seems appropriate to ask whether *P* elements really are typical insect Class II transposable elements and whether their continued use as a paradigm in transgenic insect research and development is justified and helpful?

The enormous advances in our understanding of transposable elements over the past two and a half decades have revealed a number of interesting features of *P* elements, that indicate the unusual nature of these elements. Compared to all known Class II transposable elements *P* elements form a distinct clade within transposable element phylogeny. Most insect Class II transposable elements, for example *Tc1/mariner* and *hAT* elements, are related by virtue of a distinct characteristic of the catalytic core of their transposases. *Tc1/mariner* and *hAT* elements have two aspartates and a glutamate that play key roles in catalysis.[41] This core is found in a protein pocket that displays a high degree of conservation at the structural level to retroviral integrases. These two large groups of Class II transposable elements appear to have evolved from an ancient integrase protein that predates the evolution of retrotransposons and Class II elements.[41] While *P* elements appear to be mechanistically related to the rather large superfamily of polynucleotidyltransferases, including transposons and retroviral integrases, they may have evolved quite independently. It has been suggested that they may be more related to certain polymerases and restriction endonucleases.[5]

Early interest in non-drosophilid insect transformation justifiably focused on *P* elements but this interest diminished considerably when it was realized that these elements had a highly restricted host range and appeared to only function in insects closely related to Drosophila.[123] All insect transposable elements discovered and tested since then have shown exactly the opposite characteristics. Without exception, insect Class II transposable elements have been shown to have broad host ranges within insects and in many cases we know that element mobility can extend in non-insect systems as well. So, the restricted mobility characteristics of *P* elements are highly unusual for Class II transposable elements.

The recent evolutionary history of *P* elements is also distinctly different from most known non-drosophilid insect gene vectors. *P* elements have been recently introduced into the genome of

D. melanogaster from *D. willistoni* by a mechanism that is not understood.[124] This introduction took place sometime within the last century and perhaps as recently as the mid-twentieth century. When *P* elements were discovered, isolated and cloned they had not yet completed their world-wide invasion of *D. melanogaster*.[121,125,126] While many of the transposable elements currently used as insect gene vectors belong to families of elements with historical evidence for horizontal gene transfer, none of them are currently known to be undergoing a species invasion. The impression left from the study of *P* element natural history is that horizontal gene transfer is common, swift and easy. This perspective has been very influential on those proposing to use Class II transposable elements as genetic drive systems to introduce a transgenes into natural populations. Clearly the study of transposable elements has revealed numerous examples of horizontal transfer and this appears to be a common property of these genetic elements.[127] What is not known, however, is how readily do these transfers occur and how rapid is any subsequent invasion. Because only successful horizontal transfers can be detected in nature we cannot estimate from existing natural history data how many horizontal transfers have been unsuccessful. While we know that many insect transposable elements can transpose and be used as primary gene vectors in divergent species, there are few data that tell us anything about how autonomous, natural elements would behave in these foreign hosts or whether, even under artificial conditions, the introduced elements would spread as readily as *P* elements in *D. melanogaster*. So *P* elements are unusual in that they are recent additions to the genome of *D. melanogaster* and the phylogenetic distance covered during this transfer (e.g., *D. willistoni* to *D. melanogaster*) is small relative to the transfers being performed during insect transformation experiments. Recent, albeit limited, efforts to introduce and spread transposable elements in non-drosophilid species using existing insect transposable elements have been met with low remobilization rates and spreading potential but, to date, many element/species combinations have yet to be tested.[128-130] These data suggest, however, that the powerful drive potential of *P* elements in *D. melanogaster* may not typify horizontal transfers of other elements.

P element invasion of naïve genomes of *D. melanogaster* occur not only quickly but the invasions are short-lived. That is, in laboratory populations within 20 generations altered forms of the element arise (usually as the result of internal deletions) that have transposition repression activity.[131] Not long thereafter all element activity ceases. Similar experiments performed with *hobo* elements in *D. melanogaster* yielded somewhat different results.[132,133] In addition, there are clearly host effects on element behavior. The dynamics of movement of introduced *P* elements in *D. simulans* was found to be distinctly different from that observed in *D. melanogaster*.[134] Our understanding of transposable element natural history has also provided us with reason to think that equilibrium may take a very long time to be reached, if ever.[135,136]

So, should we use *P* element invasion of *D. melanogaster* as the paradigm for element invasion in general? It might be advisable not to assume that existing gene vectors will behave in foreign species as *P* elements behave in *D. melanogaster*, thereby placing emphasis on obtaining empirical data regarding the behavior of individual elements in species and under conditions of interest.

Broadening Our Perspective

As we consider the use of transgenesis to control arthropod vectors of disease it is worth noting how perceptions play an important role within the life of an idea. Ideas within the 'mainstream' are perceived differently from 'novel' or 'exotic' ideas. Insect population modification through the introduction of transgenes seems to be perceived as somewhat exotic. Perhaps it is time to alter that perception.

Following the success of the sterile insect technique to control the new world screwworm, George Craig, in 1963, published a report in which the idea of using genetics to alter the disease transmission potential of mosquitoes was described.[119] Considerable effort during the following decades focused on developing genetic drive systems using conventional genetic manipulations as a means to spread a desirable gene into the target population.[137] The insects produced from these earlier efforts competed poorly in the field and, eventually, enthusiasm for the idea of manipulating the genetics of standing populations of insects waned. With the advent of insect genetic engineering technologies in the 1980s came a revival of the population replacement idea since new ways would be available to introduce and spread desirable genes.[3,120] The genetic engineering version of the idea of population conversion or replacement as a means to control disease transmission seems exotic and largely outside most of the paradigms under which vector biologists and insect control specialists operate. But perhaps manipulating the genotypes of populations of insect vectors of disease is less related to insect control and more related to wildlife management.

For example, rabies is a fatal encephalitis in warm-blooded vertebrates caused by viruses where infected hosts are major reservoirs for the virus.[3] Programs aimed at population reduction of animals at risk were largely unsuccessful for many of the same reasons insect vector eradication programs have been at times unsuccessful. In recent decades an oral wildlife rabies vaccine was developed that was successfully integrated into effective operational programs where it is deployed over wide areas in an edible bait.[3] Results have been dramatic, with Western Europe essentially rabies-free, and with programs in the U.S. meeting with similar success.[138,139]

Similar strategies are now being explored for *Borrelia burgdorferi* transmission.[140] Wildlife reproduction in natural populations is being controlled in some cases through the use of immuno-contraceptives.[141] While these have been administered through micro-injection, species specific viral vaccines are also now being developed. Wildlife management, therefore appears to have integrated the idea of genetically manipulating populations of wild animals (albeit somatic manipulations) into their arsenal of tools and these ideas appear to now be 'mainstream'.

A somewhat different approach to manipulation of standing populations is illustrated by an effort to combat Chestnut Blight (*Cryphonectira parasiticum*) in American Chestnut trees. Current efforts involve the use of an RNA virus to attenuate the virulence of the fungal pathogen. Recombinant fungi with virulence attenuation genes have been created that enhance the transmission dynamics of the avirulence phenotype within the populations. The objective of these efforts is to modify indigenous populations of *C. parasiticum* in situ so they are no longer a threat to American Chestnut trees.[142,143]

These large scale manipulations of indigenous populations of wildlife or plants for the purposes of interrupting disease transmission is very much analogous to proposed efforts to introduce anti-pathogen genotypes into insect vector populations. The insect programs would be self-sustaining by virtue of the germ-line integration of effector genes that would permit their vertical transmission over time.

Now that most of the components necessary for creating a system for genetically manipulating insect vector populations are in place (but certainly not optimized), moving forward toward implementation may be facilitated by placing the work within a broader context so that it does not appear to be a completely foreign idea. It seems clear that others are also considering genetic manipulation of wild populations as a means of controlling disease and in some cases outstanding success has been reported. Insect vector biologists might benefit from an awareness of these related efforts as they attempt to explain and justify their work and plans to skeptical program officers, funding agencies and the interested public.

Acknowledgements

Grateful appreciation is extended to colleagues for providing unpublished results and to the National Institutes of Health (DAO'B) and the USDA National Research Initiative-Competitive Grants Program (AMH) for support during preparation of this manuscript.

References

1. Handler AM, O'Brochta DA. Prospects for gene transformation in insects. Ann Rev Entomol 1991; 36:159-183.

2. McCrane V, Carlson JO, Miller BR et al. Microinjection of DNA into Aedes triseriatus ova and detection of integration. Amer J Trop Med Hyg 1988; 39:502-510.
3. Miller LH, Sakai RK, Romans P et al. Stable integration and expression of a bacterial gene in the mosquito Anopheles gambiae. Science 1987; 237:779-781.
4. Morris AC, Eggelston P, Crampton JM. Genetic transformation of the mosquito Aedes aegypti by micro-injection of DNA. Med Vet Entomology 1989; 3:1-7.
5. Rio DC. P transposable elements in Drosophila melanogaster. In: Craig NL, Craige R, Gellert M et al, eds. Mobile DNA II. Washington, DC: ASM Press, 2002:1204.
6. Jasinskiene N, Coates CJ, Benedict MQ et al. Stable, transposon mediated transformation of the yellow fever mosquito, Aedes aegypti, using the Hermes element from the housefly. Proc Natl Acad Sci 1998; 95:3743-3747.
7. Coates CJ, Jasinskiene N, Miyashiro L et al. Mariner transposition and transformation of the yellow fever mosquito, Aedes aegypti. Proc Natl Acad Sci 1998; 95:3748-3751.
8. Fang J, Han Q, Li J. Isolation, characterization, and functional expression of kynurenine aminotransferase cDNA from the yellow fever mosquito, Aedes aegypti. Insect Biochem Mol Biol 2002; 32:943-950.
9. Bhalla SC. White eye, a new sex-linked mutant of Aedes aegypti. Mosquito News 1968; 28:380-385.
10. Loukeris TG, Livadaras I, Arca B et al. Gene transfer into the Medfly, Ceratitis capitata, using a Drosophila hydei transposable element. Science 1995; 270:2002-2005.
11. Handler AM, McCombs SD, Fraser MJ et al. The lepidopteran transposon vector, piggyBac, mediates germ-line transformation in the Mediterranean fruit fly. Proc Natl Acad Sci USA 1998; 95:7520-7525.
12. Zwiebel LJ, Saccone G, Zacharaopoulou A et al. The white gene of Ceratitis capitata: A phenotypic marker of germline transformation. Science 1995; 270:2005-2008.
13. Horn C, Offen N, Nystedt S et al. piggyBac-based insertional mutagensis and enhancer detection as a tool for functional insect genomics. Genetics 2003; 163:647-661.
14. Higgs S, Traul D, Davis BS et al. Green fluorescent protein expressed in living mosquitoes—without the requirement of transformation. Biotechniques 1996; 21:660-664.
15. Catteruccia F, Nolan T, Loukeris TG et al. Stable germline transformation of the malaria mosquito Anopheles stephensi. Nature 2000; 405:959-962.
16. Yoshida S, Watanabe H. Robust salivary gland-specific transgene expression in Anopheles stephensi mosquito. Insect Mol Biol 2006; 15:403-410.
17. Lombardo F, Nolan N, Lycett G et al. An Anopheles gambiae salivary gland promoter analysis in Drosophila melanogaster and Anopheles stephensi. Insect Mol Biol 2005; 14:207-216.
18. Brown AE, Bugeon L, Crisanti A et al. Stable and heritable gene silencing in the malaria vector Anopheles stephensi. Nuc Acid Res 2003; 31:e85.
19. Moreira LA, Edwards MJ, Adhami F et al. Robust gut-specific gene expression in transgenic Aedes aegypti mosquitoes. Proc Natl Acad Sci USA 2000; 97:10895-10898.
20. Moreira LA, Ito J, Ghosh A et al. Bee venom phospholipase inhibits marlaria parasite development in transgenic mosquitoes. J Biol Chem 2002; 25:40839-40843.
21. Grossman GL, Rafferty CS, Clayton JR et al. Germline transformation of the malaria vector, Anopheles gambiae, with the piggyBac transposable element. Insect Molec Biol 2001; 10:597-604.
22. Kim W, Koo H, Richman AM et al. Ectopic expression of a cecropin transgene in the human malaria vector mosquito Anopheles gambiae (Diptera: Culicidae): Effects on susceptibility to Plasmodium. J Med Entomol 2004; 41:447-455.
23. Tamura T, Thibert C, Royer C et al. Germline transformation of the silkworm Bombyx mori L. using a piggyBac transposon-derived vector. Nat Biotechnol 2000; 18:81-84.
24. Adachi T, Tomita M, Shimizu K et al. Generation of hybrid transgenic silkworms that express Bombyx mori prolyl-hydroxylase alpha-subunits and human collagens in posterior silk glands: Production of cocoons that contained collagens with hydroxylated proline residues. J Biotechnol 2006; 126:205-219.
25. Imamura M, Nakahara Y, Kanda T et al. A transgenic silkworm expressing the immune-inducible cecropin B-GFP reporter gene. Insect Biochem Mol Biol 2006; 36:429-434.
26. Finnegan DJ. Eukaryotic transposable elements and genome evolution. Trends Genetic 1989; 5:103-107.
27. Rubin GM, Spradling AC. Genetic transformation of Drosophila with transposable element vectors. Science 1982; 218:348-353.
28. Rubin GM, Spradling AC. Vectors for P element gene transfer in Drosophila. Nuc Acids Res 1983; 11:6341-6351.
29. Bingham PM, Kidwell MG, Rubin GM. The molecular basis of P-M hybrid dysgenesis: The role of the P element, a P-strain-specific transposon family. Cell 1982; 29:995-1004.

30. Handler AM. An introduction to the history and methodology of insect gene transformation. In: Handler AM, James AA, eds. Transgenic Insects: Methods and Applications. Boca Raton: CRC Press LLC, 2000:397.
31. Franz G, Savakis C. Minos, a new transposable element from Drosophila hydei, is a member of the Tc1-like family of transposons. Nucleic Acids Res 1991; 19:6646.
32. Medhora MM, MacPeek AH, Hartl DL. Excision of the Drosophila transposable element mariner: Identification and characterization of the Mos factor. EMBO J 1988; 7:2185-2189.
33. Atkinson PW, Warren WD, O'Brochta DA. The hobo transposable element of Drosophila can be cross-mobilized in houseflies and excises like the Ac element of maize. Proc Natl Acad Sci USA 1993; 90:9693-9697.
34. Sundararajan P, Atkinson PW, O'Brochta DA. Transposable element interactions in insects: Crossmobilization of hobo and Hermes. Insect Molec Biol 1999; 8:359-368.
35. Handler AM, Gomez SP. The hobo transposable element excises and has related elements in tephritid species. Genetics 1996; 143:1339-1347.
36. Handler AM. Isolation and analysis of a new hopper hAT transposon from the Bactrocera dorsalis white eye strain. Genetica 2003; 118:17-24.
37. Cary LC, Goebel M, Corsaro BG et al. Transposon mutagenesis of baculoviruses: Analysis of Trichoplusia ni transposon IFP2 insertions within the FP-locus of nuclear polyhedrosis viruses. Virology 1989; 172:156-169.
38. Fraser MJ, Smith GE, Summers MD. The acquisition of host cell DNA sequences by baculoviruses: Relation between host DNA insertions and FP mutants of Autographa californica and Galleria mellonella NPVs. J Virology 1983; 47:287-300.
39. Arensburger P, Orsetti J, Kim YJ et al. A new active transposable element, Herves, from the African malaria mosquito Anopheles gambiae. Genetics 2005; 169:697-708.
40. Rowan K, Orsetti J, Atkinson PW et al. Tn5 as an insect gene vector. Insect Biochem Mol Biol 2004; 34:695-705.
41. Hickman AB, Perez ZN, Zhou L et al. Molecular architecture of a eukaryotic DNA transposase. Nat Struct Mol Biol 2005; 12:715-721.
42. Butler MG, Chakraborty SA, Lampe DJ. The N-terminus of Himar1 mariner transposase mediates multiple activities during transposition. Genetica 2006; 127:351-366.
43. Lampe DJ, Witherspoon DJ, Soto-Adames FN et al. Recent horizontal transfer of mellifera subfamily mariner transposons into insect lineages representing four different orders shows that selection acts only during horizontal transfer. Mol Biol Evol 2003; 20:554-562.
44. Hartl DL, Lohe AR, Lozovskaya ER. Modern thoughts on an ancyent marinere: Function, evolution, regulation. Annu Rev Genet 1997; 31:337-358.
45. de Almeida LM, Carareto CM. Multiple events of horizontal transfer of the Minos transposable element between Drosophila species. Mol Phylogenet Evol 2005; 35:583-594.
46. Arca B, Savakis C. Distribution of the transposable element Minos in the genus Drosophila. Genetica 2000; 108:263-267.
47. Zagoraiou L, Drabek D, Alexaki S et al. In vivo transposition of Minos, a Drosophila mobile element, in mammalian tissues. Proc Natl Acad Sci USA 2001; 98:11474-11478.
48. Sasakura Y, Awazu S, Chiba S et al. Germ-line transgenesis of the Tc1/mariner superfamily transposon Minos in Ciona intestinalis. Proc Natl Acad Sci USA 2003; 100:7726-7730.
49. Warren WD, Atkinson PW, O'Brochta DA. The Hermes transposable element from the housefly, Musca domestica, is a short inverted repeat-type element of the hobo, Ac, and Tam3 (hAT) element family. Genetical Res Camb 1994; 64:87-97.
50. Zimowska GJ, Handler AM. Highly conserved piggyBac elements in noctuid species of Lepidoptera. Insect Biochem Mol Biol 2006; 36:421-428.
51. Handler AM, McCombs SD. The piggyBac transposon mediates germ-line transformation of the Oriental fruit fly and closely related elements exist in its genome. Insect Molec Biol 2000; 9:605-612.
52. Xu HF, Xia QY, Liu C et al. Identification and characterization of piggyBac-like elements in the genome of domesticated silkworm, Bombyx mori. Mol Genet Genomics 2006; 276:31-40.
53. Wang J, Ren X, Miller TA et al. piggyBac-like elements in the tobacco budworm, Heliothis virescens (Fabricius). Insect Mol Biol 2006; 15:435-443.
54. Sarkar A, Sim C, Hong YS et al. Molecular evolutionary analysis of the widespread piggyBac transposon family and related "domesticated" sequences. Mol Genet Genomics 2003; 270:173-180.
55. Sarkar A, Collins FH. Eye color genes for selection of transgenic insects. In: Handler AM, James AA, eds. Insect Transgenesis. Boca Raton: CRC, 2000.
56. ffrench-Constant RH, Benedict MQ. Resistance genes as candidates for insect transgenesis. In: Handler AM, James AA, eds. Insect Transgenesis. Boca Raton: CRC Press, 2000.

57. Berghammer AJ, Klingler M, Wimmer EA. A universal marker for transgenic insects. Nature 1999; 402:370.
58. Horn C, Wimmer EA. A versatile vector set for animal transgenesis. Dev Genes Evol 2000; 210:630-637.
59. Handler AM, Harrell RA. Germline transformation of Drosophila melanogaster with the piggyBac transposon vector. Insect Molec Biol 1999; 4:449-458.
60. Burn TC, Vigoreaux JO, Tobin SL. Alternative 5C actin transcripts are localized in different patterns during Drosophila embryogenesis. Dev Biol 1989; 131:345-355.
61. Lee HS, Simon JA, Lis JT. Structure and expression of ubiquitin genes of Drosophila melanogaster. Mol Cell Biol 1988; 8:4727-2735.
62. Handler AM, Harrell RA. Transformation of the Caribbean fruit fly with a piggyBac transposon vector marked with polyubiquitin-regulated GFP. Insect Biochem Mol Biol 2001; 31:199-205.
63. Kokoza V, Ahmed A, Wimmer EA et al. Efficient transformation of the yellow fever mosquito Aedes aegypti using the piggyBac transposable element vector pBac[3xP3-EGFP afm]. Insect Biochem Mol Biol 2001; 31:1137-1143.
64. Lobo NF, Hua-Van A, Li X et al. Germ line transformation of the yellow fever mosquito, Aedes aegypti, mediated by transpositional insertion of a piggyBac vector. Insect Molec Biol 2002; 11:133-139.
65. Handler AM. Understanding and improving transgene stability and expression in insects for SIT and conditional lethal release programs. Insect Biochem Mol Biol 2004; 34:121-130.
66. Robinson AS, Franz G, Fisher K. Genetic sexing strains in the medfly, Ceratitis capitata: Development, mass rearing and field application. Trends in Entomol 1999; 2:81-104.
67. Wilson C, Bellen HJ, Gehring W. Position effects on eukaryotic gene expression. Ann Rev Cell Biol 1990; 6:679-714.
68. Thibault ST, Singer MA, Miyazaki WY et al. P and piggyBac transposons display a complementary insertion spectrum in Drosophila: A multifunctional toolkit to manipulate an insect genome. Nat Genet 2004; 36:283-287.
69. Catteruccia F, Godfray HC, Crisanti A. Impact of genetic manipulation on the fitness of Anopheles stephensi mosquitoes. Science 2003; 299:1225-1227.
70. Irvin N, Hoddle MS, O'Brochta DA et al. Assessing fitness costs for transgenic Aedes aegypti expressing the GFP marker and transposase genes. Proc Natl Acad Sci USA 2004; 101:891-896.
71. Allen ML, Berkebile DR, Skoda SR. Postlarval fitness of transgenic strains of Cochliomyia hominivorax (Diptera: Calliphoridae). J Econ Entomol 2004; 97:1181-1185.
72. Heinrich JC, Scott MJ. A repressible female-specific lethal genetic system for making transgenic insect strains suitable for a sterile-release program. Proc Natl Acad Sci USA 2000; 97:8229-8232.
73. Horn C, Wimmer EA. A transgene-based, embryo-specific lethality system for insect pest management. Nat Biotechnol 2003; 21:64-70.
74. Robinson AS, Franz G. The application of transgenic insect technology in the sterile insect technique. In: Handler AM, James AA, eds. Insect Transgenesis: Methods and Applications. Boca Raton: CRC Press LLC, 2000:307-319.
75. Sarkar A, Atapattu A, EJB et al. Insulated piggyBac vectors for insect transgenesis. BMC Biotechnol 2006; 6:27.
76. Groth AC, Fish M, Nusse R et al. Construction of transgenic Drosophila by using the site-specific integrase from phage phiC31. Genetics 2004; 166:1775-1782.
77. Horn C, Handler AM. Site-specific genomic targeting in Drosophila. Proc Natl Acad Sci USA 2005; 102:12483-12488.
78. Oberstein A, Pare A, Kaplan L et al. Site-specific transgenesis by Cre mediated recombination in Drosophila. Nat Methods 2005; 2:583-585.
79. Rong YS, Golic KG. Gene targeting by homologous recombination in Drosophila. Science 2000; 288:2013-2018.
80. Nimmo DD, Alphey L, Meredith JM et al. High efficiency site-specific genetic engineering of the mosquito genome. Insect Mol Biol 2006; 15:129-136.
81. Andrews BJ, Proteau GA, Beatty LG et al. The FLP recombinase of the 2 micron circle DNA of yeast: Interaction with its target sequences. Cell 1985; 40:795-803.
82. Siegal ML, Hartl DL. Transgene coplacement and high efficiency site-specific recombination with the Cre/loxP system in Drosophila. Genetics 1996; 144:715-726.
83. Golic KG, Golic MM. Engineering the Drosophila genome: Chromosome rearrangements by design. Genetics 1996; 144:1693-1711.
84. Golic MM, Rong YS, Petersen RB et al. FLP-mediated DNA mobilization to specific target sites in Drosophila chromosomes. Nuc Acid Res 1997; 25:3665-3671.
85. Handler AM, Zimowska GJ, Horn C. Post-integration stabilization of a transposon vector by terminal sequence deletion in Drosophila melanogaster. Nat Biotechnol 2004; 22:1150-1154.

86. Dafa'alla TH, Condon GC, Condon KC et al. Transposon-free insertions for insect genetic engineering. Nat Biotechnol 2006; 24:820-821.
87. Knipling EF. Possibilities of insect control or eradication through the use of sexually sterile males. J Econ Entomol 1955; 48:459-462.
88. Fackenthal JD, Turner FR, Raff EC. Tissue-specific microtubule functions in Drosophila spermatogenesis require the beta 2-tubulin isotype-specific carboxy terminus. Dev Biol 1993; 158:213-227.
89. Catteruccia F, Benton JP, Crisanti A. An Anopheles transgenic sexing strain for vector control. Nat Biotechnol 2005; 23:1414-1417.
90. Smith RC, Walter MF, Hice RH et al. Testis-specific expression of the β2 tubulin promoter of Aedes aegypti and its application as a genetic sex-separation marker Insect. Mol Biol 2007, (doi: 10.1111/j.1365-2583.2006.00701.x).
91. Furlong EE, Profitt D, Scott MP. Automated sorting of live transgenic embryos. Nat Biotechnol 2001; 19:153-156.
92. Gossen M, Bonin AL, Bujard H. Control of gene activity in higher eukaryotic cells by prokaryotic regulatory elements. Trends Biochem Sci 1993; 18:471-475.
93. Gossen M, Bujard H. Tight control of gene expresion in mammalian cells by tetracycline-responsive promoters. Proc Natl Acad Sci USA 1992; 89:5547-5551.
94. Abrams JM, White K, Fessler LI et al. Programmed cell death during Drosophila embryogenesis. Development 1993; 117:29-43.
95. Thomas DD, Donnelly RJ, Wood LS et al. Insect population control using a dominant, repressible, lethal genetic system. Science 2000; 287.
96. Gong P, Epton MJ, Fu G et al. A dominant lethal genetic system for autocidal control of the Mediterranean fruitfly. Nat Biotechnol 2005; 23:453-456.
97. Alphey L. Reengineering the sterile insect technique. Insect Biochem Mol Biol 2002; 32:1243-1247.
98. Brand AH, Manoukian AS, Perrimon N. Extopic expression in Drosophila. Methods Cell Biol 1994; 44:635-654.
99. McGuire SE, Le PT, Osborn AJ et al. Spatiotemporal rescue of memory dysfunction in Drosophila. Science 2003; 302:1765-1768.
100. Bellen HJ, D'Evelyn D, Harvey M et al. Isolation of temperature-sensitive diptheria toxins in yeast and their effects on Drosophila cells. Development 1992; 114:787-796.
101. Moffat KG, Gould JH, Smith HK et al. Inducible cell ablation in Drosophila by cold-densitve ricin A chain. Development 1992; 114:681-687.
102. Fryxell KJ, Miller TA. Autocidal biological control: A general strategy for insect control based on genetic transformation with a highly conserved gene. J Econ Entomol 1994; 88:1221-1232.
103. Saville KJ, Belote JM. Identification of an essential gene, l(3)73Ai, with a dominant temperature-sensitive lethal allele, encoding a Drosophila proteasome subunit. Proc Natl Acad Sci USA 1993; 90:8842-8846.
104. Covi JA, Belote JM, Mykles DL. Subunit compositions and catalytic properties of proteasomes from developmental temperature sensitive mutants of Drosophila melanogaster. Arch Biochem Biophys 1999; 368:85-97.
105. Beard CB, Mason PW, Aksoy S et al. Transformation of an insect symbiont and expression of a foreign gene in the Chagas's disease vector Rhodnius. Am J Trop Med Hyg 1992; 46:195-200.
106. Durvasula RV, Gumbs A, Panackal A et al. Prevention of insect-borne disease: An approach using transgenic symbiotic bacteria. Proc Natl Acad Sci USA 1997; 94:3274-3278.
107. Dotson EM, Plikaytis B, Shinnick TM et al. Transformation of Rhodococcus rhodnii, a symbiont of the Chagas disease vector Rhodnius prolixus, with integrative elements of the L1 mycobacteriophage. Infect Genet Evol 2003; 3:103-109.
108. Stouthammer R, Breeuwer JA, Hurst GD. Wolbachia pipientis: Microbial manipulator of arthropod reproduction. Ann Rev Microbiol 1999; 53:71-102.
109. Sinkins SP, O'Neill SL. Wolbachia as a vehicle to modify insect populations. In: Handler AM, James AA, eds. Insect Transgenesis: Methods and Applications. Boca Raton: CRC, 2000:271-288.
110. Xi Z, Khoo CC, Dobson SL. Interspecific transfer of Wolbachia into the mosquito disease vector Aedes albopictus. Proc Biol Sci 2006; 273:1317-1322.
111. Zabalou S, Riegler M, Theodorakopoulou M et al. Wolbachia-induced cytoplasmic incompatibility as a means for insect pest population control. Proc Natl Acad Sci USA 2004; 101:15042-15045.
112. Foy BD, Myles KM, Pierro DJ et al. Development of a new Sindbis virus transducing system and its characterization in three Culicine mosquitoes and two Lepidopteran species. Insect Mol Biol 2004; 13:89-100.
113. Ando T, Fujiyuke T, Kawashima T et al. In vivo gene transfer into the honeybee using a nucleopolyhedrovirus vector. Biochem Biophys Res Commun 2007; 352:335-340.

114. Carlson JO, Suchman E, Buchatsky L. Densoviruses for control and genetic manipulation of mosquitoes. Adv Virus Res 2006; 68:361-392.
115. Jordan TV, Shike V, Boulo V et al. Pantropic retroviral vectors mediate somatic cell transformation and expression of foreign genes in dipteran insects. Insect Mol Biol 1998; 7:215-222.
116. Anonymous. Prospects for malaria control by genetic manipulation of its vectors (TDR/BCV/MAL-ENT/91.3). Geneva: World Health Organization, 1991.
117. James AA. Mosquito molecular genetics: The hands that feed bite back. Science 1992; 257:37-38.
118. Miller LH. The challenge of Malaria. Science 1992; 257:36-37.
119. Craig GB. Prospects for vector control through manipulation of populations. Bull World Health Organ 1963; 29:89-97.
120. Collins FH, Sakai RK, Vernick KD et al. Genetic selection of a Plasmodium-refractory strain of the malaria vector Anopheles gambiae. Science 1986; 234:607-610.
121. Anxolabehere D, Kai H, Nouaud D et al. The geographical distribution of P-M hybrid dysgenesis in Drosophila melanogaster. Genet Sel Evol 1984; 16:15-26.
122. Quesneville H, Anxolabehere D. Dynamics of transposable elements in metapopulations: A model of P element invasion in Drosophila. Theor Pop Biol 1998; 54:175-193.
123. O'Brochta DA, Handler AM. Mobility of P elements in drosophilids and non-drosophilids. Proc Natl Acad sci USA 1988; 85:6052-6056.
124. Kidwell MG. Horizontal transfer of P elements and other short inverted repeat transposons. Genetica 1992; 86:275-286.
125. Anxolabehere D, Kidwell MG, Periquet G. Molecular characteristics of diverse populations are consistent with the hypothesis of a recent invasion of Drosophila melanogaster by mobile P elements. Mol Biol Evol 1988; 5:252-269.
126. Kidwell MG, Frydryk T, Novy JB. The hybrid dysgenesis potential of Drosophila melanogaster strains of diverse temporal and geographical natural origins. Drosophila Inform. Serv 1983; 51:97-100.
127. Sanchez-Gracia A, Maside X, Charlesworth B. High rate of horizontal transfer of transposable elements in Drosophila. Trends in Genetics 2005; 21:200-2003.
128. Guimond N, Bideshi DK, Pinkerton AC et al. Patterns of Hermes Transposition in Drosophila melanogaster. Molec Gen Genet 2003; 268:779-790.
129. Wilson R, Orsetti J, Klocko AD et al. Post-integration behavior of a Mos1 gene vector in Aedes aegypti. Insect Biochem Mol Biol 2003; 33:853-863.
130. O'Brochta DA, Sethuraman N, Wilson R et al. Gene vector and transposable element behavior in mosquitoes. J Exp Biol 2003; 206:3823-3834.
131. Preston CR, Engels WR. Spread of P transposable elements in inbred lines of Drosophila melanogaster. Prog Nuc Acid Res Molec Biol 1989; 36:71-85.
132. Galindo MI, Ladeveze V, Lemeunier F et al. Spread of the autonomous transposable element hobo in the genome of Drosophila melanogaster. Mol Biol Evol 1995; 12:723-734.
133. Ladeveze V, Galindo I, Chaminade N et al. Transmission pattern of hobo transposable element in transgenic lines of Drosophila melanogaster. Genet Res Camb 1998; 71:97-107.
134. Kimura K, Kidwell MG. Differences in P element population dynamics between the sibling species Drosophila melanogaster and Drosophila simulans. Genet Res Camb 1994; 63:27-38.
135. Biemont C, Tsitrone A, Vieira C et al. Transposable element distribution in Drosophila. Genetics 1997; 147:1997-1999.
136. Deceliere G, Charles S, Biemont C. The dynamics of transposable elements in structured populations. Genetics 2005; 169:467-474.
137. Curtis CF. Possible use of translocations to fix desirable gene in insect pest populations. Nature 1968; 218:368-369.
138. Rupprecht CE, Hanlon CA, Slate D. Oral vaccination of wildlife against rabies: Opportunities and challenges in prevention and control. Developments in Biologicals 2004; 119:173-184.
139. Rupprecht CE, Hanlon CA, Slate D. Control and prevention of rabies in animals: Paradigm shifts. Dev Biol (Basel) 2006; 125:103-111.
140. Tsao JI, Wootton JT, Bunikis J et al. An ecological approach to preventing human infection: Vaccinating wild mouse reservoirs intervenes in the Lyme disease cycle. Proc Natl Acad Sci USA 2004; 101:18159-18164.
141. Barfield JP, Nieschlag E, Cooper TG. Fertility control in wildlife: Humans as a model. Contraception 2006; 73:6-22.
142. Dawe AL, Nuss DL. Hypoviruses and chestnut blight: Exploiting viruses to understand and modulate fungal pathogenesis. Ann Rev Genet 2001; 35:1-29.
143. Milgroom MG, Cortesi P. Biological control of chestnut blight with hypovirulence: A critical analysis. Annu Rev Phytopathol 2004; 42:311-338.

Chapter 2

Alphavirus Transducing Systems

Brian D. Foy and Ken E. Olson*

Abstract

Alphavirus transducing systems (ATSs) are important tools for expressing genes of interest (GOI) in mosquitoes and nonvector insects. ATSs are derived from infectious cDNA clones of mosquito-borne RNA viruses (family *Togaviridae*). The most common ATSs in use are derived from Sindbis viruses; however, ATSs have been derived from other alphaviruses as well. ATSs generate viruses with genomes that contain GOI's that can be expressed from additional viral subgenomic promoters. ATSs in which an exogenous gene sequence is positioned 5' to the viral structural genes is used for stable protein expression in insects. ATSs in which a gene sequence is positioned 3' to the structural genes is used to trigger RNAi and silence expression of that gene in the insect. ATSs are proving to be invaluable tools for understanding vector-pathogen interactions, vector competence, and other components of vector-pathogen amplification and maintenance cycles in nature. These virus-based expression systems also facilitate the researcher's ability to decide which gene-based disease control strategies merit a further investment in time and resources in transgenic mosquitoes.

Introduction

Alphavirus transducing systems (ATSs) are infectious recombinant viruses derived from mosquito-borne RNA viruses of the genus *Alphavirus* (family: *Togaviridae*). ATSs are used for rapid and efficient gene transcription and expression in a variety of mosquitoes and nonvector insect species. Originally developed as nonheritable expression systems to complement transposon-based DNA transformation systems in mosquitoes, ATSs have been developed from Sindbis (SIN), O'nyong-nyong (ONN), and Chikungunya (CHIK) viruses, but they can readily be developed for other alphaviruses as well.[1-5] Different ATSs have been designed to infect specific taxonomic groups of mosquitoes. The ONNV, CHIKV, SINV ATSs infect and express genes of interest (GOI) in *Anopheles gambiae*, *Aedes aegypti*, and *Aedes* and *Culex* species respectively. ATSs have been engineered to express reporter genes, anti-sporozoite/single chain antibodies, anti-mosquito neurotoxins, and anti-bacterial peptides in mosquitoes.[3,6-9] Expression of a GOI is only limited to the tropism of the alphavirus to specific insect cells. Following parenteral injection, virus can spread throughout hemocoelic tissues, including the fat body, neural tissues, muscle cells, Malphigian tubules, and salivary glands.[1,3,6,10-14] When ATSs are mixed into bloodmeals and fed to mosquitoes, efficient infections of the midgut and foregut epithelium can be also be achieved.[2-5] The power of ATSs as a tool in vector and insect research is enhanced by their ability to also stimulate RNA interference using RNA fragments

*Corresponding Author: Ken E. Olson—Arthropod-Borne and Infectious Diseases Laboratory, Department of Microbiology, Immunology and Pathology, 3185 Rampart Road, Mail Delivery 1692, Foothills Research Campus, Colorado State University, Fort Collins, Colorado 80523, USA. Email: kolson@colostate.edu

Transgenesis and the Management of Vector-Borne Disease, edited by Serap Aksoy.

cloned into the ATS expression cassette. They have been used to transcribe pieces of RNA from genetically unrelated arboviruses, such as flaviviruses (dengue or yellow fever viruses), to inhibit flavivirus replication through RNAi or a similar mechanism.[15-20] ATSs are also powerful tools for posttranscriptionally silencing endogenous genes and have been used successfully to silence reporter (luciferase) and endogenous genes (phenoloxidase; GATA repressor) following injection of ATSs into adult mosquitoes, and Broad complex transcriptional activator gene in nonvector insects.[13,14,21-23] Lastly, ATSs are very important tools in identifying virus determinants of mosquito infection and in understanding alphavirus-vector interactions. In this review, we will briefly describe the molecular biology of these viruses, discuss how ATSs are generated, examine their role in understanding alphavirus-vector interactions, and describe which ATSs are most useful for protein expression and which are best for gene silencing in a given species. Lastly, we will address biosafety considerations for working with these recombinant viruses.

Alphaviruses

The Alphavirus genus consists of approximately 30 distinct mosquito-borne viruses found both in the New World [Eastern, Venezuelan, and Western equine encephalitis viruses (EEEV, VEEV, and WEEV)] and the Old World (e.g., Sindbis, Semliki Forest, O'nyong-nyong, Chikungunya viruses). For excellent reviews of alphavirus biology see Strauss and Strauss,[24] and Schlesinger and Schlesinger.[25] Alphaviruses are enveloped, RNA viruses that are approximately 70 nm in diameter and replicate exclusively in the cytoplasm of infected cells. Alphavirus genomes are positive-sense, single-stranded, nonsegmented RNA of about 11.7 kb. The 5' end is capped with 7-methylguanosine and the 3' end contains a poly(A) tail. The 5' two-thirds of the alphavirus genome is translated to form polyproteins (P123 or P1234) that contain the viral nonstructural proteins (nsP1-nsP4). Posttranslational processing of the polyproteins form different replicase complexes that synthesize positive or negative RNAs for genome replication.[24,26] Replication occurs at intracellular membranes in infected cells that are induced by the viral nsP1 protein and requires the formation of double stranded RNA (dsRNA) replicative intermediates.[27] The replication complexes include nsP4, the virus-encoded RNA-dependent RNA polymerase.

The subgenomic (26S) mRNA, colinear with the 3' one-third of the genome, is translated into a structural polyprotein. Capsid (C) protein subunits are autocatalytically cleaved from the structural polyprotein cotranslationally; the remaining polyprotein molecule is translocated into the cell's endoplasmic reticulum (ER). Cleavage and modification of the envelope glycoproteins (E1 and PE2) is a multi-step process that takes place during vesicular transport through the ER/Golgi complex. E1 association with PE2 through disulfide bridges is required for oligomer formation and the export of E1 from the endoplasmic reticulum.[28,29] Glycoprotein spikes (triplets of E2/E1 dimers) and nucleocapsids assemble at the plasma membrane to produce progeny virions. Two smaller, unpackaged polypeptides (E3 and 6K) are produced during SIN virus infections as cleavage products during glycoprotein processing. A noncoding region (NCR) at the 3' end of genomic and subgenomic RNAs, contiguous with a poly (A) tail, contains characteristic repeated sequence elements[24] and plays a role in host specificity, possibly through interactions with cellular proteins.[30]

The development of infectious cDNA clone technology has been important for analyzing the genomes of many postive-sense RNA viruses.[31-36] Infectious cDNA clones of alphaviruses are generated by reverse transcription of the viral RNA genome and the cDNA copies are assembled and inserted into a bacterial plasmid containing an origin of replication and an antibiotic resistance marker.[37] The infectious clone essentially becomes a large plasmid that can be manipulated with modern cloning techniques much like any other plasmid. An RNA polymerase promoter sequence, such as that recognized by T7 or SP6 polymerases, is inserted into the infectious clone plasmid immediately upstream of the first nucleotide in the virus genome and the first nucleotide is modified to allow for insertion of a capped nucleotide analogue. Furthermore, a unique restriction site is added to the infectious clone plasmid immediately following the poly dT tail sequence. Once modifications to the virus genome

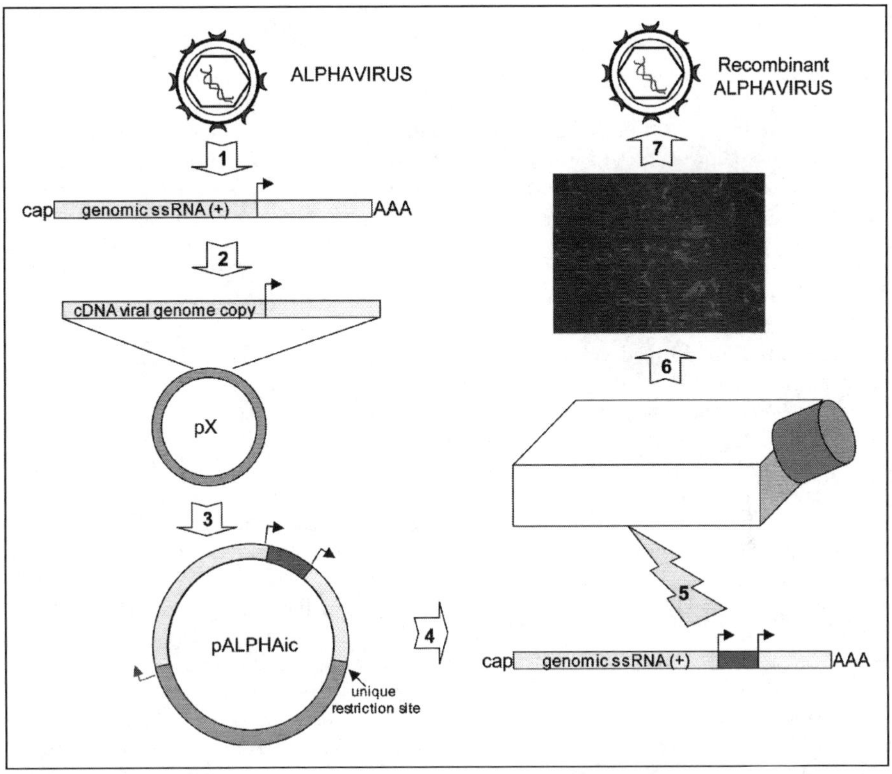

Figure 1. Generation of an Alphavirus Tranducing System. 1) Viral RNA is harvested from an Alphavirus (ALPHAVIRUS). The RNA contains a 5' capped nucleotide, a poly A tail, and a subgenomic promotor for transcribing the 28S RNA (black arrow). 2) The ssRNA(+) viral genome is reverse transcribed with specific primers and reverse transcriptase. The resulting cDNA fragments are amplified with PCR and ligated at novel restriction sites to reconstruct a cDNA clone of the entire viral genome (see ref. 37 for details). This dsDNA virus copy is then ligated into a plasmid (pX) that contains an origin of replication and an antibiotic selection marker. 3) The resulting plasmid is modified to make an infectious clone (pALPHAic): (a) an RNA polymerase promotor sequence (usually for SP6 polymerase) is inserted upstream of the virus genome (red arrow), (b) a unique restriction site is added following the virus genome sequence, and (c) the subgenomic promotor is duplicated with a downstream multiple cloning site and this cassette is reinserted into the virus genome. Here we depict a 5′dsATS construction and the fluorescent green segment denotes that the green fluorescent protein gene is inserted into the double subgenomic multiple cloning site. 4) Infectious viral RNA is generated from the infectious clone by an in vitro transcription reaction. The plasmid is linearized at the unique restriction site, and incubated with the RNA polymerase, rNTPs, and a capped rNTP analogue corresponding to the first nucleotide in the viral genome. 5) The infectious ssRNA (+) is electroporated (or transfected with liposomes) into cultured mammalian or insect cells. 6) RNA that enters the cells will replicate in the same fashion as normal virus. Additionally, GFP RNA will be transcribed from the replicating virus genome and so will be expressed only in virus-infected cells. Infectious virus will be released into the cell culture supernatant from transfected cells within 12hrs. and go on to infect the rest of the cell culture monolayer. 7) Infectious recombinant virus (Recombinant ALPHAVIRUS) is then harvested from the cell and the supernatant for later studies.

are incorporated into the infectious clone, new recombinant virus is generated by linearizing the plasmid at the unique restriction site, and then performing a transcription reaction using the plasmid as a template, the proper RNA polymerase and free ribonucleotides. The viral RNA that is generated is transfected into cultured cells and resulting viral genome replication eventually produces recombinant infectious virions from the transfected cells (Fig. 1). This technology has allowed researchers to genetically manipulate RNA genomes and identify viral deteminants of host range, virulence, replication, disassembly, assembly and packaging. Full-length infectious clones have been generated for SIN, CHICK, ONN, Semliki Forest (SF), Ross River (RR), Sagiyama (SAG), and VEE viruses.[4,5,37-43]

ATS Construction

Some of the aforentioned alphavirus infectious clones have also been made into ATSs. An ATS is made by duplicating the virus subgenomic promoter sequence from the infectious clone and reinserting it into the virus genome as a cassette containing a multiple cloning site for GOI insertion.[3] When this cassette is inserted in the 3'UTR of the virus genome, it is commonly referred to as a 3' double subgenomic ATS (3'dsATS). Alternatively, the cassette can be inserted immediately upstream of the original subgenomic promoter and the resulting ATS is commonly referred to as a 5' double subgenomic ATS (5'dsATS). There are genome stability differences between these two ATSs, the latter configuration has been shown to be the most stable after multiple virus passages,[2,9] but they may also differ in their ability to induce RNAi (see Fig. 5). The subgenomic promoter of alphaviruses has been well-characterized[44] and occurs in a stretch of approximately 112 nucleotides at the end of the nsP4 gene and prior to the capsid start codon. The alphavirus genome has some plasticity in its length as demonstrated by the variable length of the nonconserved C-terminus of the nsP3 gene among different alphavirus species.[45] This plasticity allows the virion to package viral RNA that contains up to 2 kb of additional sequence. Therefore, manipulation of the virus genome, as occurs with an ATS, only minimally compromises alphavirus replication if the GOI is less than 1kb in size. For example, we have observed that a 5'dsSINV ATS expressing enhanced green fluorescent protein (~800bp) can infect all the same mosquito tissues as the wild type parent virus and can pass between mosquito and mammalian cells several times without noticeably losing GFP expression.[3] However, the speed at which this ATS disseminates through mosquitoes is slower than the wild type parent virus, probably due to a compromise in virus packaging efficiency.

Currently Used ATSs

Sindbis Virus ATSs

TE/3'2J ATS and Derivatives

One of the first ATSs, TE/3'2J (3' designates that this ATS expressed the GOI from a duplicated subgenomic promotor cassette in the 3'UTR), was constructed by Hahn et al in 1992.[38] The origin of the glycoproteins of this ATS is a mouse neurovirulent SINV strain.[46] Researchers at the Arthropod Borne Infectious Diseases Lab first used this ATS to express chloramphenicol acetyltransferase (CAT) in mosquitoes and mosquito cells. TE/3'2J was found to readily replicate in many insect species and nonmidgut tissues when infected by the parenteral route. For instance, 4 days after intrathoracic inoculation of *Aedes triseriatus* with 4.0 $\log_{10}$ $TCID_{50}$ of TE/3'2J/CAT virus, we readily detected virus titers >6.0 $\log_{10}$ $TCID_{50}$ and approximately 7.2×10^{11} CAT polypeptides.[1,47] Immunofluorescence and CAT activity assays were used to localize expression in infected mosquitoes and demonstrated that CAT was efficiently expressed in neural, midgut, ovarian, and salivary gland tissues. The TE/3'2J ATS has since been used to express protein products in *Ae. aegypti, Ae. triseriatus, Cx. pipiens, An. gambiae* and nonvector insects.[7,8,10,15,48,49] These initial studies clearly established TE/3'2J-based ATSs as important tools for expressing heterologous genes in mosquito cells and in adult mosquitoes.

A recombinant 5'dsATS clone of TE/3'2J has also been constructed in which the duplicated subgenomic promoter cassette was placed upstream of the SINV structural genes.[50,51] This design has been found to be much more resistant to virus recombination and therefore replicating viruses in mosquito tissues will continue to express the GOI (such as GFP) as they disseminate in vivo or are passed through cultured cells.[2,3]

MRE16 ATS and Derivatives

The TE ATSs replicate well in mosquito hemocoelic tissues after intrathoracic injection, but these viruses have lost the ability to efficiently infect mosquito midgut cells because the envelope glycoproteins are derived from a SIN virus that was selected to be neurovirulent in mice. Consequently, they are not useful in studies that examine alphavirus infection following an oral bloodmeal. This problem was alleviated by ATSs constructed based on genomic sequence of the MRE16 strain of SIN virus. Chimeric ATSs based on TE/5'2J nonstructural and MRE16 structural coding sequences have been developed that infect the midgut epithelium of *Aedes aegypti* when fed per os with a bloodmeal.[2,12,52] These chimeric ATSs have been used to show that particular residues in the the E2 glycoprotein of MRE16 virus are critical for establishing midgut epithelium infections.[37,53] A full length infectious clone of the MRE16 SIN virus has also been made and characterized[37] and a 5'dsATS and a 3'dsATS were developed from this infectious clone.[3] The 5'dsMRE16ic-GFP ATS was characterized by its infection and dissemination patterns in three culicine mosquitoes and in two lepidopterans through visualizing GFP expression. The data showed that foregut/midgut junction infections can be observed in some *Aedes aegypti* mosquitoes within 1-2 days after a bloodmeal and that initial midgut infection is a dose dependent process. Additionally, the data showed that midgut tissues of certain species, such as *Culex tritaeniorhynchus*, are initially infected at high rates but that these infections are quickly resolved, ultimately leading to few mosquitoes having disseminated infections.

ONNV-Based ATSs for Expression of Proteins in Anopheline Mosquitoes

Arthropod-borne alphaviruses are transmitted almost exclusively by culicine mosquitoes; however, the O'nyong-nyong alphavirus (ONNV) has the unusual characteristic of being transmitted by anopheline mosquitoes. ONNV also has an unusual epidemiology in that very large outbreaks have occurred in between decades of inactivity.[54] Therefore, the ONNV ATS is a valuable tool for exploring this unusual system and for expressing GOIs in these malaria-transmitting mosquito species. The ONNV ATS was made from a full-length infectious cDNA clone of the SG650 strain, isolated in a Ugandan epidemic in 1996.[55] There are now several ATSs using the SG650 strain as the backbone alphavirus genome, including two different 5'dsATSs and a 3'dsATS.[4,6] We have inserted the GFP gene into the polylinker of this ATS and used the construct to characterize infection and dissemination patterns of ONNV

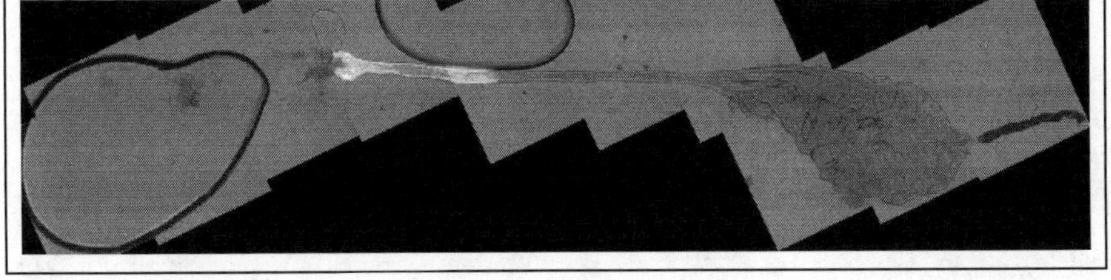

Figure 2. *Anopheles gambiae* midgut infected with 5'dsONNic- GFP. The mosquito was fed a bloodmeal with a titer of $1x10^5$ pfu/ml and its alimentary canal was dissected 7 days later. Image was captured under UV and low intensity visible light. GFP expression demonstrates that the virus is infecting anterior midgut cells and cell of the cardia.

within *An. gambiae* mosquitoes. These experiments have allowed us to identify atypical sites of initial infection and dissemination patterns in this mosquito species not frequently observed in comparable culicine infections (Fig. 2). Furthermore, we used the 5'dsONNV-GFP construct to first prove that RNAi acts as a natural antagonist to alphavirus replication in mosquitoes.[6] The utility of ONNV ATSs for studies in anopheline mosquitoes includes the potential for identifying vector infection determinants and to serve as tools for anti-malaria studies.

CHIK-Based ATS for Expression in *Aedes aegypti*

ATSs based on the Chickungunya alphavirus (CHIKV) have now been constructed. These constructs infected *Aedes aegypti* at high levels and expressed GFP in the midgut, salivary glands, and nervous tissue of these mosquitoes following oral infection.[5] *Aedes aegypti* dissemination rates of this new viral vector exceed that of previous SINV systems, thus expanding the repertoire and potential for gene expression studies in this important vector species.

Alphavirus/Mosquito Interactions

Wild type alphavirus interactions with mosquitoes have been the subject of many studies.[53,56-64] The sequence of events beginning with the mosquito's intake of an infected blood meal and ending with the mosquito's ability to transmit a pathogen is termed the extrinsic incubation period (EIP).[65] The EIP of alphaviruses usually ranges between 3 and 10 days,[66] and this factor varies with both temperature[67] and specific alphavirus/mosquito combinations. During the EIP, the virus interacts with receptors on mosquito midgut epithelial cells (or a subset of these cells), replicates within the midgut cells, disseminates through these cells, enters the hemocoel, and disseminates to other secondary target organs. Infection of salivary glands permits horizontal transmission of the virus to the next susceptible vertebrate host, when the female feeds again.[68] Vector competence refers to the intrinsic permissiveness of a vector to infection, replication, and transmission of a virus.[65,68] At the macro level, vector competence for an alphavirus can be divided into commonly recognized barriers that ultimately influence successful transmission; these are midgut infection barriers, midgut escape barriers, and salivary gland infection barriers. These barriers can be subdivided into different processes, known and unknown, that are influenced by the molecular biology of both the alphavirus and the mosquito. While extensive work using wild type alphaviruses has defined the existence of these barriers, ATSs are increasingly important tools for understanding the molecular intricacies of vector competence in alphavirus/mosquito interactions.

Midgut Infection

Midgut infection in compatible alphavirus-mosquito combinations is a dose dependent process, where usually a threshold titer of $\geq 1 \times 10^6$ pfu/ml is needed to achieve infection rates above 50% of bloodfed mosquitoes.[3] There is evidence that artificial bloodmeals (commonly used to feed ATSs to mosquitoes) result in artificial virus associations with the mosquito midgut epithelium that can affect infection rates and patterns,[59] but feeding on viremic animals is costly and requires rigorous animal approval protocols. There is also evidence that ATS virions produced from mammalian cell culture are slightly more infectious to midgut epithial cells than virus produced from insect cells (Pierro, Olson, unpublished data), suggesting glycosylation or codon usage differences. However higher virus titers are usually acquired from virus grown in insect cells and we have always noticed that a higher dose can overcome this reduced infectivity. An ATS expressing GFP will most often infect posterior midgut epithelial cells within 8-24hrs after ingestion, and the infected cells will express GFP within this time frame. No evidence has yet been found of a cell type in the posterior midgut that is preferentially infected by alphaviruses. In *Aedes aegypti*, individual cells are infected and then surrounding cells are subsequently infected as the ATS spreads laterally into an infection focus (Fig. 3). In the SINV/*Culex tritaeniorhynchus* system, many cells of the posterior midgut are initially infected but most seem to clear the infection within 1-3 days by an as yet unknown mechanism. Less often, virus is observed infecting the distal-most portion of the posterior midgut and spreading into

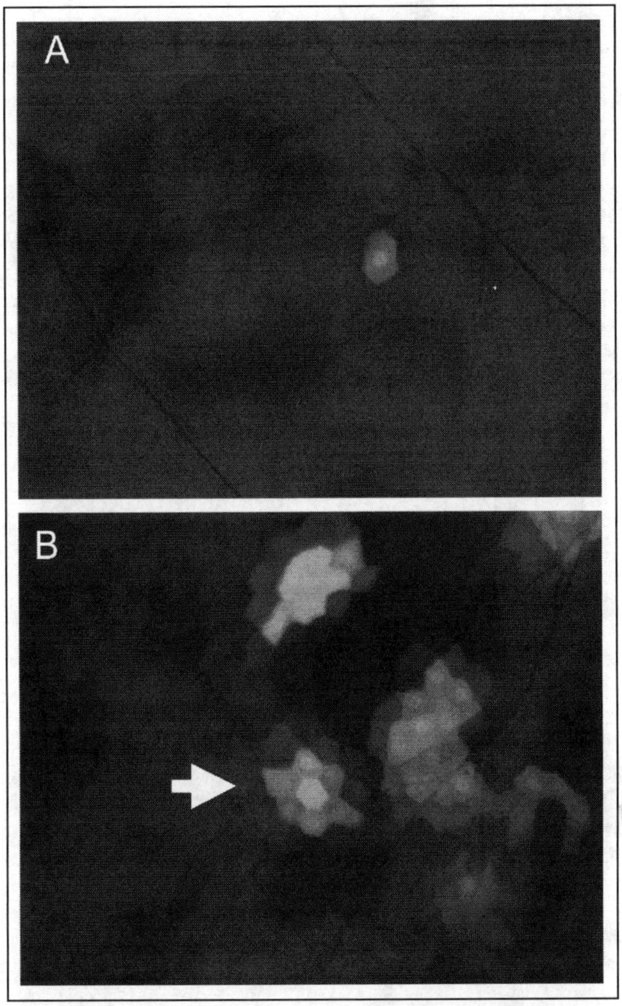

Figure 3. *Aedes aegypti* midgut cells infected with 5'dsMRE16ic-GFP. The mosquito was fed a bloodmeal with a titer of 1×10^7 pfu/ml and its midgut was dissected 24hrs. later. Images were captured under UV and low intensity visible light. Undigested blood in the midgut colors the images deep red, while infected cells are delineated by GFP expression. A) An individual infected midgut epithelial cell. Virus has not yet spread to surrounding cells. B) Infection foci of midgut epithelial cells. The focus highlighted by the arrow demonstrated how virus spreads; the brightest fluorescence occurs in the earliest infected cells, likely because the virus has been replicating longer in that cell and so it has accumulated more GFP. The center most cell of this focus was initially infected from the midgut lumen, the six cells directly adjacent to this cell were subsequently infected, and newly infected cells on the outer edge of the focus are just beginning to express GFP. A color version of this figure is available online at www.eurekah.com.

the base of one or more Malphigian tubules. ONNV in *Anopheles gambiae* and SINV (MRE16 strain) in *Aedes aegypti* will also often be observed infecting cells of the anterior midgut, just posterior to the intussuscepted foregut (Fig. 2). It is not clear if these are primary infections that occur as the bloodmeal passes through the intussuscepted foregut or if they are secondary infections that occur from replicating virus passing into the midgut lumen and moving into the anterior midgut to infect these cells.

Midgut Escape

The process of viral escape from the alimentary canal is not fully understood, despite the fact that this is often the most important barrier to successful virus transmission, but ATS are helping to illuminate this important process. Following intrathoracic inoculation of SINV ATSs, the gut-associated musculature and trachioles are infected at an early stage in the infection but viral antigen is not generally detected in the epithelial cell layer that lines the midgut lumen.[11,58] Thus, dissemination of alphaviruses in competent mosquitoes following per os infection seems to be mostly a unidirectional event, whereby virus can leave the alimentary canal but cannot reenter. It has been assumed that arboviral dissemination occurs out of the alimentary canal through the posterior midgut because this is the final destination of ingested blood[69] and because posterior midgut cells are often infected.[57] Indeed, confocal imaging of SINV in *Ae. aegypti* posterior midgut cells clearly shows that viral antigen accumulates at the basal labyrinth (hemocoel side) of these cells.[37,53] However, it is not clear how viruses pass through the basal lamina surrounding the mosquito midgut.[70] The pore size of *Ae. dorsalis* basal lamina was measured to be only 10 nm in diameter,[71] and particles larger than 8 nm could not permeate the basal lamina of *Cx. tarsalis* midguts from the hemocoel side.[72] While the pore-size of the basal lamina nearly doubles during the immediate stretching that occurs from ingestion of a bloodmeal,[71] alphavirus dissemination does not occur within the short time frame that the basal lamina is stretched. Furthermore, the diameter of an alphavirus is 69 nm[73] and other arboviruses range from 50-60 nm in diameter, so it is unclear how they can pass through this seemingly impenetrable barrier.

It has been demonstrated that Rift Valley fever virus (RVFV; Bunyaviridae) can disseminate from the alimentary canal of *Cx. pipiens* through the cardial junction.[74-76] The observation of ATSs infecting the proximal neck and cardial epithelium of the anterior midgut[3] opens the possibility that this is a common route of midgut escape for arboviruses in general. Indeed, the very fast speed of dissemination that has been observed in some alphavirus (3 days) does not fit well with the slower posterior midgut epithelium infection and escape model. The salivary glands lay in the thorax immediately next to the foregut/midgut junction, thus mosquitoes that have been recorded as infectious in 3-4 days might have had virus disseminating through this route and quickly into the salivary glands.

Dissemination and Salivary Gland Infection

ATSs disseminate throughout many tissues of the hemocoel, but they usually have the greatest tropism for nervous tissue, salivary glands, the fat body, and the circular musculature that surrounds the alimentary canal. They will randomly infect one or more legs and more rarely, they infect Malphigian tubules. Almost never are they seen infecting ovarian tissues or follicles, keeping with the observation that alphaviruses generally are not transoviarally transmitted. SINV ATSs will disseminate in hemocoelic tissues of mosquitoes that they cannot infect orally (such as *Culex pipiens* or *Anopheles gambiae*) if they are inoculated parenterally. This is also true with many other insect species and arthropods in general.[3,23,49] Salivary gland infection barriers are normally not common with alphaviruses. One exception seems to be *An. gambiae* and the ONNV ATS. If this ATS is injected into the *An. gambiae* hemocoel, a strong barrier is observed despite the fact that fat body surrounding the salivary glands can be infected and expressing GFP. However, when virus is fed, salivary glands infection rates correlates with the low rate of anterior midgut infections (Tompkins, Olson, Foy; unpublished data). This phenomenon suggests that only virus infecting the anterior midgut is infecting the salivary glands.

Alphavirus Pathology to Mosquito Cells and Mosquitoes

Pathology has been observed in several mosquito species resulting from arbovirus infections.[65,77,78] Studies have demonstrated that both EEEV and WEEV can cause cytopathic effects in midgut cells of natural vectors.[62,66,77] And decreases in the number of virions of

EEEV and SLEV have been observed over time in mosquito midguts.[79,80] We have observed a similar phenomenon whereby SINV (MRE16 strain) midgut infections are resolved quickly in most *Culex tritaeniorhynchus* mosquitoes and TR339 strain infection is similarly resolved in *Aedes aegypti* mosquitoes (Foy, Campbell, Keene, Olson; unpublished data). Furthermore, we are discovering that RNAi mechanisms play a role in this mosquito defense. And this is likely a broad insect phenomenon; recent evidence from Drosophila mutants shows that RNA viruses efficiently kill infected flies that have key RNAi defense proteins disabled.[81,82]

Gene Expression with ATSs

Any understanding of vector-pathogen interactions must include in-depth knowledge of the molecular biology of the vector. Only recently have the molecular tools been available to study vector species. Four mosquito transformation systems, based on Mos1, Hermes, Minos, and piggyBac transposable elements, have been shown to integrate exogenous genes into the mosquito genome in a predictable manner.[83-87] This is an important accomplishment that should greatly facilitate gene manipulation in mosquitoes. However, transformation remains an intensely laborious and time-consuming procedure for characterization, mutagenesis, and expression of genes in mosquitoes. The propagation and maintenance of multiple transformed mosquito lines is difficult in even the most spacious, well-equipped insectaries. Additionally, the complex life cycle of some medically important mosquitoes make routine transgenesis difficult, whereas other transient expression systems may more easily and rapidly answer biological questions posed by researchers. ATSs can transiently express heterologous RNAs and proteins in mosquito cell culture and in mosquito larvae and adults. ATSs benefit vector biology for additional reasons by allowing rapid, long-term gene expression in mosquitoes that can facilitate understanding basic biology and vector-pathogen interactions and give researchers the ability to rapidly test the merits a given gene-based anti-pathogen strategy prior to use in transformed mosquitoes. ATSs by no means lessen the importance of arthropod vector transformation, rather they complement transformation studies by allowing researchers an opportunity to rationally decide what gene products or effector molecules merit further analysis in transgenic mosquitoes.

Scorpion toxin was one of the first effector molecules expressed in mosquito tissues via the TE/3'2J ATS following intrathoracic injection.[8] The toxin was efficiently expressed in vivo and killed all injected mosquitoes, but had no effect on ATS-challenged houseflies and ticks. TE/3'2J was also used to express an anti-circumsporozoite antibody fragment in *Ae. aegypti* that successfully bound to circulating *Plasmodium gallinaceum* sporozoites in the same mosquitoes and inhibited their entry into salivary glands.[7] And TE/3'2J was used to abnormally express ultrabithorax protein in butterfly pupae and beetle embryos.[49] An orally infectious ATS expressing *Ae. aegypti* defensin was fed to *Ae. aegypti* larvae and the defensin protein was continually expressed as the larvae molted into adults.[9]

While gene expression has proven fruitful in mosquitoes and nonvector insects, perhaps even more promising is the ability of ATSs to transcribe noncoding RNAs that serve to induce RNAi within mosquito and other insect cells.

ATSs for Induction of RNA Interference

DsRNA has been shown to trigger sequence-specific RNAi in nematodes, trypanosomes, fruit flies and planaria.[88-91] RNAi acts posttranscriptionally and silences mRNAs of high sequence identity with the dsRNA trigger. In Drosophila and mosquitoes, dsRNAs of >200 nt are efficiently digested into small, 21-22 nt dsRNAs by a RNase III protein termed Dicer-2. These small dsRNAs are thought to unwind, associate with the RNAi silencing complex (RISC) and act as guide sequences to target and destroy mRNAs of similar sequence. Excellent reviews of the RNAi pathway in Drosophila and in mosquitoes are available.[19,92]

Synthetic dsRNA has been used to silence expression of endogenous genes in many vector and nonvector insects, but the use of synthesized dsRNA to target specific genes in adult insects has had mixed results.[23,93-95] Despite the promise of RNAi as a tool for functional genomics, injection of dsRNA interferes with gene expression over a short time frame which may not be

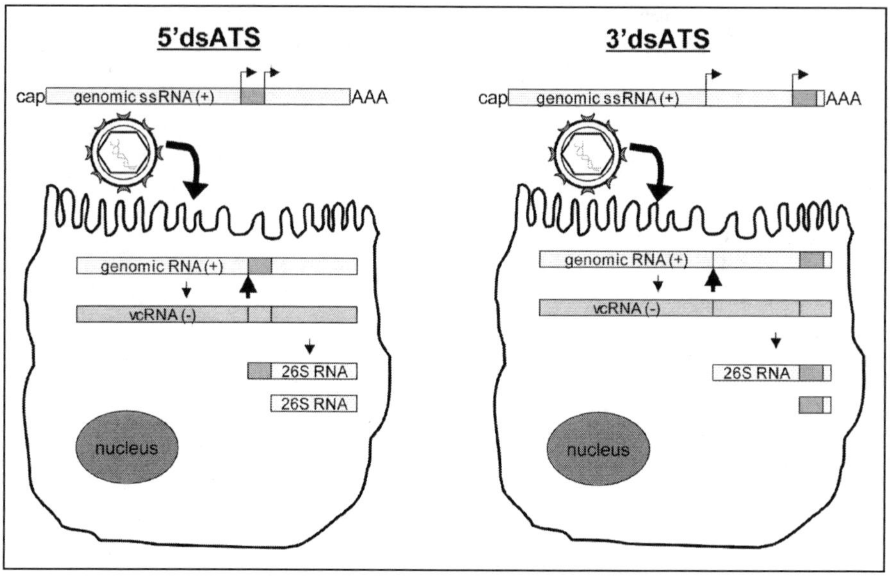

Figure 4. Hypothetical ATS function in silencing endogenous mRNAs. A) An effector sequence (pink) complementary to an endogenous mRNA in the cell is transcribed from the duplicated subgenomic promotor in either a 5'dsATS or a 3'dsATS. When the alphavirus genome replicates, the viral complementary RNA [vcRNA(-), orange] is transiently produced and serves as a template to form more genomic RNA (+) as well as subgenomic RNA. Both a 5'dsATS and a 3'dsATS yield two subgenomic RNAs, but only the 3'dsATS yields two subgenomic RNAs with the effector sequence. A color version of this figure is available online at www.eurekah.com.

ideal for knocking down gene expression involved with virus-vector interactions. Moreover, this dsRNA must be injected into adult mosquitoes using pulled capillary needles, which compromises the fitness of mosquitoes that later must be used for experiments. Consequently, use of synthesized dsRNA to initiate RNAi and study gene function in the late stages of insect development can be limited. It is particularly problematic for development of disease models that rely on post-natal individuals and long extrinsic incubation periods for pathogen replication or development.

In mosquitoes, in addition to GOI expression studies, ATSs were first used to demonstrate RNA cross protection between recombinant SINV and dengue viruses.[15,16,47,96,97] These studies led to the initial descriptions of the RNAi pathway in mosquitoes. As an initial proof of concept that ATSs could silence endogenous genes, we constructed the TE/3'2J ATS to express an antisense RNA from the gene encoding luciferase to silence expression of luciferase in a transgenic line of *Ae. aegypti.*[21] Within 5 days post injection luciferase was significantly depleted and by 9 days less than 10% of the enzyme activity was detected. A number of groups have now used ATSs for inducing RNAi to targeted genes in adult mosquitoes.[13,14,18,98]

Several lines of evidence indicate that ATSs are excellent tools for selectively silencing genes for long periods of time in adult mosquitoes. First, plant RNA viruses have been used for a number of years to silence endogenous genes by a process known as virus-induced gene silencing (VIGS). Recombinant PVX containing a portion of the phytoene desaturase gene in sense or antisense orientation silenced the targeted gene in tobacco plants.[99] Second, ATSs express genes over days or weeks in the adult mosquito.[1,3,9,14] Third, ATSs may form dsRNAs by (a) generation of viral dsRNA replication intermediates, (b) expression of antisense RNAs that form dsRNAs after annealing to the target mRNA, and (c) expression of inverted repeat RNAs that form dsRNA after transcription. Current evidence suggests that dsRNA intermediates are

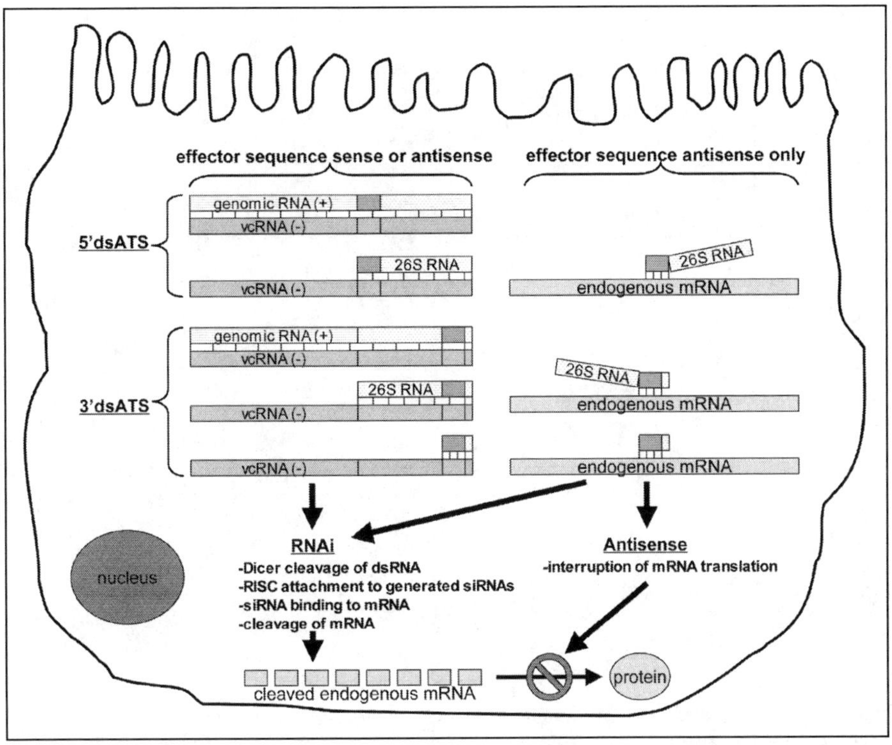

Figure 5. The effector sequence can either be a sense or an antisense match to the endogenous mRNA. Sense sequence can only work to silence the endogenous transcript through the RNAi pathway stimulated by transient dsRNA intermediates between the genomic vRNAs(+) and the vcRNA(-). On the other hand, antisense effector sequence can work either through this pathway or by binding to the endogenous mRNA. Hypothetically, a 3'dsATS has more possibilities to silence the endogenous mRNA because it transcribes two species of effector sequence that could act through either the RNAi or the antisense pathway, or both.

the primary stimulus for RNAi, but questions remain how an ATS specifically induces RNAi and what is the best ATS configuration to induce RNAi (see Figs. 4, 5). Tamang et al[14] have studied the ATS effector sequence length for RNAi induction capability against endogenous mosquito mRNAs and demonstrated that small effector sequences (<50bp) can induce RNAi but that longer sequences (>500bp) are the most effective.

Tissue tropism of synthetic dsRNAs is currently unresolved and can be problematic for some vector species. Some tissues and cells, such as those in the antennae or maxillary palps are generally refractory to RNA silencing by dsRNAs, and there are no current studies that systematically have examined the dsRNA/siRNA entry and silencing efficiency in specific mosquito tissues. ATSs that are inducing RNAi can be tracked in tissues and individual cells by either immunofluorescence assays or by expressing marker genes such as GFP. We have created a series of ATSs that express GFP but that also contain a secondary multiple cloning site immediately after the stop codon of the GFP coding sequence (Fig. 6). With these ATSs, one can insert a small noncoding RNA for RNA silencing into this secondary site and also track the cells in which the ATS replicates in vivo, thus marker gene expression is linked to RNAi.

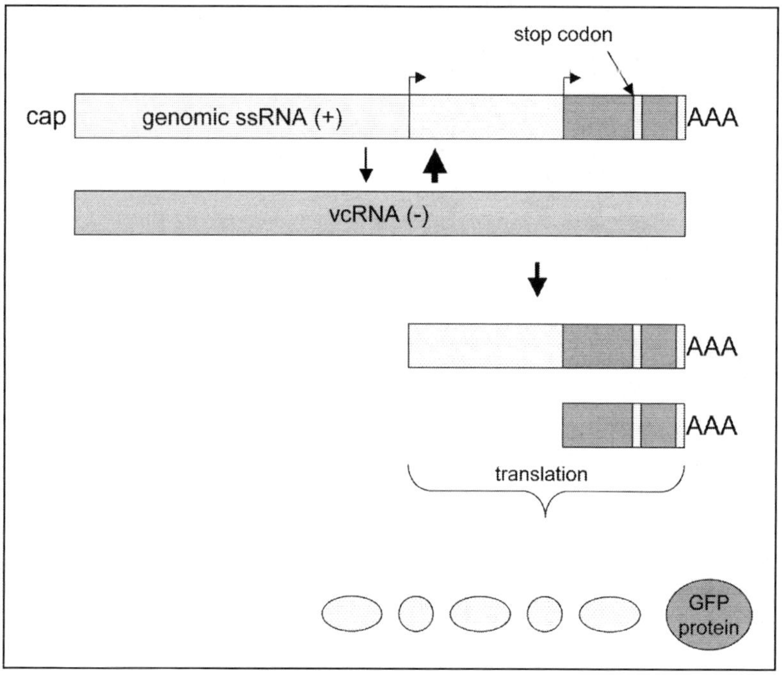

Figure 6. A new 3′dsATS for simultaneous GFP expression and gene silencing. The ATS was constructed with 3 unique restriction sites in its polylinker. The full length GFP gene (green), including stop codon, is inserted into the upstream restriction sites. The effector sequence for RNAi (pink) is cloned into the downstream restriction site. Upon replication, the subgenomic RNAs will contain the RNAi effector sequence, but translation of these subgenomic RNAs will only produce the 5 structural genes and GFP. Infected cells will express GFP and also have their target mRNA silenced via the effector sequence. A color version of this figure is available online at www.eurekah.com.

Biosafety Considerations for Using ATSs

All ATSs designed and constructed to date have the potential to infect researchers in the laboratory. This is problematic when using ATSs in adult arthropod vectors that readily seek blood meals from humans and increase the risk of laboratory infections. While the recombinant ATS most likely will be attenuated with respect to the wild-type parental virus, recombinant viruses may possess an etiology that is unknown or incompletely understood. The minimum biosafety level for arbovirus infected vectors is the biosafety level designation of the infectious agent. It is up to the researcher to know the appropriate biosafety level for a given alphavirus. For instance, Sindbis viruses are biosafety level 2 (BSL2) agents and CHIKV is a BSL3 agent. There are several excellent references to determine the appropriate biosafety conditions for the ATS used in vectors[100] (www.cdc.gov/od/ohs/biosfty/bmbl4/bmbl4s2.htm) (www.astmh.org/SIC/files/ACGv31.pdf). The researcher should discuss with, and seek approval of, the Institutional Biosafety Committee and Institutional Animal Care and Use Committee before conducting any experiments with ATSs.

Conclusion

Mosquito-borne pathogens are having increasingly negative impacts on human and animal health. Our present knowledge of arbovirus morphogenesis and molecular biology of virus-vector interactions is to a large extent derived from the pathogen's interactions with their vertebrate hosts. Although we have made considerable progress, much is still to be learned about the virus molecular determinants of infection in the invertebrate host and the invertebrate's immune response against these viruses. ATSs are essential molecular biology tools that can facilitate research in vector biology, especially research requiring expression of exogenous and endogenous genes in the vector and silencing of mRNA for functional analyses of genes.

References

1. Olson KE, Higgs S, Hahn CS et al. The expression of chloramphenicol acetyltransferase in Aedes albopictus (C6/36) cells and Aedes triseriatus mosquitoes using a double subgenomic recombinant Sindbis virus. Insect Biochem Mol Biol 1994; 24(1):39-48.
2. Pierro DJ, Myles KM, Foy BD et al. Development of an orally infectious Sindbis virus transducing system that efficiently disseminates and expresses green fluorescent protein in Aedes aegypti. Insect Mol Biol 2003; 12(2):107-16.
3. Foy BD, Myles KM, Pierro DJ et al. Development of a new Sindbis virus transducing system and its characterization in three Culicine mosquitoes and two Lepidopteran species. Insect Mol Biol 2004; 13(1):89-100.
4. Brault AC, Foy BD, Myles KM et al. Infection patterns of o'nyong nyong virus in the malaria-transmitting mosquito, Anopheles gambiae. Insect Mol Biol 2004; 13(6):625-35.
5. Vanlandingham DL, Tsetsarkin K, Hong C et al. Development and characterization of a double subgenomic chikungunya virus infectious clone to express heterologous genes in Aedes aegypti mosquitoes. Insect Biochem Mol Biol 2005; 35(10):1162-70.
6. Keene KM, Foy BD, Sanchez-Vargas I et al. RNA interference acts as a natural antiviral response to O'nyong-nyong virus (Alphavirus; Togaviridae) infection of Anopheles gambiae. Proc Natl Acad Sci USA 2004; 101(49):17240-5.
7. de Lara Capurro M, Coleman J, Beerntsen BT et al. Virus-expressed, recombinant single-chain antibody blocks sporozoite infection of salivary glands in Plasmodium gallinaceum-infected Aedes aegypti. Am J Trop Med Hyg 2000; 62(4):427-33.
8. Higgs S, Olson KE, Klimowski L et al. Mosquito sensitivity to a scorpion neurotoxin expressed using an infectious Sindbis virus vector. Insect Mol Biol 1995; 4(2):97-103.
9. Cheng LL, Bartholomay LC, Olson KE et al. Characterization of an endogenous gene expressed in Aedes aegypti using an orally infectious recombinant Sindbis virus. J Insect Sci 2001; 1(10).
10. Kamrud KI, Olson KE, Higgs S et al. Detection of expressed chloramphenicol acetyltransferase in the saliva of Culex pipiens mosquitoes. Insect Biochem Mol Biol 1997; 27(5):423-9.
11. Olson KE. Sindbis virus expression systems in mosquitoes: Background, methods, and applications. In: Handler AM, James AA, eds. Insect Transgenesis: Methods and Applications. Boca Raton, FL: CRC Press, 2000:161-90.
12. Olson KE, Myles KM, Seabaugh RC et al. Development of a Sindbis virus expression system that efficiently expresses green fluorescent protein in midguts of Aedes aegypti following per os infection. Insect Mol Biol 2000; 9(1):57-65.
13. Shiao SH, Higgs S, Adelman Z et al. Effect of prophenoloxidase expression knockout on the melanization of microfilariae in the mosquito Armigeres subalbatus. Insect Mol Biol 2001; 10(4):315-21.
14. Tamang D, Tseng SM, Huang CY et al. The use of a double subgenomic Sindbis virus expression system to study mosquito gene function: Effects of antisense nucleotide number and duration of viral infection on gene silencing efficiency. Insect Mol Biol 2004; 13(6):595-602.
15. Olson KE, Higgs S, Gaines PJ et al. Genetically engineered resistance to dengue-2 virus transmission in mosquitoes. Science 1996; 272(5263):884-6.
16. Higgs S, Rayner JO, Olson KE et al. Engineered resistance in Aedes aegypti to a west african and a south american strain of yellow fever virus. Am J Trop Med Hyg 1998; 58(5):663-70.
17. Blair CD, Adelman ZN, Olson KE. Molecular strategies for interrupting arthropod-borne virus transmission by mosquitoes. Clin Microbiol Rev 2000; 13(4):651-61.
18. Adelman ZN, Blair CD, Carlson JO et al. Sindbis virus-induced silencing of dengue viruses in mosquitoes. Insect Mol Biol 2001; 10(3):265-73.
19. Sanchez-Vargas I, Travanty EA, Keene KM et al. RNA interference, arthropod-borne viruses, and mosquitoes. Virus Res 2004; 102(1):65-74.
20. Franz AW, Sanchez-Vargas I, Adelman ZN et al. Engineering RNA interference-based resistance to dengue virus type 2 in genetically modified Aedes aegypti. Proc Natl Acad Sci USA 2006; 103(11):4198-203.
21. Johnson BW, Olson KE, Allen-Miura T et al. Inhibition of luciferase expression in transgenic Aedes aegypti mosquitoes by Sindbis virus expression of antisense luciferase RNA. Proc Natl Acad Sci USA 1999; 96(23):13399-403.

22. Raikhel AS, Kokoza VA, Zhu J et al. Molecular biology of mosquito vitellogenesis: From basic studies to genetic engineering of antipathogen immunity. Insect Biochem Mol Biol 2002; 32(10):1275-86.
23. Uhlirova M, Foy BD, Beaty BJ et al. Use of Sindbis virus-mediated RNA interference to demonstrate a conserved role of Broad-Complex in insect metamorphosis. Proc Natl Acad Sci USA 2003; 100(26):15607-12.
24. Strauss JH, Strauss EG. The alphaviruses: Gene expression, replication, and evolution. Microbiol Rev 1994; 58(3):491-562.
25. Schlesinger S, Schlesinger MJ. Togaviridae: The viruses and their replication. In: Fields BN, Knipe DM, Howley PM, eds. Fields Virology. 3rd ed. Philadelphia: Lippincott-Raven, 1996:825-41.
26. Sawicki DL, Sawicki SG. Alphavirus positive and negative strand RNA synthesis and the role of polyproteins in formation of viral replication complexes. Arch Virol Suppl 1994; 9:393-405.
27. Barton DJ, Sawicki SG, Sawicki DL. Solubilization and immunoprecipitation of alphavirus replication complexes. J Virol 1991; 65(3):1496-506.
28. Carleton M, Brown DT. Disulfide bridge-mediated folding of Sindbis virus glycoproteins. J Virol 1996; 70(8):5541-7.
29. Carleton M, Lee H, Mulvey M et al. Role of glycoprotein PE2 in formation and maturation of the Sindbis virus spike. J Virol 1997; 71(2):1558-66.
30. Kuhn RJ, Hong Z, Strauss JH. Mutagenesis of the 3' nontranslated region of Sindbis virus RNA. J Virol 1990; 64(4):1465-76.
31. Taniguchi T, Palmieri M, Weissmann C. QB DNA-containing hybrid plasmids giving rise to QB phage formation in the bacterial host. Nature 1978; 274(5668):223-8.
32. Racaniello VR, Baltimore D. Cloned poliovirus complementary DNA is infectious in mammalian cells. Science 1981; 214(4523):916-9.
33. Ahlquist P, Janda M. cDNA cloning and in vitro transcription of the complete brome mosaic virus genome. Mol Cell Biol 1984; 4(12):2876-82.
34. Rice CM, Grakoui A, Galler R et al. Transcription of infectious yellow fever RNA from full-length cDNA templates produced by in vitro ligation. New Biol 1989; 1(3):285-96.
35. Rice CM, Levis R, Strauss JH et al. Production of infectious RNA transcripts from Sindbis virus cDNA clones: Mapping of lethal mutations, rescue of a temperature-sensitive marker, and in vitro mutagenesis to generate defined mutants. J Virol 1987; 61(12):3809-19.
36. Kinney RM, Butrapet S, Chang GJ et al. Construction of infectious cDNA clones for dengue 2 virus: Strain 16681 and its attenuated vaccine derivative, strain PDK-53. Virology 1997; 230(2):300-8.
37. Myles KM, Pierro DJ, Olson KE. Deletions in the putative cell receptor-binding domain of Sindbis virus strain MRE16 E2 glycoprotein reduce midgut infectivity in Aedes aegypti. J Virol 2003; 77(16):8872-81.
38. Hahn CS, Hahn YS, Braciale TJ et al. Infectious Sindbis virus transient expression vectors for studying antigen processing and presentation. Proc Natl Acad Sci USA 1992; 89(7):2679-83.
39. Simpson DA, Davis NL, Lin SC et al. Complete nucleotide sequence and full-length cDNA clone of S.A.AR86 a South African alphavirus related to Sindbis. Virology 1996; 222(2):464-9.
40. Davis NL, Willis LV, Smith JF et al. In vitro synthesis of infectious venezuelan equine encephalitis virus RNA from a cDNA clone: Analysis of a viable deletion mutant. Virology 1989; 171(1):189-204.
41. Kuhn RJ, Niesters HG, Hong Z et al. Infectious RNA transcripts from Ross River virus cDNA clones and the construction and characterization of defined chimeras with Sindbis virus. Virology 1991; 182(2):430-41.
42. Yamaguchi Y, Shirako Y. Engineering of a Sagiyama alphavirus RNA-based transient expression vector. Microbiol Immunol 2002; 46(2):119-29.
43. Liljestrom P, Lusa S, Huylebroeck D et al. In vitro mutagenesis of a full-length cDNA clone of Semliki Forest virus: The small 6,000-molecular-weight membrane protein modulates virus release. J Virol 1991; 65(8):4107-13.
44. Wielgosz MM, Raju R, Huang HV. Sequence requirements for Sindbis virus subgenomic mRNA promoter function in cultured cells. J Virol 2001; 75(8):3509-19.
45. Lastarza MW, Grakoui A, Rice CM. Deletion and duplication mutations in the C-terminal nonconserved region of Sindbis virus nsP3: Effects on phosphorylation and on virus replication in vertebrate and invertebrate cells. Virology 1994; 202(1):224-32.
46. Lustig S, Jackson AC, Hahn CS et al. Molecular basis of Sindbis virus neurovirulence in mice. J Virol 1988; 62(7):2329-36.
47. Higgs S, Powers AM, Olson KE. Alphavirus expression systems: Applications to mosquito vector studies. Parasitol Today 1993; 9(12):444-52.

48. Higgs S, Traul D, Davis BS et al. Green fluorescent protein expressed in living mosquitoes—without the requirement of transformation. Biotechniques 1996; 21(4):660-4.
49. Lewis DL, DeCamillis MA, Brunetti CR et al. Ectopic gene expression and homeotic transformations in arthropods using recombinant Sindbis viruses. Curr Biol 1999; 9(22):1279-87.
50. Raju R, Huang HV. Analysis of Sindbis virus promoter recognition in vivo, using novel vectors with two subgenomic mRNA promoters. J Virol 1991; 65(5):2501-10.
51. Pugachev KV, Mason PW, Shope RE et al. Double-subgenomic Sindbis virus recombinants expressing immunogenic proteins of Japanese encephalitis virus induce significant protection in mice against lethal JEV infection. Virology 1995; 212(2):587-94.
52. Seabaugh RC, Olson KE, Higgs S et al. Development of a chimeric sindbis virus with enhanced per Os infection of Aedes aegypti. Virology 1998; 243(1):99-112.
53. Myles KM, Pierro DJ, Olson KE. Comparison of the transmission potential of two genetically distinct Sindbis viruses after oral infection of Aedes aegypti (Diptera: Culicidae). J Med Entomol 2004; 41(1):95-106.
54. Lanciotti RS, Ludwig ML, Rwaguma EB et al. Emergence of epidemic O'nyong-nyong fever in Uganda after a 35-year absence: Genetic characterization of the virus. Virology 1998; 252(1):258-68.
55. Rwaguma EB, Lutwama JJ, Sempala SD et al. Emergence of epidemic O'nyong-nyong fever in southwestern Uganda, after an absence of 35 years. Emerg Infect Dis 1997; 3(1):77.
56. Scott TW, Burrage TG. Rapid infection of salivary glands in Culiseta melanura with eastern equine encephalitis virus: An electron microscopic study. Am J Trop Med Hyg 1984; 33(5):961-4.
57. Scott TW, Hildreth SW, Beaty BJ. The distribution and development of eastern equine encephalitis virus in its enzootic mosquito vector, Culiseta melanura. Am J Trop Med Hyg 1984; 33(2):300-10.
58. Bowers DF, Abell BA, Brown DT. Replication and tissue tropism of the alphavirus Sindbis in the mosquito Aedes albopictus. Virology 1995; 212(1):1-12.
59. Weaver SC, Lorenz LH, Scott TW. Distribution of western equine encephalomyelitis virus in the alimentary tract of Culex tarsalis (Diptera: Culicidae) following natural and artificial blood meals. J Med Entomol 1993; 30(2):391-7.
60. Weaver SC, Scherer WF, Cupp EW et al. Barriers to dissemination of Venezuelan encephalitis viruses in the Middle American enzootic vector mosquito, Culex (Melanoconion) taeniopus. Am J Trop Med Hyg 1984; 33(5):953-60.
61. Weaver SC, Scott TW, Lorenz LH. Patterns of eastern equine encephalomyelitis virus infection in Culiseta melanura (Diptera: Culicidae). J Med Entomol 1990; 27(5):878-91.
62. Weaver SC, Scott TW, Lorenz LH et al. Togavirus-associated pathologic changes in the midgut of a natural mosquito vector. J Virol 1988; 62(6):2083-90.
63. Weaver SC, Scott TW, Lorenz LH et al. Detection of eastern equine encephalomyelitis virus deposition in Culiseta melanura following ingestion of radiolabeled virus in blood meals. Am J Trop Med Hyg 1991; 44(3):250-9.
64. Vanlandingham DL, Hong C, Klingler K et al. Differential infectivities of o'nyong-nyong and chikungunya virus isolates in Anopheles gambiae and Aedes aegypti mosquitoes. Am J Trop Med Hyg 2005; 72(5):616-21.
65. Hardy JL. Susceptibility and resistance of vector mosquitoes. In: Monath TP, ed. The Arboviruses: Ecology and Epidemiology. Boca Raton, FL: CRC Press, 1988:87.
66. Weaver SC. Vector biology in virus pathogenesis. In: Nathanson, ed. Viral Pathogenesis. New York: Raven Press, 1996.
67. Chamberlain RW. Epidemiology of arthropod-borne togaviruses: The role of arthropods as hosts and vectors and of vertebrate hosts in natural transmission cycles. In: Schlesinger RW, ed. The Togaviruses: Biology, Structure, Replication. New York: Academic Press Inc., 1980:175.
68. Woodring JI, Higgs S, Beaty BJ. Natural cycles of vector-borne pathogens. In: Beaty BJ, Marquardt WC, eds. The Biology of Disease Vectors. Niwot, CO: University of Colorado Press, 1996:51.
69. Hardy JL, Houk EJ, Kramer LD et al. Intrinsic factors affecting vector competence of mosquitoes for arboviruses. Annu Rev Entomol 1983; 28:229-62.
70. Reinhardt C, Hecker H. Structure and function of the basal lamina and of the cell junctions in the midgut epithelium (stomach) of female Aedes aegypti L.(Insecta, Diptera). Acta Trop 1973; 30(4):213-36.
71. Houk EJ, Chiles RE, Hardy JL. Unique midgut lamina in the mosquito, Aedes dorsalis (Meigan) (Insecta:Diptera). Int J Insect Morphol Embryol 1980; 9:161.
72. Houk EJ, Hardy JL, Chiles RE. Permeability of the midgut basal lamina in the mosquito Culex tarsalis Coquillett (Insecta, Diptera). Acta Tropica 1981; 38:163-171.
73. Paredes AM, Simon MN, Brown DT. The mass of the Sindbis virus nucleocapsid suggests it has T = 4 icosahedral symmetry. Virology 1992; 187(1):329-32.

74. Lerdthusnee K, Romoser WS, Faran ME et al. Rift Valley fever virus in the cardia of Culex pipiens: An immunocytochemical and ultrastructural study. Am J Trop Med Hyg 1995; 53(4):331-7.
75. Romoser WS, Faran ME, Bailey CL. Newly recognized route of arbovirus dissemination from the mosquito (Diptera: Culicidae) midgut. J Med Entomol 1987; 24(4):431-2.
76. Romoser WS, Faran ME, Bailey CL et al. An immunocytochemical study of the distribution of Rift Valley fever virus in the mosquito Culex pipiens. Am J Trop Med Hyg 1992; 46(4):489-501.
77. Scott TW, Lorenz LH. Reduction of Culiseta melanura fitness by eastern equine encephalomyelitis virus. Am J Trop Med Hyg 1998; 59(2):341-6.
78. Girard YA, Popov V, Wen J et al. Ultrastructural study of West Nile virus pathogenesis in Culex pipiens quinquefasciatus (Diptera: Culicidae). J Med Entomol 2005; 42(3):429-44.
79. Whitfield SG, Murphy FA, Sudia WD. St. Louis encephalitis virus: An ultrastructural study of infection in a mosquito vector. Virology 1973; 56(1):70-87.
80. Murphy FA. Cellular resistance to arbovirus infection. Ann NY Acad Sci 1975; 266:197-203.
81. Galiana-Arnoux D, Dostert C, Schneemann A et al. Essential function in vivo for Dicer-2 in host defense against RNA viruses in drosophila. Nat Immunol 2006.
82. Wang XH, Aliyari R, Li WX et al. RNA interference directs innate immunity against viruses in adult Drosophila. Science 2006; 312(5772):452-4.
83. Coates CJ, Jasinskiene N, Miyashiro L et al. Mariner transposition and transformation of the yellow fever mosquito, Aedes aegypti. Proc Natl Acad Sci USA 1998; 95(7):3748-51.
84. Jasinskiene N, Coates CJ, Benedict MQ et al. Stable transformation of the yellow fever mosquito, Aedes aegypti, with the Hermes element from the housefly. Proc Natl Acad Sci USA 1998; 95(7):3743-7.
85. Kokoza V, Ahmed A, Wimmer EA et al. Efficient transformation of the yellow fever mosquito Aedes aegypti using the piggyBac transposable element vector pBac[3xP3-EGFP afm]. Insect Biochem Mol Biol 2001; 31(12):1137-43.
86. Catteruccia F, Nolan T, Blass C et al. Toward Anopheles transformation: Minos element activity in anopheline cells and embryos. Proc Natl Acad Sci USA 2000; 97(5):2157-62.
87. Grossman GL, Rafferty CS, Clayton JR et al. Germline transformation of the malaria vector, Anopheles gambiae, with the piggyBac transposable element. Insect Mol Biol 2001; 10(6):597-604.
88. Ngo H, Tschudi C, Gull K et al. Double-stranded RNA induces mRNA degradation in Trypanosoma brucei. Proc Natl Acad Sci USA 1998; 95(25):14687-92.
89. Fire A, Xu S, Montgomery MK et al. Potent and specific genetic interference by double-stranded RNA in Caenorhabditis elegans. Nature 1998; 391(6669):806-11.
90. Fire A. RNA-triggered gene silencing. Trends Genet 1999; 15(9):358-63.
91. Kennerdell JR, Carthew RW. Use of dsRNA-mediated genetic interference to demonstrate that frizzled and frizzled 2 act in the wingless pathway. Cell 1998; 95(7):1017-26.
92. Kavi HH, Fernandez HR, Xie W et al. RNA silencing in Drosophila. FEBS Lett 2005; 579(26):5940-9.
93. Kennerdell JR, Carthew RW. Heritable gene silencing in Drosophila using double-stranded RNA. Nat Biotechnol 2000; 18(8):896-8.
94. Blandin S, Moita LF, Kocher T et al. Reverse genetics in the mosquito Anopheles gambiae: Targeted disruption of the Defensin gene. EMBO Rep 2002; 3(9):852-6.
95. Osta MA, Christophides GK, Kafatos FC. Effects of mosquito genes on Plasmodium development. Science 2004; 303(5666):2030-2.
96. Powers AM, Olson KE, Higgs S et al. Intracellular immunization of mosquito cells to LaCrosse virus using a recombinant Sindbis virus vector. Virus Res 1994; 32(1):57-67.
97. Powers AM, Kamrud KI, Olson KE et al. Molecularly engineered resistance to California serogroup virus replication in mosquito cells and mosquitoes. Proc Natl Acad Sci USA 1996; 93(9):4187-91.
98. Attardo GM, Higgs S, Klingler KA et al. RNA interference-mediated knockdown of a GATA factor reveals a link to anautogeny in the mosquito Aedes aegypti. Proc Natl Acad Sci USA 2003; 100(23):13374-9.
99. Ruiz MT, Voinnet O, Baulcombe DC. Initiation and maintenance of virus-induced gene silencing. Plant Cell 1998; 10(6):937-46.
100. Powers AM, Olson KE. Working safely with recombinant viruses and vectors. In: Richmond JY, ed. Anthology of Biosafety: VI. Arthropod Borne Diseases. Mundelein, IL: American Biological Safety Association, 2003:39-52.

CHAPTER 3

Paratransgenesis Applied for Control of Tsetse Transmitted Sleeping Sickness

Serap Aksoy,* Brian Weiss and Geoffrey Attardo

Abstract

African trypanosomiasis (sleeping sickness) is a major cause of morbidity and mortality in Subsaharan Africa for human and animal health. In the absence of effective vaccines and efficacious drugs, vector control is an alternative intervention tool to break the disease cycle. This chapter describes the vectorial and symbiotic biology of tsetse with emphasis on the current knowledge on tsetse symbiont genomics and functional biology, and tsetse's trypanosome transmission capability. The ability to culture one of tsetse's commensal symbiotic microbes, *Sodalis* in vitro has allowed for the development of a genetic transformation system for this organism. Tsetse can be repopulated with the modified *Sodalis* symbiont, which can express foreign gene products (an approach we refer to as paratransgenic expression system). Expanding knowledge on tsetse immunity effectors, on genomics of tsetse symbionts and on tsetse's parasite transmission biology stands to enhance the development and potential application of paratransgenesis as a new vector-control strategy. We describe the hallmarks of the paratransgenic transformation technology where the modified symbionts expressing trypanocidal compounds can be used to manipulate host functions and lead to the control of trypanosomiasis by blocking trypanosome transmission in the tsetse vector.

Introduction

There is an urgent need for cost-effective strategies for the sustainable control of human sleeping sickness (HAT), which is a fatal zoonotic disease that has caused devastating epidemics during the past century.[1] The disease agent is a unicellular flagellated protozoan, *Trypanosoma brucei* spp. that spends parts of its life cycle in the mammalian host bloodstream and the rest in the tsetse fly's midgut and salivary gland tissues. In addition to its medical significance, diseases in animals (nagana) caused by closely related parasite species are among the major deterrents for agricultural practices in subSahara with profound economic effect on much of the continent (over $3.5 billion annually).[2,3] In East Africa, HAT caused by *Trypanosoma brucei rhodesiense* is a zoonosis, cycling between wild game and the mammalian/domestic hosts. Although disease control can be effectively realized by targeting parasite's viability in either its vertebrate or invertebrate host, given the extensive antigenic variation the parasite displays in its mammalian host, efforts to develop mammalian vaccines have been unsuccessful. The available drugs used for chemotherapy in humans are highly toxic and expensive and have been made available mostly through international efforts during epidemics.[4] Additionally, the emergence of drug resistance in parasites threatens their efficacy.[5-8] The use of vector control for sleeping sickness

*Corresponding Author: Serap Aksoy—Yale University School of Medicine, Department of Epidemiology and Public Health, New Haven, CT 06520 USA. Email: serap.aksoy@yale.edu

Transgenesis and the Management of Vector-Borne Disease, edited by Serap Aksoy.

control has been met with varying success efforts given the problems associated with the maintenance of these activities and lack of their sustainability over time.[9,10] Control activities mostly occur during the major epidemics, and farmers and communities must fend for themselves during the intervening periods.

There have been a number of technological advancements that stand to enhance traditional disease control efforts and bring about new strategies. In particular, the genome sequence of *T. b. brucei* has been completed and the genomes of the related parasites *T. vivax* and *T. congolense* are near completion.[11] Various genomics resources are also available for tsetse, including several gene discovery projects representing tissue specific transcriptomes, in addition to the ongoing *Glossina morsitans morsitans* whole genome sequence project.[12,13] The genomes of the mutualistic symbiotic bacteria that are vital for tsetse physiology, genus *Wigglesworthia glossinidae* and genus *Sodalis glossinidius*, have also been completed.[14,15] These resources can now form the foundation for discovery of new drug targets, insecticides and vector control tools.[16] Below we describe one potential vector control strategy based on paratransgenic expression technology. This approach exploits tsetse's symbiotic biology and aims to manipulate tsetse's vectorial capacity to block parasite transmission.

Symbiosis in Tsetse

Many animals, but insects in particular, have entered into symbiotic associations with microorganisms to expand their niche into ecologically difficult terrains. Among these is the tsetse fly, which survives strictly on vertebrate blood during all developmental stages. To date three distinct vertically transmitted microorganisms have been characterized from tsetse populations. Two of these symbionts, primary symbiont *Wigglesworthia glossinidia*[17] and the secondary symbiont *Sodalis glossinidius*,[18] belong to the family Enterobacteriaceae. The obligate *Wigglesworthia* resides exclusively within the cytoplasm of specialized cells called 'bacteriocytes'. These cells comprise an organ called the 'bacteriome' which is associated with tsetse's midgut (shown in Fig. 1). *Sodalis*, on the other hand, is found inter- and intracellularly in several tsetse tissues, including hemolymph, midgut and milk gland.[19] Also present in some tsetse populations are parasitic members of the genus *Wolbachia*.[20] *Wolbachia* resides primarily in reproductive tissues, and is transovarially transmitted from mothers to their offspring while the enteric gut symbionts are acquired into the intrauterine larva via the mother's milk gland secretions.

Many obligate intracellular endosymbionts, including *Wigglesworthia* have extraordinary genome features. These may include an extremely small genome size, high AT bias and reduced coding capacity in comparison to their free-living relatives. *Wigglesworthia glossinidia brevipalpis* has a genome size of about 700 kb and cannot be cultivated in-vitro, presumably due to its dependence of eukaryotic host cell functions. Phylogenetic reconstruction studies indicate that the initial association between the free-living *Wigglesworthia* bacterium and the ancestral tsetse host species is ancient, going back 50-100 million years ago. *Wigglesworthia* has apparently been vertically transmitted from host generation to generation, giving rise to concordant evolution that we now observe between the tsetse host and *Wigglesworthia* species.[21] Analysis of the *Wigglesworthia* transcriptome has indicated that a large proportion of this genome is devoted to the biosynthesis of cofactors and vitamins. Accordingly, one of *Wigglesworthia*'s functional roles in host biology may to be to supplement its single diet with vitamin products, which are known to be limited in the vertebrate blood. Similar phylogenetic reconstruction studies of *Sodalis* strains characterized from different tsetse species has not revealed extensive differences, implying a recent origin for this symbiosis, during which the different *Sodalis* species has not apparently had enough evolutionary time to diverge.[21,22] In support of its recent establishment, *Sodalis*' genome was found to be about 4.5 kb, much closer in size to related free-living enteric microbes such as *E. coli* and *Yersinia* and *Salmonella* than the obligate *Wigglesworthia*.[15] The sequencing and subsequent annotation of *Sodalis*'s genome has revealed an ongoing functional streamlining process, where only 51% of the genome has retained coding capacity with an unusually large number of pseudogenes present.[15] *Sodalis*, like

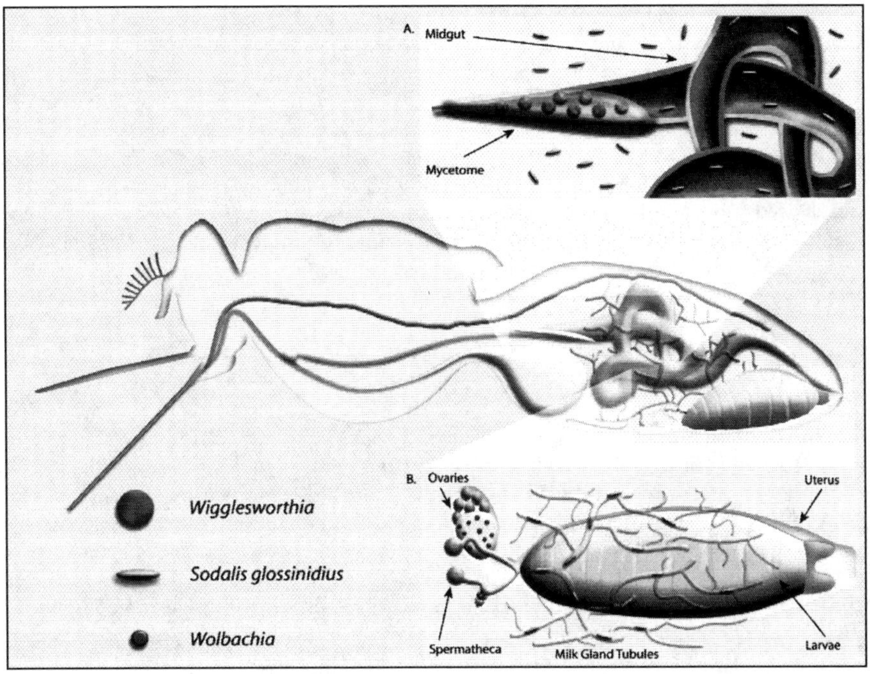

Figure 1. Symbiont localization within the tsetse fly. A) The *Wigglesworthia* endosymbiont is exclusively intracellular and found in specialized epithelial cells, forming the organ called the mycetome. The mycetome is found at the anterior end of the midgut. The *Sodalis* symbiont is extracellular and can be found in the midgut and throughout the hemolymph of the fly. B) Symbiont localization in the reproductive tract. The *Wolbachia* bacteria can be found in the germ cells of the ovary and in the cytoplasm of developing oocytes in the ovaries. *Sodalis* and *Wigglesworthia* can be found in the milk gland tubules, which function to deliver nourishment to the developing larvae. It is thought that the gut enteric symbionts are transmitted into the developing larvae through the mother's milk gland secretions.

Wigglesworthia, is transmitted vertically between host generations. Hence, Sodalis may have lost genes encoding many under-utilized functions during its evolution to a symbiotic lifestyle within the tsetse host since a large gene inventory and corresponding genomic plasticity are no longer required. On the whole, *Sodalis* has a relatively inactive biochemical profile, with the preferential loss of genes involved in carbohydrate transport and metabolism. *Sodalis* is able to import *N*-acetyl-glucosamine, mannose and mannitol.[15,18] Given the low carbohydrate content of vertebrate blood, it is not surprising that genes encoding glycolytic enzymes, such as glucosidase and galactosidase are not present, while genes encoding components of amino acid biosynthesis pathways have largely escaped erosion. In addition, lack of drug resistance genes and O-antigen in the lipopolysaccharide (LPS) protein further supports the transitioning state of *Sodalis* from a free-living to a symbiotic state, where defense functions may no longer be as necessary.

Symbiont-Host Interactions During Development

Studies on symbiont density regulation during host development provide further insight into the mutualistic nature of *Sodalis* and *Wigglesworthia* associations. Using a real-time quantitative PCR assay, genome density of the two different symbionts harbored within the same individual host was studied throughout development (i.e., spanning from first instar larva (L1) to 30 days into adulthood). Symbiont infection prevalence among the individuals analyzed was 100%, in that every fly examined was found infected with *Wigglesworthia* and *Sodalis*. Although both enteric symbionts proliferated in parallel to host cell replication during larval development, there were significant differences in both *Wigglesworthia* and *Sodalis* densities between early and late pupal stages. There was a 3-fold increase in *Sodalis* and a 30-fold increase in *Wigglesworthia* in late in pupal development. *Sodalis* proliferation continued for about 48 hours post-eclosion, before being regulated by possibly host immune responses. The two-week adults showed only a two-fold increase in *Sodalis* density relative to newly eclosed teneral adults. In contrast, in fertile females, *Wigglesworthia* continued to proliferate during the whole age spectrum analyzed, while in males proliferation was limited to the first few weeks, possibly reflecting its mutualistic functions in female fecundity.

Tsetse's Reproductive Biology and Symbiont Transmission

Tsetse flies have developed a complex viviparous reproductive biology, whereby they are able to give birth to live young (viviparity), diagrammatically depicted in (Fig. 1). Tsetse flies have two ovaries with only two ovarian follicles per ovary. This is a significant reduction from what is observed in other Diptera, which typically contain many follicles per ovary. In tsetse a single oocyte is developed at a time and the active ovary alternates between the left and the right at each cycle of oogenesis. Upon completion of development, the oocyte undergoes ovulation from the ovary into the uterus. During ovulation the oocyte passes the spermathecal ducts and is fertilized with sperm that is stored within the spermathecae. The fertilized oocyte begins embryonic development within the uterus. Larvae proceed through three instars, all entirely within the mother. Pregnant females give birth to mature third instar larvae, which quickly pupate after parturition (birth). Larval nourishment is provided by a modified accessory gland (the milk gland), which connects to the dorsal side of the uterus where the larval mouthparts are positioned. The gland tubules extend from the uterus, bifurcate many times and radiates throughout the abdominal cavity. The milk secretion generated by this gland contains essential nutrients, such as protein and fat necessary for larval development.

In addition to essential nutrients, PCR-based evidence suggests that the enteric symbionts are also introduced into the larvae via the milk gland secretions.[23] Electron microscopic analysis has also shown the presence of endosymbionts in the lumen of the milk gland tubules.[24] The transmission of symbionts from mother to offspring via this system suggests that the symbionts have adapted to tsetse's reproductive biology to ensure their transmission to the next generation of offspring. Once in the larva, the symbionts have to establish infections in the larval tissues. Interestingly, the Type III secretion systems (TTSSs) encoded on *Sodalis*' chromosome are apparently expressed preferentially in the milk gland tissue and during early larval development, indicating that they may function in the infection establishment process.[15]

Biology of African Trypanosome Transmission in Tsetse

The life cycle of the African trypanosome begins when a tsetse fly feeds from an infected mammalian host. In the gut lumen the stumpy-form parasites that are preadapted for life in the fly rapidly undergo structural and metabolic transformation to insect-stage cells (procyclic form, PF). They activate their mitochondria and adapt to use a proline-based energy metabolism, which is also the major energy source that tsetse flies use for flight. During their development in the fly, trypanosomes of the *Trypanosoma brucei* spp. complex face several obstacles. Within several days of digestion, replicating trypanosomes can be seen in the ectoperitrophic space of the midgut, between the gut epithelia and the continuously synthesized chitinous peritrophic matrix, which forms a sleeve-like barrier throughout the tsetse midgut (shown in Fig. 2). Around day 3, the parasites undergo a massive attrition—a process that completely eliminates infections in a large proportion of the parasite exposed flies.[25] In only those few permissive flies, parasites continue to proliferate and establish midgut infections. In the ectoperitrophic space of the midgut, trypanosomes are in close proximity to *Sodalis* and *Wigglesworthia* cells. Successful colonization of the salivary glands, where the parasites mature to mammalian infective cells

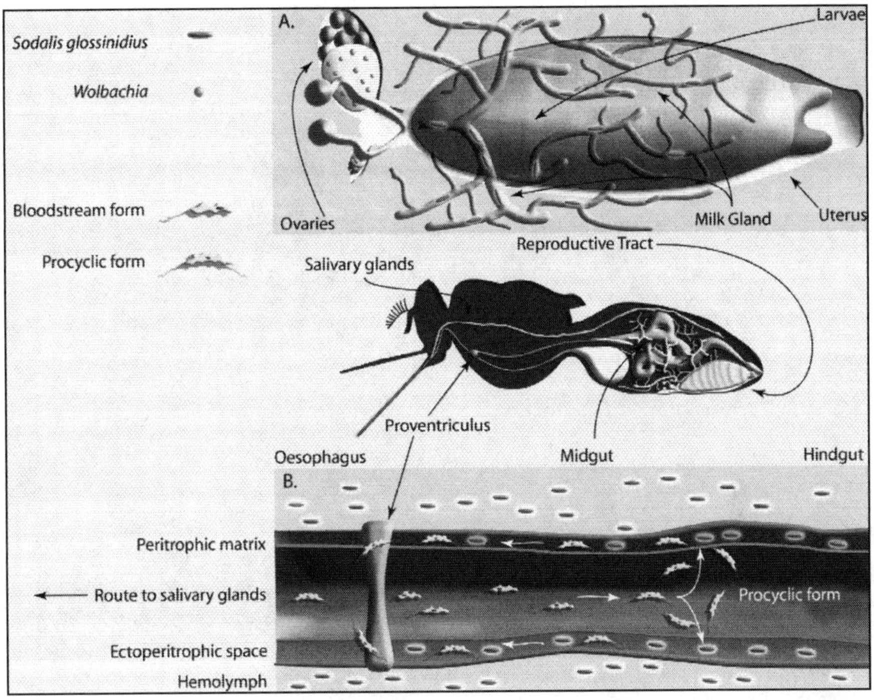

Figure 2. Localization of parasites and symbionts in Tsetse. A) Reproductive tract of the tsetse fly. Tsetse endosymbionts can be found in developing oocytes (*Wolbachia*) and the milk gland tubules (*Sodalis glossinidius*). B) Parasite and symbiont co localization in the midgut. Parasites entering via an infected blood meal are colocalized with the symbiont *Sodalis* in the ectoperitrophic space between the peritrophic matrix surrounding the blood meal and the midgut epithelium.

capable of infecting the next mammalian host, is again restricted to only a subset of those flies with midgut infections. Once infected, the fly remains so during its 2-3 month adult life and can transmit the parasite to a new host as it takes a blood meal approximately every 48 hours.

Paratransgenic Gene Expression in Tsetse

The viviparous reproductive biology of tsetse makes germline transformation via embryo injections difficult, thus a paratransgenic strategy has been developed (schematically depicted in Fig. 3).[26-28] With this approach, genes are not inserted into tsetse's chromosome, but instead into the chromosome of a tsetse symbiont. Subsequently, the transgenic bacterium is introduced back into its host. *Sodalis* is well suited for paratransgenesis because: a) it resides in the gut in close proximity to pathogenic trypanosomes, b) a system for culturing *Sodalis in vitro* (both in liquid media and on agar plates) has been developed[29] and can be used in conjunction with standard molecular biology techniques to insert and express foreign genes of interest in this bacterium,[29,30] c) *Sodalis* is highly resistant to many trypanocidal peptides,[31,32] d) recombinant *Sodalis* (rec*Sodalis*) can be reintroduced into tsetse by thoracic microinjection and passed on to future progeny where they successfully express the marker gene product,[19] and e) its genome is completely sequenced and annotated, and this information will serve as a valuable resource that can be exploited to improve the efficiency of this expression system. Because these commensal bacteria live naturally in close proximity to where trypanosomes develop and replicate in the tsetse midgut, expression of trypanocidal products in *Sodalis* has the potential to block parasite development in the fly as we describe below.

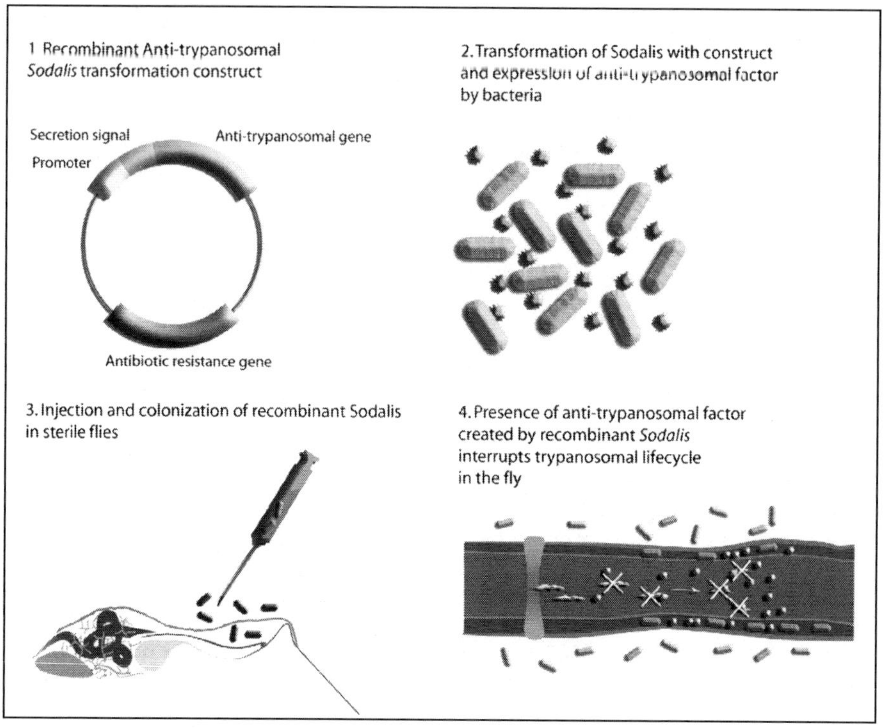

Figure 3. Principles of paratransgenesis. 1) A plasmid construct is engineered containing an anti-trypanosomal gene (effector) carrying a 5'-end localized secretion signal and a promoter to ensure correct temporal expression and secretion of the product in vivo. 2) The symbiont is transformed with the plasmid and expression of the anti-trypanosomal factor is confirmed. 3) Natural symbionts are cleared from flies with antibiotic treatment and flies are recolonized with recombinant symbionts via intrathoracic injection. 4) Recombinant symbionts expressing anti-trypanosomal factor create a hostile environment in which the trypanosomes cannot survive, thereby blocking the disease transmission cycle.

Sodalis *in Vitro Cultivation Method*

In vitro growth of *Sodalis*, both in liquid and on solid media, requires the initiation of a primary culture. To achieve this, pupae are collected mid-way through their 30-day developmental period, and repeatedly surface sterilized in consecutively lower concentrations of ethanol. Immediately proceeding surface sterilization, the pupae are placed into a culture dish containing a semi-confluent *Aedes albopictus* C6/36 cells grown in Mitsuhashi-Maramorosch media (M-M media) supplemented with 10% fetal bovine serum (FBS). Pupal contents are sucked out using a 30.5 gauge needle and subsequently deposited onto the C6/36 cells.[30] The symbiont can also be directly obtained from extracted hemolymph prior to cultivation on the C6/36 cells.[33] Within approximately 7 days at 28°C, *Sodalis* cells begin to establish an infection in the C6/36 cells and can be split and thereafter maintained in cell-free media where they double about every 24 hours and reach $OD_{600} \geq 0.8$.

Certain applications such as transformation may require the use of clonal populations of *Sodalis*. In this case, cells must be plated onto a solid medium and the desired clones isolated. The solid medium is prepared by combining M-M media with 1% molten agar (and the appropriate selective antibiotic if desired) so that both are at a final 1x concentration. Because *Sodalis* is a microaerophilic organism, cultures grown on solid media are incubated under microaerobic conditions.[18] We use disposable hydrogen carbon dioxide generating envelopes that produce a microaerophilic atmosphere in culture jars and incubate the *Sodalis* spread plates in the sealed jars for colony formation. Colonies which are typically visible in 5-7 days, are then placed into liquid culture and maintained as described above.

Sodalis *Genetic Transformation*

Efficient transformation of *Sodalis* is critical to the development of a successful paratransgenic system. Currently, heat shock and electroporation are the two methods used to perform this operation. The former technique involves preparing transformation competent *Sodalis* cells, which are incubated with target DNA prior to heat shock briefly at 37°C.[30] This procedure typically requires large quantities of target DNA (~1-5µg). This problem becomes especially troublesome when plasmid constructs of interest encode gene products toxic to *E. coli*, such as some of the trypanolytic antimicrobial peptides, which readily kill *E. coli* but not *Sodalis*. In this case, the construct cannot be propagated in *E. coli* using a shuttle vector via the conventional cloning techniques prior to transforming *Sodalis*. An alternative transformation method is electroporation.[34] Electroporation requires several orders of magnitude less target DNA (~10-50ng) than does heat shock, and this facilitates transformation using ligation reactions, thus rendering the entire *E. coli* propagation step superfluous.

Transformation of *Sodalis* has been achieved by using broad-host-range plasmids. The use of extrachromosomal DNA is useful for laboratory investigations with paratransgenic studies during which flies can be fed an antibiotic-supplemented diet. However, maintenance of plasmids generally requires continuous selection and is of concern for eventual applications of such modified symbionts in the field. Foreign DNA has been directly inserted into *Sodalis*'s chromosome via homologous recombination (Aksoy, unpublished). This is accomplished by transforming *Sodalis* with a nonreplicating piece of circular DNA, the sequence of which is homologous to the desired chromosomal loci. An antibiotic resistance gene expressed under the control of a *Sodalis*-specific promoter is typically inserted within the homologous DNA. Following electroporation, and isolation of transformants using the antibiotic selection process, transgenic individuals can be maintained in the absence of selection. Furthermore, the risk of a double-crossover event taking place is unlikely due to the absence of an original exogenous plasmid (this can occur when using transposons or allelic exchange plasmids that replicate and remain in the host following recombination).

Reconstitution of Tsetse with Modified Symbiont Flora

The final step to producing paratransgenic tsetse is the reconstitution of flies with rec*Sodalis*, which has been possible via thoracic microinjection.[19] Injections, which contain a physiological concentration of *Sodalis*, are made into tsetse's thorax with a glass needle attached to a microinjector that regulates the quantity of liquid dispensed. The symbionts are introduced into fertile females, which are subsequently maintained on an ampicillin-supplemented diet. Progeny deposited by mothers receiving this treatment are collected and shown to have acquired the rec*Sodalis* by PCR-amplification. To date we have been able to use this antibiotic enrichment regiment with amplicillin, which clears native *Sodalis* flora of tsetse without eliminating the obligate *Wigglesworthia*. If an antibiotic such as tetracycline is used that clears intracellular infections including *Wigglesworthia*, the mother's diet needs to be supplemented with a battery of vitamins to maintain fertility (unpublished data Aksoy).[35]

Given the close phylogenetic relationship and similar restriction fragment-length polymorphism (RFLP) analysis of different species of *Sodalis*, transinfection experiments were used as a means of revealing if functional differences exist between *Sodalis* species and if host-symbiont relationships have been subject to coevolution.[22] *Sodalis* species isolated from *Glossina morsitans morsitans* and *G. fuscipes* were transformed with a plasmid (pGFPuv) that encodes the enhanced green fluorescent protein (eGFP). These recombinant *Sodalis* (rec*Sodalis*) were then microinjected into the other nonnative tsetse species, which had its wild-type *Sodalis* cleared via ampicillin treatment. The 'transinfected' flies were then monitored to determine whether the rec*Sodalis* were able to successfully colonize a nonnative host, if transinfected hosts experienced a reduction in fitness compared to wild-type individuals, and if transinfected hosts maintained their rec*Sodalis* at densities comparable to wild-type individuals. From a functional perspective, flies with transinfected *Sodalis* exhibited no significant reductions in fitness (as measured by fecundity and longevity) when compared to wild-type individuals. These transinfected individuals also maintained their nonnative symbionts at densities similar to their wild-type counterparts. When put into the context of controlling HAT by modifying the tsetse vector, our success in interspecies specific transinfection experiments may serve as a means of streamlining paratransgenesis. Specifically, one strain of *Sodalis* may be developed for the paratransgenic strategy and subsequently transinfected into different species of tsetse that vector the disease instead of custom transforming *Sodalis* for each tsetse species.

Effector Genes with Trypanocidal Activity

A crucial step in transgenic technology is the identification of gene products that can have an adverse effect on pathogen development (effector genes) when expressed in insect hosts. The identification of monoclonal antibodies (mABs) with parasite-transmission blocking characteristics, and their subsequent expression as single-chain antibody gene fragments, provides a vast array of potential antipathogenic effectors. Towards this goal, transmission-blocking antibodies affecting the major surface protein procyclin of the insect-stage procyclic trypanosomes have already been reported.[36] The feasibility of expressing in bacterial symbionts single chain antibody fragments, which retain their functional activities, has been demonstrated in *Rhodoccocus*[37] as well as in human commensal flora.[38-40]

Alternatively, insect immunity genes that naturally confer resistance to pathogens can be pursued as potential effectors. Because most insects mount a significant immune response to the presence of pathogens and can effectively clear these infections, immunity genes are good candidates to be further explored. In tsetse, parasite transmission is also restricted by a natural phenomenon of refractoriness. In the laboratory, only a proportion of flies exposed to a parasite infected blood meal establish midgut infections, and only a subset of those midgut infected flies can give rise to salivary gland infections. These are then transmitted to the next mammalian host.[41] During an infection process, the natural immune responses may be elicited either too late to combat the parasites, or may not be expressed in the correct compartments where parasites reside. In tsetse, the complex lifecycle of the parasite may take from a few days for *Trypanosoma vivax* to three-four weeks for *T. brucei* subspecies.[12] In experiments where the immune system of the newly emerged adult tsetse was artificially stimulated prior to providing the parasite infected blood meal, the prevalence of trypanosome infection rates in tsetse midgut was substantially lowered.[31] This indicated that tsetse immune components could interfere with parasite development. Among the induced responses in such immune challenged tsetse flies are a battery of small antimicrobial peptides. In particular, immune products synthesized via the Immune deficiency (Imd) pathway have been implicated in greater parasite-clearance in tsetse.[42,43] Among these products is the effector Attacin, which has been well characterized in *Drosophila melanogaster* as an inducible immune peptide with specificity against some Gram-negative bacteria and protozoa.[44] Recombinant *Glossina* Attacin also exhibits trypanocidal activity in vitro and in vivo, while not interfering with *Sodalis* viability.[32] In fact, the sensitivity of *Sodalis* to another immune peptide Diptericin, with known antimicrobial activity, was also found to be significantly less than that of *E. coli*.[31] The greater resistance of *Sodalis* to its host immune molecules may reflect its constant exposure to these products. During their development in the fly, trypanosomes differentiate and replicate in the midgut in close proximity to where the *Sodalis* symbionts reside. Thus, constitutive and abundant expression of immunity molecules such as Attacin can lead to parasite clearance. It will be important to evaluate the fitness of both the recombinant symbionts as well as the insects harboring these modified symbionts in order to assess any potential fitness costs. Such potential fitness loss would not be desirable, as these insects will not be able to compete with their natural counterparts, and be eliminated in nature over time. Equally vital, no matter which transformation approach is used, is to consider whether over time the pathogens being targeted could develop resistance to the transgene product(s). With transmission blocking mABs, it is possible to have a multitude of potential antiparasitic effectors, which can be sequentially expressed in bacteria to inhibit the development of resistance. Alternatively, several transgenes can potentially be simultaneously expressed in the symbiont expression system to deter resistance against any one individual target.

Gene Driver System

The ability to interfere with disease transmission via insect manipulations requires the spread of desirable engineered phenotypes into natural populations.[45] To this end, *Wolbachia* infections detected in different tsetse populations provide a potential opportunity. This is because the functional presence of *Wolbachia* has been shown to result in a variety of reproductive abnormalities in the various invertebrate hosts they infect. One of these abnormalities is termed cytoplasmic incompatibility (CI), and when expressed, commonly results in embryonic death due to disruptions in early fertilization events.[46] In an incompatible cross, the sperm enters the egg but does not successfully contribute its genetic material to the potential zygote and in most species this results in none or very few hatching eggs. In crosses between *Wolbachia*-infected and uninfected individuals, the infected females have a reproductive advantage over their uninfected counterparts as they can produce successful progeny with both the imprinted and normal sperm, eventually allowing the *Wolbachia*-infected insects to spread into populations. *Wolbachia* population invasion has been directly observed in nature in *Drosophila simulans*. The most important parameters for *Wolbachia* mediated spread are the relative hatch rates that result from incompatible versus compatible crosses, the fecundity of infected females and the maternal transmission efficiency. Based on laboratory data, all symbionts including *Wolbachia* in tsetse exhibit 100% maternal transmission efficacy into the intrauterine larva.[47]

Phylogenetic analysis of *Wolbachia* infections in natural tsetse populations have indicated that different tsetse species are infected with different strains of *Wolbachia* belonging to A and B types, based on their outer surface protein *wsp* DNA sequence.[20] The *Wolbachia* prevalence in laboratory colonies and the maternal transmission efficiency of *Wolbachia* infections by individual mothers to their multiple progeny is 100%.[47] However, the infection prevalence in different field populations shows variability.[20] Analysis of some of these polymorphic field populations for infection prevalence at multiple times are needed to understand *Wolbachia* infection dynamics in the field. The potential extent of CI expression and relative hatch rates from incompatible versus compatible crosses needs evaluation to assess the utility of this gene-drive system for spreading modified tsetse. Most functional studies have involved curing insects of their *Wolbachia* infections by administering antibiotics, such as tetracyline, in their diet. The *Wolbachia* then cured lines are used to test for CI expression with the original infected individuals. This approach, however, has not been feasible in tsetse since treatment of flies with the antibiotic tetracycline results in the clearing of the obligate *Wigglesworthia*, which in turn results in fly sterility. Perhaps *Wolbachia* infected and uninfected lines can be developed from the polymorphic field populations and used to elucidate the functional role of this organism in tsetse biology. Given the perfect maternal transmission rate observed in colony flies, it should be possible to capture field females, collect their deposited progeny, identify the mother's *Wolbachia* infection status and subsequently develop lines from the progeny of infected and uninfected females.

Applications with Parasite Resistant Tsetse

There are two potential applications where parasite resistant tsetse lines can be used for trypanosomiasis control. The first approach involves a population replacement strategy where the engineered refractory flies can be driven into natural populations. The second application stands to enhance ongoing SIT programs, and involves the use of the engineered lines for field releases.

The ability to replace susceptible flies with their modified counterparts requires the presence of a strong genetic drive system. *Wolbachia* infections described above can provide one potential approach pending demonstration of a strong CI outcome for infections in tsetse. As *Wolbachia* spreads itself into populations, it will also spread other maternally linked traits and organelles such as mitochondria[48] and maternally linked symbionts such as *Sodalis*. The refractoriness conferring products expressed in modified *Sodalis* hence can lead to parasite resistance in natural populations. Furthermore, the potential use of bi-directional incompatibility promises to replace populations already infected with a strain of *Wolbachia* with that of another.

An alternative use of the parasite resistant tsetse is the Sterile Insect Technique (SIT), a genetic population suppression approach. SIT which involves sustained systematic releases of irradiated sterile male insects into the wild population. The male insects to be released are typically sterilized by irradiation prior to their release in a selected area. Releasing sterile males in high numbers over a period of 3-4 generations, after having reduced population density by other techniques (trapping, insecticide spraying, etc.), can lower the reproductive capacity of the target population, eventually leading to its eradication.[49,50] The recent successful eradication of *G. austeni* from the island of Zanzibar by SIT has demonstrated the feasibility and applicability of this technology in integrated tsetse control programs.[50] Further plans to apply this approach to mainland tsetse control are underway. The release of parasite refractory flies in the SIT programs would enhance the efficacy of releases in disease endemic areas. In addition, if release of only male flies can be guaranteed, the chemical sterility dose can be reduced in conjunction with *Wolbachia* mediated genetic sterility and result in better survival of the released males. This in turn will necessitate fewer released flies and result in lower cost of the overall program.

Application of Paratransgenesis for Control of Other Insect Transmitted Diseases

Symbiosis in animals, in particular in arthropods, is widespread. Similar infections with obligate and facultative microbes have been reported in various insect systems, that have restricted nutrients and rely on their symbiotic flora to supplement their dietary needs. In particular, aphids and carpenter ants have established infections with a bacterium in Enterobacteriacia that shares an evolutionary lineage with Wigglesworthia.[51,52] Similarly weevils and tsetse harbor facultative symbionts that form of a closely related lineage, Sodalis and SOPE.[53] In addition to these maternally transmitted microbes, other insects have symbionts, which they acquire as adults from their environment by coprophagy. Among these are the reduviid bugs Rhodnius prolixus, vectors of Trypanosoma cruzi, agent of Chagas disease. Reduviid bugs rely on their extracellular gut symbiont Rhodoccocus rhodnii for fertility and without it can not go through their full developmental cycles. Rhodoccus has been genetically transformed and bugs have been reconstituted with modified symbiont flora.[54] Expression of the antimicrobial peptide cecropin in Rhodoccocus reduces the development of T. cruzi infections in hindgut.[55] In addition, expression of functional monoclonal antibody gene fragments have also been demonstrated in the Rhodoccus microbe.[37] The modified symbiont expressing trypanocidal products could be applied in the field for blocking parasite transmission in the gut using an approach that simulates the natural transmission mode of the organism. Spread of Rhodococcus within populations of Rhodnius occurs via coprophagy, the ingestion of fecal deposits. Newly emerging first-instar nymphs are transiently aposymbiotic (devoid of symbiotic bacteria). Probing the fecal droplets deposited by adult bugs either, on the eggshell or in the immediate environment, permits nymphs to ingest Rhodococcus and establish gut infections. In studies where the symbiont was provided to the newly hatched first-instar larva in a simulated fecal paste, the modified Rhodococcus established gut infections in these bugs, and furthermore, competed successfully with the unmodified counterparts (reviewed in Durvasula et al).[56] Hence, it may be possible to deliver and spread the horizontally transmitted genetically modified symbionts within the domestic insect populations. Recently bacteria related to insect

Box I. Features that will enable the application of a paratransgenic expression strategy for control of vector-transmitted pathogens

1. The availability of symbionts that can be isolated and cultured.
2. Availability of knowledge on the transmission mode as well as the population dynamics of the symbiont throughout the host lifecycle.
3. Development of symbiont transformation system that produces stable phenotypes.
4. Availability of effective anti-pathogenic products (transgenes) that will block pathogen transmission when expressed in the insects.
5. Expression of transgene products by symbionts without compromising either the symbiont or host fitness.
6. Ability to reintroduce and populate hosts with modified symbionts.
7. Synthesis of the transgene products in the tissues or insect compartments that can interrupt the pathogen lifecycle.
8. Lack of potential risk for emergence of resistance by target pathogens to the transgene products.
9. Ability to express multiple transgene products to combat potential resis tance emergence.
10. Availability of effective gene driving mechanisms to replace field populations with engineered pathogen refractory lines.
11. An ecological assessment of potential spread of modified symbionts and barriers of dispersal to nontarget organisms.
12. In-depth analysis of the potential harmful effects of symbionts or transgene products to humans and animals.
13. An environmentally-sound implementation project for delivering insects repopulated with modified symbionts into the field.

symbionts have also been identified in mosquitoes, in particular a *Nocardia* species closely related to *Rhodococcus rhodnii* as well as several *Anaplasma* and *Spiroplasma* species have been detected in *Anopholes funestus*.[57] The feasibility of a paratransgenic Anopheles mosquito and its utility for parasite transmission inhibition can now be entertained (Chapter 4).

Future Directions

Symbiotic flora are widespread in insects and can be exploited to block other organisms vectored in their hosts via a Trojan-horse approach. Our molecular knowledge on the symbiotic interactions is in its infancy, but one that is expanding quickly. The paratransgenic technology will surely be refined as knowledge on the molecular and functional aspects of the symbiotic organisms, on their transmission biology and on the pathogen inhibitory products advance. Expanded knowledge is also required on the ecology of the symbiotic interactions in order to effectively evaluate the feasibility of the paratransgenic applications for disease control. This knowledgebase is especially needed to develop a set of criteria, which will need to be met for ecological, and environmental safety requirements before field applications can be entertained.

Acknowledgements

We are grateful to past members of our group, Drs. Xiao-ai Chen, Song Li, Quiying Cheng, Jian Yan, Leyla Akman, Zhengrong Hao, Patricia M. Strickler, Irene Kasumba, Dana Nayduch, Rita V.M. Rio, Youjia Hu, Changyun Hu, Nurper Guz, Sarah Perkin and the current members Amy Savage, Jingwen Yang, Yineng Wu, Guangxiao Yang, Uzma Alam and Roshan Pais for their contributions to this work.

References

1. Ekwanzala M, Pepin J, Khonde N et al. In the heart of darkness: Sleeping sickness in Zaire. Lancet 1996; 348:1427-1430.
2. Welburn SC, Coleman PG, Maudlin I et al. Crisis, what crisis? Control of Rhodesian sleeping sickness. Trends Parasitol Mar 2006; 22(3):123-128.
3. Fevre EM, Picozzi K, Jannin J et al. Human African trypanosomiasis: Epidemiology and control. Adv Parasitol 2006; 61:167-221.
4. Etchegorry MG, Helenport JP, Pecoul B et al. Availability and affordability of treatment for Human African Trypanosomiasis. Trop Med Int Health 2001; 6(11):957-959.
5. Mamman M, Katende J, Moloo SK et al. Variation in sensitivity of Trypanosoma congolense to diminazene during the early phase of tsetse-transmitted infection in goats. Vet Parasitol 1993; 50(1-2):1-14.
6. Moloo SK, Kutuza SB. Expression of resistance to isometamidium and diminazene in Trypanosoma congolense in Boran cattle infected by Glossina morsitans centralis. Acta Trop 1990; 47(2):79-89.
7. Nok AJ. Arsenicals (melarsoprol), pentamidine and suramin in the treatment of human African trypanosomiasis. Parasitol Res 2003; 90(1):71-79.
8. Matovu E, Enyaru JC, Legros D et al. Melarsoprol refractory T. b. gambiense from Omugo, north-western Uganda. Trop Med Int Health 2001; 6(5):407-411.
9. Gouteux JP, Blanc F, Pounekrozou E et al. Tsetse and livestock in Central African Republic: Retreat of Glossina morsitans submorsitans (Diptera, Glossinidae). Bulletin de la Societe de Pathologie Exotique 1994; 87:52-56, [French].
10. Joja LL, Okoli UA. Trapping the vector: Community action to curb sleeping sickness in southern Sudan. Am J Public Health 2001; 91(10):1583-1585.
11. Berriman M, Ghedin E, Hertz-Fowler C et al. The genome of the African trypanosome Trypanosoma brucei. Science 2005; 309(5733):416-422.
12. Lehane MJ, Aksoy S, Gibson W et al. Adult midgut expressed sequence tags from the tsetse fly Glossina morsitans morsitans and expression analysis of putative immune response genes. Genome Biol 2003; 4(10):R63.
13. Attardo G, Strickler-Dinglasan P, Perkin S et al. Analysis of fat body transcriptome from the adult tsetse fly, Glossina morsitans morsitans. Insect Mol Biology 2006; 15(4):411-424.
14. Akman L, Yamashita A, Watanabe H et al. Genome sequence of the endocellular obligate symbiont of tsetse, Wigglesworthia glossinidia. Nat Genet 2002; 32(2):402-407.
15. Toh H, Weiss BL, Perkin SA et al. Massive genome erosion and functional adaptations provide insights into the symbiotic lifestyle of Sodalis glossinidius in the tsetse host. Genome Res 2006; 16(2):149-156.
16. Aksoy S, Berriman M, Hall N et al. A case for a Glossina genome project. Trends in Parasitology 2005; 21(3):107-111.
17. Aksoy S. Wigglesworthia gen. nov. and Wigglesworthia glossinidia sp. nov., taxa consisting of the mycetocyte-associated, primary endosymbionts of tsetse flies. Int J Syst Bacteriol 1995; 45(4):848-851.
18. Dale C, Maudlin I. Sodalis gen. nov. and Sodalis glossinidius sp. nov., a microaerophilic secondary endosymbiont of the tsetse fly Glossina morsitans morsitans. Int J Syst Bacteriol 1999; 49(Pt 1):267-275.
19. Cheng Q, Aksoy S. Tissue tropism, transmission and expression of foreign genes in vivo in midgut symbionts of tsetse flies. Insect Molecular Biology 1999; 8(1):125-132.
20. Cheng Q, Ruel TD, Zhou W et al. Tissue distribution and prevalence of Wolbachia infections in tsetse flies, Glossina spp. Med Vet Entomol 2000; 14(1):44-50.
21. Chen XA, Li S, Aksoy S. Concordant evolution of a symbiont with its host insect species: Molecular phylogeny of genus Glossina and its bacteriome-associated endosymbiont, Wigglesworthia glossinidia. J Mol Evolution 1999; 48(1):49-58.
22. Weiss BL, Mouchotte R, Rio RVM et al. Inter-specific transfer of bacterial endosymbionts between tsetse species: Infection establishment and effect on host fitness. Applied and Environmental Microbiology, 2006; 72(11):7013-7021..
23. Aksoy S, Chen X, Hypsa V. Phylogeny and potential transmission routes of midgut- associated endosymbionts of tsetse (Diptera: Glossinidae). Insect Molecular Biology 1997; 6(2):183-190.
24. Ma WC, Denlinger DL. Secretory discharge and microflora of milk gland in tsetse flies. Nature 1974; 247:301-303.
25. Gibson W, Bailey M. The development of Trypanosoma brucei within the tsetse fly midgut observed using green fluorescent trypanosomes. Kinetoplastid Biol Dis 2003; 2(1):1.

26. Rio RV, Aksoy S. Interactions among multiple genomes: Tsetse, its symbionts and trypanosomes. Insect Biochem and Molecular Biology 2005; (35):691-698.
27. Beard C, O'Neill S, Tesh R et al. Modification of arthropod vector competence via symbiotic bacteria. Parasitology Today 1993; 9(5):179-183.
28. Beard C, Durvasula R, Richards F. Bacterial symbiosis in arthropods and the control of disease transmission. Emerg Infect Dis 1998; 4(4):581-591.
29. Welburn SC, Maudlin I, Ellis DS. In vitro cultivation of rickettsia-like-organisms from Glossina spp. Annals of Tropical Medicine and Parasitology 1987; 81(3):331-335.
30. Beard CB, O'Neill SL, Mason P et al. Genetic transformation and phylogeny of bacterial symbionts from tsetse. Insect Mol Biology 1993; 1:123-131.
31. Hao Z, Kasumba I, Lehane MJ et al. Tsetse immune responses and trypanosome transmission: Implications for the development of tsetse-based strategies to reduce trypanosomiasis. Proc Natl Acad Sci USA 2001; 98(22):12648-12653.
32. Hu Y, Aksoy S. An antimicrobial peptide with trypanocidal activity characterized from Glossina morsitans morsitans. Insect Biochem Mol Biol 2005; 35:105-115.
33. Welburn SC, Gibson WC. Isolation and cloning of a repetitive element from the rickettsia-like organisms of tsetse (Glossina). Parasitology 1989; 98:81-84.
34. Dale C, Young SA, Haydon DT et al. The insect endosymbiont Sodalis glossinidius utilizes a type III secretion system for cell invasion. Proc Natl Acad Sci USA 2001; 98(4):1883-1888.
35. Nogge G. Significance of symbionts for the maintenance of an optimal nutritional state for successful reproduction in haematophagous arthropods. Parasitology 1981; 82(4):101-104.
36. Nantulya VM, Moloo SK. Suppression of cyclical development of Trypanosoma brucei brucei in Glossina morsitans centralis by an anti-procyclics monoclonal antibody. Acta Trop 1988; 45(2):137-144.
37. Durvasula R, Gumbs A, Panackal A et al. Expression of a functional antibody fragment in the gut of Rhodnius prolixus via transgenic bacterial symbiont Rhodococcus rhodnii. Med Vet Entomology 1999; 13(2):115-119.
38. Oggioni MR, Beninati C, Boccanera M et al. Recombinant Streptococcus gordonii for mucosal delivery of a scFv microbicidal antibody. Int Rev Immunol 2001; 20(2):275-287.
39. Beninati C, Oggioni MR, Boccanera M et al. Therapy of mucosal candidiasis by expression of an anti-idiotype in human commensal bacteria. Nat Biotechnol 2000; 18(10):1060-1064.
40. Kruger C, Hu Y, Pan Q et al. In situ delivery of passive immunity by lactobacilli producing single-chain antibodies. Nat Biotechnol 2002; 20(7):702-706.
41. Gibson WC, Bailey M. The development of Trypanosoma brucei within the tsetse fly midgut observed using green fluorescent trypanosomes. Kinetoplastid Biol Dis 2003; 1-3.
42. Hu C, Aksoy S. Innate immune responses regulate trypanosome parasite infection of the tsetse fly Glossina morsitans morsitans. Mol Microbiol 2006; 60(5):1194-1204.
43. Hao Z, Kasumba I, Aksoy S. Proventriculus (cardia) plays a crucial role in immunity in tsetse fly (Diptera: Glossinidiae). Insect Bichem and Molecular Biology 2003; 33(11):1155-64.
44. Asling B, Dushay M, Hultmark D. Identification of early genes in the Drosophila immune response by PCR-based differential display: The Attacin A gene and the evolution of attacin-like proteins. Insect Biochem Mol Biol 1995; 25(4):511-518.
45. Sinkins SP, Gould F. Gene drive systems for insect disease vectors. Nat Rev Genet 2006; 7(6):427-435.
46. Bourtzis K, O'Neill S. Wolbachia infections and arthropod reproduction. BioScience 1998; 48(4):287-293.
47. Rio RV, Wu YN, Filardo G et al. Dynamics of multiple symbiont density regulation during host development: Tsetse fly and its microbial flora. Proc Biol Sci 2006; 273(1588):805-814.
48. Turelli M, Hoffmann AA, McKechnie SW. Dynamics of cytoplasmic incompatibility and mtDNA variation in natural Drosophila simulans populations. Genetics 1992; 132(3):713-723.
49. Politzar H, Cuisance D. An integrated campaign against riverine tsetse Glossina palpalis gambiensis and G. tachinoides by trapping and the release of sterile males. Insect Sci 1984; 5(60):439-442.
50. Vreysen MJ, Saleh KM, Ali MY et al. Glossina austeni (Diptera: Glossinidae) eradicated on the Island of Unguga, Zanzibar, using the sterile insect technique. J Econ Entomology 2000; 93:123-135.
51. Douglas AE. Buchnera bacteria and Other Symbionts of Aphids. In: Bourtzis K, Miller TA, eds. Insect Symbiosis. New York: CRC Press, 2003.
52. Wernegreen JJ. Genome evolution in bacterial endosymbionts of insects. Nat Rev Genet 2002; 3(11):850-861.
53. Weiss BL, Mouchotte R, Rio RV et al. Inter-specific transfer of bacterial endosymbionts between tsetse species: Infection establishment and effect on host fitness. Appl Environ Microbiol 2006; 72(11):7013-7021.

54. Beard C, Mason P, Aksoy S et al. Transformation of an insect symbiont and expression of a foreign gene in the Chagas' disease vector Rhodnius prolixus. Am J Trop Med Hyg 1992; 46(2):195-200.
55. Durvasula RV, Gumbs A, Panackal A et al. Prevention of insect-borne disease: An approach using transgenic symbiotic bacteria. Proceedings of the National Academy of Sciences of the United States of America 1997; 94(7):3274-3278.
56. Durvasala R, Sundarum KR, Cordon-Rosales C et al. Rhodnius prolixus and its symbiont, Rhodococcus rhodnii, A model for Paratransgenic control of Disease Transmission. In: Bourtzis K, Thomas AM, eds. Insect Symbiosis. New York: CRC Press, 2003.
57. Lindh JM, Terenius O, Faye I. 16S rRNA gene-based identification of midgut bacteria from field-caught Anopheles gambiae sensu lato and A. funestus mosquitoes reveals new species related to known insect symbionts. Appl Environ Microbiol 2005; 71(11):7217-7223.

Chapter 4

Bacteria of the Genus *Asaia*:

A Potential Paratransgenic Weapon Against Malaria

Guido Favia,* Irene Ricci, Massimo Marzorati, Ilaria Negri, Alberto Alma, Luciano Sacchi, Claudio Bandi, and Daniele Daffonchio

Abstract

Symbiotic bacteria have been proposed as tools for control of insect-borne diseases. Primary requirements for such symbionts are dominance, prevalence and stability within the insect body. Most of the bacterial symbionts described to date in *Anopheles* mosquitoes, the vector of malaria in humans, have lacked these features. We describe an α-Proteobacterium of the genus *Asaia*, which stably associates with several *Anopheles* species and dominates within the body of *An. stephensi*. *Asaia* exhibits all the required ecological characteristics making it the best candidate, available to date, for the development of a paratransgenic approach for manipulation of mosquito vector competence. Key features of *Asaia* are: (i) dominance within the mosquito-associated microflora, as shown by clone prevalence in 16S rRNA gene libraries and quantitative real time Polymerase Chain Reaction (qRT-PCR); (ii) cultivability in cell-free media; (iii) ease of transformation with foreign DNA and iv) wide distribution in the larvae and adult mosquito body, as revealed by transmission electron microscopy, and in situ-hybridization experiments. Using a green fluorescent protein (GFP)-tagged *Asaia* strain, it has been possible to show that it effectively colonizes all mosquito body organs necessary for malaria parasite development and transmission, including female gut and salivary glands. *Asaia* was also found to massively colonize the larval gut and the male reproductive system of adult mosquitoes. Moreover, mating experiments showed an additional key feature necessary for symbiotic control, the high transmission potential of the symbiont to progeny by multiple mechanisms. *Asaia* is capable of horizontal infection through an oral route during feeding both in preadult and adult stages and through a venereal pattern during mating in adults. Furthermore, *Asaia* is vertically transmitted from mother to progeny indicating that it could quickly spread in natural mosquito populations.

Introduction

Symbiotic control is a novel method for controlling human and plant insect-borne diseases. This approach harnesses the symbiotic microbes to provide anti-disease strategies in the insect hosts. Microbes and insects have co-evolved from 10 to several hundred million years, and these associations often reflect extensive cooperation between the partners. The study of the microbes' role in insects, using molecular techniques, has opened a previously unknown world of possible interactions, but much still remains to be explored.[1]

Recently, several reviews documented the role of microrganisms in insects as well as the potential use of these microbes and their metabolic capabilities as biocontrol agents against

*Corresponding Author: Guido Favia—Dipartimento di Medicina Sperimentale e Sanità Pubblica, Università degli Studi di Camerino, 62032 Camerino, Italy. Email: guido.favia@unicam.it

Transgenesis and the Management of Vector-Borne Disease, edited by Serap Aksoy.

important diseases.[2-6] Some of these symbiotic microbes in insects can have a nutritional role, as in the case of aphids where the symbiont *Buchnera* recycles nitrogen, or a protective function that enables the insect host to resist to parasitoids or to environmental stresses.[7] Other bacterial symbionts such as the α-Proteobacterium *Wolbachia* and the Bacteroidetes *Cardinium* strongly influence the sexual behavior of the host.[8-10] Insect symbionts have been proposed for disease control based either on their ability to manipulate insect populations by affecting sex balance in natural populations and/or on paratransgenic approaches where bacterial symbionts are genetically modified to express toxins against disease causing microorganisms transmitted by the insect. Recently, paratransgenic symbiont-based protection approaches have been proposed both in medicine, for controlling African and American Trypanosomiasis[11,12] and HIV infection,[13] and in agriculture, for controlling the Pierce's disease of grapevine caused by the bacterium *Xylella fastidiosa.*[14] Decreasing the infestation of Mediterranean fruit fly *Ceratitis capitata* through the use of *Wolbachia*-infected lines have also been porposed.[15] Hence, insect associated microorganisms could exert a direct pathogenic effect against the host,[16] or interfere with their reproduction,[15,17] or reduce their vector competence.[18,19]

Malaria and Symbiotic Control Strategies

Malaria is a vector-borne infectious disease, caused by parasites of the genus *Plasmodium* transmitted by female *Anopheles* mosquitoes, widespread in tropical and subtropical regions. It threatens 300 to 500 million people[20] and kills around 2 million person per year, mostly children at pre-scholar age. The control strategies currently available, mostly represented by the use of antimalarial drugs and insecticides, are becoming less effective due to resistance developed by parasites and vectors, thus new strategies are urgently required. Accordingly, one of the major objectives of mosquito-based malaria control strategies has been to interfere with parasite transmission mechanisms in the mosquito vectors.[21] Towards this goal, at the beginning of this millennium genetic transformation systems have been developed for *Anopheles gambiae*[22] and *An. stephensi*,[23] the main malaria vectors in Africa and Asia, respectively. Subsequently, transgenic Anopheline mosquitoes impaired in transmission of malaria parasites have been produced.[24] However, it has been shown that such genetic manipulation can result in reduced mosquito fitness.[25] Recently, Marrelli and coworkers[26] for the first time have shown that transgenic Anopheline mosquitoes expressing the SM1 peptide are impaired for transmission of *Plasmodium berghei* (a rodent malaria parasite) and when fed on infected mice, are more fit and display higher fecundity and lower mortality than the corresponding non-transgenic mosquitoes. However, the transgenic malaria-resistant mosquitoes have a selective advantage over nontransgenic insects only when fed on the *Plasmodium*-infected blood meal.

Symbiotic control could represent an alternative strategy to control malaria in an environmentally friendly way. To further these studies, the identification of microorganisms potentially useful in symbiotic control is a critical prerequisite. Symbionts resident in the mosquito body would be especially of interest since they could interfere with the *Plasmodium* life cycle in different organs. About 24 hours after gaining entry into the *Anopheles* mosquito midgut through feeding, *Plasmodium* ookinetes cross the midgut epithelium and, once they reach the basal lamina, develop into oocysts. At this stage the parasite is particularly "weak" and vulnerable to control methods aimed to interrupt the disease transmission, since in natural infections only few oocysts are present. From thousands of gametocytes ingested by the insect vector in an infected blood meal, typically less than ten oocysts develop. The use of natural or genetically modified (GM) bacteria to deliver anti-parasite molecules presents some advantages over the use of GM mosquito vectors, since the release of non-pathogenic bacteria has been already monitored in different *Anopheles* species and the production of appropriate numbers of bacteria is easily achievable.[5] In contrast, the release of GM malaria vectors involves more serious technical issues, because a straightforward generalized technology suitable for different vector models, designed for replacing a wild vector population with a parasite transmission-refractory one is not yet available.[5]

Despite the great potential that symbiotic control holds, few studies have been performed on the microbiota associated with malaria mosquito vectors. Thus its potential use in biocontrol of parasite transmission represents a field of mosquito biology that should be explored further.[5,27,28,29] Khampang and colleagues[28] pioneered the field of mosquito control by symbiotic bacteria in a study describing the isolation of *Enterobacter amnigenus* from the gut of *An. dirus* larvae and its genetic modification to express the cryIVB gene of *B. thuringiensis* subsp. *israelensis* and the binary toxin genes of *B. sphaericus*. After a break of six years, Lindh and coworkers[29] screened the midgut bacterial community of field collected *An. gambiae* sensu lato and *An. funestus* by sequence analysis of bacterial 16S rRNA genes, directly amplified by PCR from the total DNA extracted from the mosquito or from isolates cultured in different growth media. They identified 16 bacterial species belonging to 14 different genera. These microbes were phylogenetically related to *Acidovorax* sp., *Spiroplasma* sp., *Aeromonas hydrophila, Bacillus simplex, Serratia odorifera* and *Nocardia corynebacterioides,* a relative of *Rhodococcus rhodnii* (Nocardiaceae) which is a symbiont found in the Chagas' disease insect vector, that has been proposed as a candidate for paratransgenic control of *Trypanosoma* transmission.[30] In a recent report, *Escherichia coli* and *Enterobacter agglomerans* isolated from the midgut of *An. stephensi* were genetically engineered to display two anti-*Plasmodium* effector proteins (SM1 and phospholipase-A2). Both engineered bacteria resulted in inhibition of *P. berghei* development. However, the fitness of the transgenic bacteria in the mosquito body was a limiting factor. *E. coli* could only survive for a short time in the mosquito body, while *E. agglomerans* survived for two weeks but its presence was restricted to the mosquitoes' midgut organ only.[27]

α-Proteobacteria of the Genus *Asaia* Dominate the Microflora of *An. stephensi*

Recently, Riehle and Jacobs-Lorena[5] reviewed the current knowledge on the use of bacteria to express anti-parasite molecules in mosquitoes and summarized the general concepts necessary to evaluate the feasibility of paratransgenic approach in vectors. An ideal microbial symbiont for malaria biocontrol should have the following characteristics:

i) dominance within the insect-associated microflora to efficiently outcompete other microbial symbionts and display maximum effect against the target parasite to be controlled;
ii) cultivatable in cell-free media;
iii) readily applicable genetic transformation system in order to introduce parasite resistance traits;
iv) exhibit stable expression and maintenance of the newly acquired anti-pathogen function;
v) have wide distribution in the larvae and adult insect body;
vi) co-localize with pathogenic parasites in the relevant insect organs (i.e.: gut, salivary glands).

Symbiotic bacteria can also be engineered to express multiple effector molecules capable of killing the parasite by different mechanisms, maximizing in this way the final effect. The ideal molecule should be small, soluble, stable and resistant to midgut digestive enzymes.

To search for mosquito symbionts that fulfill these requirements, a range of molecular and ultrastructural analytical tools, widely used in environmental microbial ecology, are available. Using some of these versatile tools, we have recently implemented a multidisciplinary approach to characterize the bacterial community associated with different *Anopheline* species (*An. stephensi, An. maculipennis,* and *An. gambiae*),[31] in order to identify a potential candidate to be used as a biocontrol agent. The initial characterization of the microbiota associated with these *Anopheline* species was performed through the analysis of 16S rRNA gene libraries and revealed the presence of microorganisms belonging to the order of Acetobacteriales, Enterobacteriales, Bacillales and Sphingomonadales. Within the order of Acetobacteriales almost all the sequences were related to the genus *Asaia*, an α-Proteobacterium phylogenetically close to acetic acid bacteria.[32,33] The sequence obtained showed over 99% nucleotide identity with those of *Asaia bogorensis* and *Asaia siamensis,* two species previously isolated from tropical flowers, likely associated with the so-called phytotelmata, structures formed by non-aquatic plants for water recovery where mosquito larvae commonly live. Within the 16S rRNA gene libraries determined from DNA extracted from the gut of *An. stephensi*, sequences related to *Asaia* represented 90% of the clones examined. This percentage decreased to 20% in *An. maculipennis* collected in the field in Italy. *Asaia* was sporadically (around 5% of the clones) found in 16S rRNA gene libraries constructed from total DNA of *An. gambiae* individuals collected in Burkina Faso. The percentage of *Asaia* clones found in *An. gambiae* was rather low with respect to the other species. Finding only 5% of the clones attributable to *Asaia* in *An. gambiae* could be associated with an analytical bias due to mosquito preservation, i.e., prolonged freezing and storage may decrease the amount of amplifiable DNA from *Asaia*. Independent sets of experiments performed on *An. stephensi* specimens, revealed that the numbers of *Asaia* colonies isolated from mosquito individuals stored for several days at low temperatures, was much lower than those obtained from freshly-processed individuals. Most likely *Asaia* is easily lost in dead individuals when stored at low temperatures if they are not properly conserved in suitable cryoprotectants. Indeed, it has been recently shown that interference with *An. gambiae* innate immune system by a transient silencing of *AgDscam* leads to a massive proliferation of *Asaia bogorensis* in the mosquito hemolymph,[34] indicating that this bacterium also coexists in this main malaria vector species. The prevalence of *Asaia* in the 16S rRNA gene clone libraries from *An. gambiae* populations should be further investigated using freshly collected insects.

By using an *Asaia*-specific PCR test, we found that its prevalence was 100% both in a lab-reared population of *An. stephensi* (300 individuals tested) and in a natural field population of *An. maculipennis* (60 individuals tested). The dominance of *Asaia* in the *An. stephensi* microbiota was analyzed by quantitative real time (qRT) PCR using total DNA extracted from dissected mosquito organs including gut, female salivary glands and the reproductive system. To minimize quantification biases due to DNA losses during DNA extraction process from these small organs, we established a relative quantification assay consisting of two separate qRT-PCR reactions. The first was performed with *Asaia*-specific primers while the second with universal bacterial primers. On the basis of the numbers obtained from the two reactions, the ratio of the *Asaia* to total bacterial 16S rRNA gene copy (ABR) was determined. The analysis showed that *Asaia* is the prevalent bacterium in all the organs analyzed (copies of *Asaia* 16S rRNA gene in the ranges of: 1.3×10^4 to 8.7×10^7 in the intestine; 5.3×10^2 to 4.0×10^6 in the salivary glands; 7.5×10^1 to 3.4×10^6 in the female reproductive system) with ABR up to 0.4 in the gut, indicating that *Asaia* can account for up to 40% of the bacteria in the gut, thus representing the dominant symbiont within the bacterial community of *An. stephensi*. High numbers of *Asaia* were also found in both salivary glands and male reproductive system, which have rarely been reported to be colonized by bacteria in other insects[35] and in the female reproductive system, where vertically transmitted endosymbionts are typically found.

The genome sequences of bacterial obligate endosymbionts have revealed a common feature for these microbes and shows the highly integrative nature of these associations with their insect host biology. These bacteria have the smallest known bacterial genomes and, as a result of the extensive genome reductions they have undergone, many genes commonly found in the phylogenetically closely related free-living bacteria are absent. The 'genome reduction' theory explains the loss of genes as a process that leads to an increase in the fitness of the obligate in the host's stable and nutrient-rich environment. Evidently, genes coding for functions no longer required while living in the cytoplasm of the host cell, are progressively lost during evolutionary time, since they constitute a metabolic load for the bacterium, for example for the transcription machinery or during DNA replication.[36] The 'genome reduction' process however has restricted these obligate endosymbiotic species to a strict intracellular lifestyle without the ability to surviving outside the host cell.[37] *Asaia* does not share this feature with obligate symbionts, and instead can be cultured in vitro in artificial media. A pre-enrichment step in liquid medium at pH 3.5, followed by plating on carbonate-rich medium, allowed the isolation of single, pink colonies capable of dissolving carbonate with the generation of dissolution haloes around the colonies.[32,33] The phylogenetic position of a bacterial isolate from *An. stephensi*, based on 16S-23S rRNA gene intergenic transcribed spacer (Marzorati and Daffonchio, unpublished results), agrees with the phylogeny deduced from 16S rRNA gene sequencing[31] and has placed the isolate close to the species *Asaia bogorensis* previously isolated from plant tissues.[32]

Asaia is Localized in Different Organs of *An. stephensi*

Transmission electron microscopy (TEM), PCR and in situ hybridization analyses with specific 16S rRNA gene-based probes revealed that *Asaia* is associated with multiple organs of *Anopheles*, including guts, salivary glands and male and female reproductive organs (Fig. 1). TEM analysis of mosquito midgut contents indicated that *Asaia* is capable of producing a thick slime matrix presumably made of exopolysaccharides that are not apparently produced when the cells are cultured in cell-free media. This observation opens the hypothesis that *Asaia*

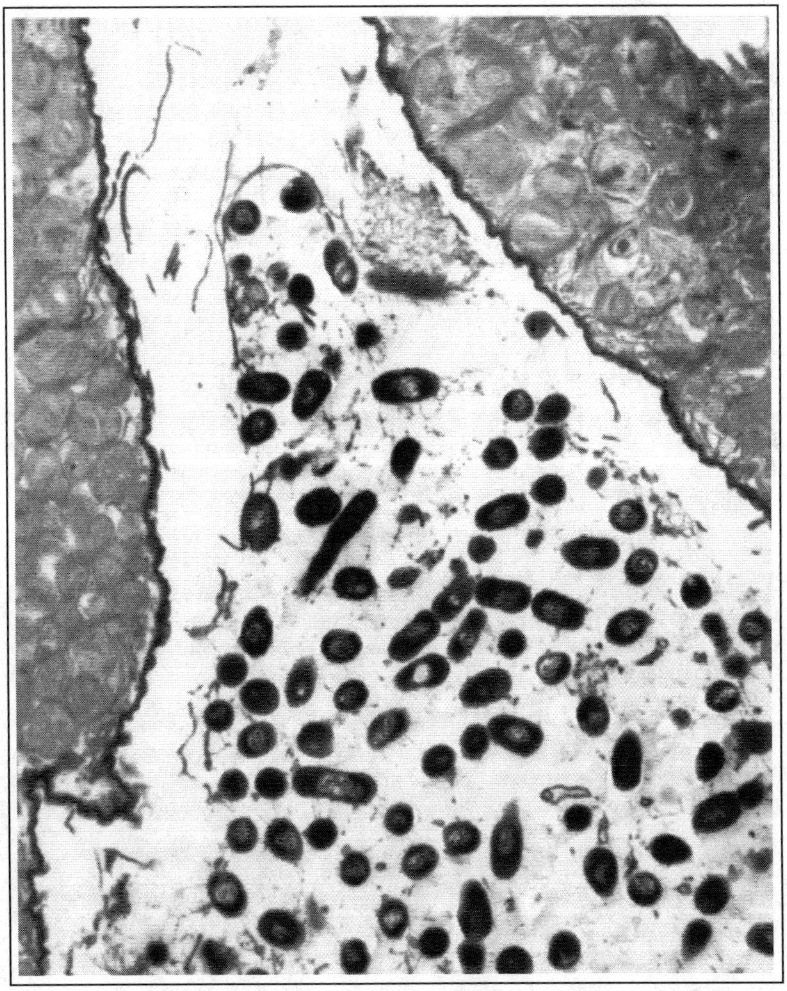

Figure 1. TEM micrograph of an *An. stephensi* adult male deferent duct showing high numbers of bacteria of the genus *Asaia*.

could establish a chemical communication with the insect tissues, which in turn trigger the expression of factors (e.g., exopolysaccharides implicated in biofilm formation) presumably important for the body colonization. The localization of *Asaia* not only in the mosquito gut but also in the salivary glands overlaps with that of *Plasmodium* and further supports the utility of *Asaia* for paratransgenic applications. As already mentioned, in the mosquito's midgut the parasite should be particularly vulnerable to symbiotic control to interrupt disease transmission, due to the relative low numbers of oocysts.

An important characteristic of a symbiotic control microorganism is the ability to transform cells with ease to allow for their genetic manipulation for expression of anti-parasite molecules. Transformation of *Asaia* isolated from *An. stephensi* was attempted by using different plasmid vectors, including a broad host range plasmid and two plasmids previously developed as shuttle vectors between *Escherichia coli* and the genera *Acetobacter* and *Gluconobacter*, closely related to the genus *Asaia*.[38] Among these, the plasmid pHM2 was the most efficient and transformed *Asaia* with an efficiency of 4.7×10^5 transformants per µg of DNA. The gene cassette coding for the green fluorescent protein (Gfp) was cloned into the plasmid vector pHM2 and provided a valuable optical marker to trace mosquito body colonization. Transformed *Asaia* cells were found to efficiently express the protein and showed bright fluorescence useful for localization of the symbiotic cells in the insect body (Fig. 2a). The Gfp-tagged bacterium is able to colonize larvae bodies when bacterial cells are suspended in the breeding water. The colonization pattern can be observed in dissected mosquitoes by fluorescent confocal laser scanning microscopy (CLSM). In our experience a relevant percentage typically around 50% of the larvae examined by fluorescence microscopy and CLSM showed a massive colonization by the Gfp-tagged strain along the gut (Fig. 3). The colonization experiments indicate that the horizontal route of infection through feeding is an efficient way of acquisition of the bacterium by the larvae. This has important practical implications for the delivery of transgenic symbionts capable of interfering with *Plasmodium* transmission. The bacterium could be easily sprayed in environments where the larvae reside, potentially allowing for a high colonization rate.

By including the Gfp-tagged bacterial cells in the feed (blood or cotton pad soaked with sugar solution) of *An. stephensi* adults, it was shown that *Asaia* can also efficiently colonize the insect body. In adult mosquitoes *Asaia* cells have been confirmed to efficiently colonize the gut (Fig. 2b) where they are localized in large fluorescent cell aggregates both in males and females, the latter fed either on sugar solution or blood. While the gut colonization by fluorescent cells

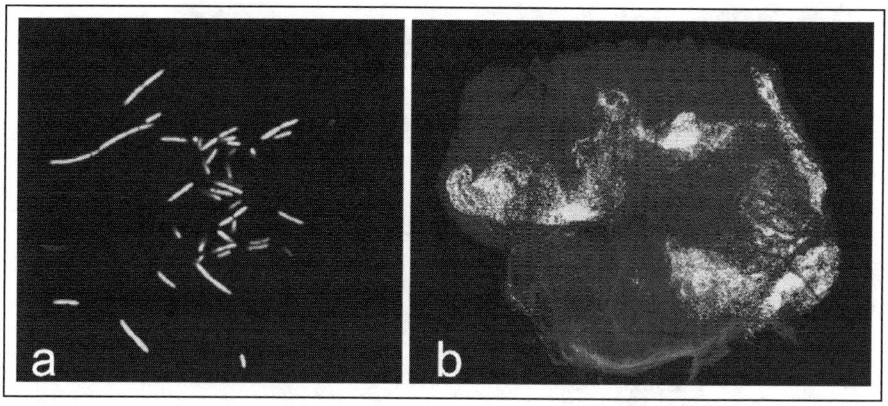

Figure 2. Laser scanning confocal microscopy images of *Asaia* sp. strain SF2.1 (Gfp) bacteria expressing a bright green fluorescence (a) and an adult mosquito gut (b) showing high concentration of the transformed bacteria.

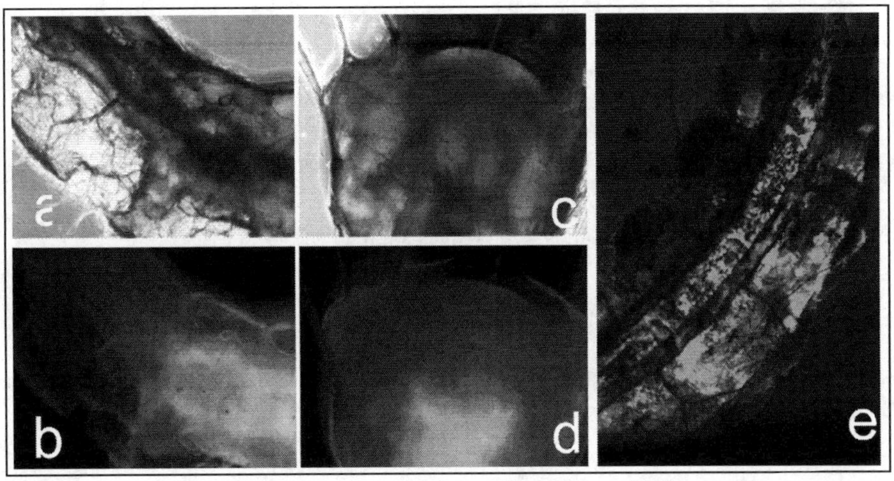

Figure 3. Fluorescence microscope images of larvae bred in medium enriched with cells of *Asaia* sp. strain SF2.1(Gfp). Phase contrast (a, c) and fluorescence (b, d) microscope images of the terminal (a, b) and apical region (c, d) of the larval gut. (e) Laser scanning confocal microscopy image of the middle portion of the abdomen showing intense bacterial fluorescence in the larvae.

occurs in about 48 h after a sucrose meal, bacterial colonization could already be observed 24 h following a blood meal. This difference between sugar and blood meal in cell dispersion in the midgut can be explained by a delay of a sugar meal in reaching the midgut since it first has to pass through the crop. After colonization, fluorescent *Asaia* cells are also detected in the salivary glands, an organ that is invaded by sporozoites, the infective stage of the malaria parasite for the vertebrate host. The localization of *Asaia* in the two mosquito organs, gut and salivary glands, both critical for completion of the *Plasmodium* life cycle in the insect, supports *Asaia's* potential role as a Trojan horse for in situ delivery of antiparasite effectors to the appropriate organs.

The colonization experiments show that *Asaia* cells are able to colonize the adult mosquito body and establish a massive infection when they are inoculated into the sugar-containing medium at concentrations in the range of 10^3-10^8 cells ml^{-1}. Furthermore colonization lasts during the full life span of adults as evidenced by periodic observation of mosquitoes by fluorescence microscopy and CLSM.[31] These experiments show the growth of *Asaia* within the mosquito body in large microcolonies found in all the organs examined.

Asaia: A Self-Spreading Carrier of Potential Antagonistic Factors in Mosquitoes

Analyses of the genital systems of *An. stephensi* adults fed with Gfp-tagged *Asaia*, revealed that the bacterium is also capable of reaching the male and female gonoducts within a few hours post acquisition. CLSM showed large microcolonies formed along the male gonoduct epithelium.[31] This observation was further confirmed by TEM analysis where a plug of bacterial cells was observed within the gonoduct in wild type males, indicating that the gonoducts are the natural niches for *Asaia* colonization in the mosquito. The presence of *Asaia* in the genital ducts opens the intriguing possibilities that this microorganism may be vertically transmitted from the mother to her progeny and that transmission can also occur paternally. The latter route of transmission has recently been demonstrated by Moran and Dunbar[39] for the secondary symbionts of aphids. It is possible that the paternal transfer might constitute an alternative route for introducing the symbionts into natural populations.

Vertical and venereal transmission of *Asaia* has been demonstrated by crossing males fed with the Gfp-tagged *Asaia* with normal females of *An. stephensi* in the laboratory. After mating, fluorescent bacteria can be detected in the spermatheca and in the terminal portions of the gastrointestinal tract, thus indicating the transmission of the bacterium along with sperm. Furthermore, the vertical transmission of the bacterium to the progeny has also been observed. Progeny resulting from matings between females and males fed with and without Gfp-tagged *Asaia,* respectively were observed to be colonized by fluorescent cells (vertical or maternal transmission). Control experiment, in which adult mosquitoes were not previously fed with the Gfp-tagged bacterium, gave negative results, i.e., the progeny did not show any fluorescent cells in the gut.

In the *Asaia/Anopheles* symbiotic system, the transmission routes among different members of a population and from parents to progeny represent a complex situation. It resembles both the obligate intracellular symbionts, which are vertically transmitted following egg cytoplasm colonization, as well as the extracellular midgut bacteria that are frequently acquired from the environment where the insects live. In this respect *Asaia* resembles the symbionts living in the gut of wood-feeding cockroaches and termites[40] for which a clear-cut distinction between environmental acquisition and vertical transmission has been difficult to establish. In the case of *Asaia,* acquisition through the environment probably represents the most common source of bacterial infections, given the ecological distribution of the bacterium, and its wide association with different plants. However the mother to offspring route of transmission and even the paternal route would increase the capacity of the bacterium to colonize the insect body. The multiple patterns of colonization mechanisms suggest that *Asaia* is evolving toward an efficient exploitation of the insect niche.[31]

Conclusions and Perspectives

Paratrangenesis is an innovative technology aimed to interfere with the host insect's biology. It aims to interfere with insects' capabilities to transmit pathogenic microorganisms by utilizing the microbial symbionts as carriers of antagonistic factors within the insect body. One of the major challenges in developing an efficient paratransgenic system is the stability of the symbiotic system that strictly depends on the type and strength of host-symbiont associations. The extent of the strength is determined by multiple factors including the relative abundance of the symbiont in the host, the localization of the symbiont in single or multiple host body organs and its capacity to efficiently colonize the insect body and spread within host populations. In relation to the latter point, sexual obligate endosymbionts such as *Wolbachia,* could be excellent paratransgenic vectors since they have the ability to rapidly spread through natural insect populations by virtue of the Cytoplasmic Incompatibility phenomenon they confirm on infected hosts. However, *Wolbachia* has major limitations for a paratransgenic approach. It is largely restricted in its localization to host tissues, which do not harbor parasites and in addition given the absence of in vitro cultures available for *Wolbachia,* its genetic manipulation has been difficult. Furthermore *Wolbachia* infections have not been reported in any natural *Anopheline* mosquito populations.[41] There are very few studies on the associated symbiotic microbiota of the Anopheline malaria vectors. The search for an 'ecologically suitable' microbial candidate, to be used as a carrier of *Plasmodium* antagonistic factors, seems to be one of the major 'technological bottlenecks' for the development of an efficient paratransgenic approach for malaria control. Indeed, several effector molecules capable of impairing transmission of *Plasmodium* have been already described and, in the absence of suitable microbial carriers, they have been engineered directly in the mosquitoes, even though at the expenses of the insect fitness that cannot compete with the natural vector populations. In such a context, bacteria of the genus *Asaia* seem to have the potential to address a paratransgenic approach for malaria control. Their ecology, localization and transmission routes within *Anopheles* positively respond to the major features such a symbiont should have.

In a genetically engineered mosquito or a microbial symbiont of a paratransgenic insect, the selected anti-pathogen molecule should be expressed during the midgut phases of the malaria parasite development. This stage represents a remarkable bottleneck in the malaria cycle, in which the parasite cell numbers are very low with respect to other parasite lifestages in the insect. Due to the low number, the interference with parasite cells would be reasonably more effective and can lead to significant reduction of transmission rates. *Asaia* is present at high densities in the mosquito midgut lumen, so the bacteria could express effector molecules directly in corsivo. Another important localization of *Asaia* is in the salivary glands, which are also invaded by the sporozoites, the human infective stages of the malaria parasites. As already shown for transgenic mosquito, the synergistic expression of antiplasmodial effectors in both the midgut and salivary gland organs, could strongly impair the transmission vector competence.[24] The double localization of *Asaia* in the midgut and the salivary glands could efficiently amplify the impact of the expression of antagonistic molecules in paratransgenic mosquitoes.

For an efficient paratransgenic approach the localization of *Asaia* in key organs of the mosquito relevant for malaria cycle should be coupled with an effective antiparasite molecule and an adequate expression system. Several potential molecules and corresponding gene cassettes have already been selected in the recent past and tested for their ability to impair *Plasmodium* transmission. By feeding *An. stephensi* with *E. coli* expressing a fusion protein of ricin and a single chain antibody against an ookinete surface protein of *P. berghei* (Pbs 21), Yoshida and coworkers[42] obtained a consistent inhibition of oocysts formation (up to 95%). Another interesting effector molecule is SM1, a synthetic peptide molecule able to interfere with parasite development. SM1 is a short peptide identified by screening a phage display library for protein domain binding salivary gland and midgut epithelia.[43] Transgenic *An. stephensi* mosquitoes that have been transformed to express a SM1 tetramer under the control of a carboxypeptidase promoter, showed a dramatic reduction of parasite trasmission.[24] In some of these experiments, the transmission was totally blocked most probably due to SM1 binding onto epithelial cell surface receptors that mediate parasite invasion. Several other molecules have also been described as potential 'bullets' to impair malaria transmission, and although their modes of action for parasite killing or interference with parasite development are unknown, their effect(s) are well documented. For example in the case of the phospolipase A2 (PLA2), the expression of this molecule strongly inhibited the ookinete invasion of mosquito midgut epithelium.[44] Although several effector molecules have been described to date, the incessant search for new molecules is strongly needed considering the widely described capacity of malaria parasites to acquire drug resistance and to increase the potential to block parasite development.[5]

Due to its ability to propagate within insect populations using both vertical and horizontal routes, and to infect both pre-adult and adult stages, *Asaia* is an exceptionally attractive candidate to drive the antiplasmodial molecules into vector populations. *Asaia* has been shown to be present in the male genital ducts,[31] which has a major impact on the transmission routes by which the bacterium can ensure its propagation to the progeny. Indeed, a male to female transmission during mating as well as a passage from the mother to the offspring have been already reported,[31] in addition to acquisition from the environment, which appears to be the main source of infection for both preadult and adult stages. The high level of mosquito colonization reached by *Asaia* after feeding indicates that horizontal transmission by the oral route can efficiently lead to the infection of the mosquito body and could possibly be exploited in the field. A potential delivery system could involve the placement of sucrose sources supplemented with recombinant bacteria carrying effector molecules for impairing *Plasmodium* transmission close to or in mosquito oviposition sites. In this way mosquitoes would be repeatedly exposed to the recombinant bacteria. Obviously, many ethical and practical issues will have to be addressed before field applications can be entertained, including for example the consequences related to the local peridomestic increase in vector populations and to the introduction of the desired phenotype(s) by paratransgenic mosquitoes in selected areas, since this has to be achieved on a wide scale (even entire countries) for efficacy. In summary, paratransgenesis has the potential to develop a very effective and innovative malaria control strategy, which can be integrated with the currently applied methodologies. In this context, *Asaia* could exert a very important rôle.

Acknowledgements

Irene Ricci was funded by 'Compagnia di San Paolo' in the context of the Italian Malaria Network.

References

1. Ishikawa H. Insect in symbiosis: An introduction. In: Bourtzis K, Miller TA, eds. Insect Symbiosis Boca Raton: Crc Press Llc, 2003:33487.
2. Dillon RJ, Dillon VM. The gut bacteria of insects: Nonpathogenic interactions. Annu Rev Entomol 2004; 49:71-92.
3. Gil R, Latorre A, Moya A. Bacterial endosymbionts of insects: Insights from comparative genomics. Environ Microbiol 2004; 6:1109-22.
4. Hoffmeister M, Martin W. Interspecific evolution: Microbial symbiosis, endosymbiosis and gene transfer. Environ Microbiol 2003; 5:641-49.
5. Riehle MA, Jacobs-Lorena M. Using bacteria to express and display anti-parasite molecules in mosquitoes: Current and future strategies. Insect Biochem Mol Biol 2005; 35:699-707.
6. Zientz E, Silva FJ, Gross R. Genome interdependence in insect-bacterium symbioses. Genome Biol 2001; 2:1032.1-32.6.
7. Douglas AE. Nutritional interactions in insect-microbial symbioses: Aphids and their symbiotic bacteria Buchnera. Annu Rev Entomol 1998; 43:17-37.
8. Sinkins SP. Wolbachia and cytoplasmic incompatibility in mosquitoes. Insect Biochem Mol Biol 2004; 34:723-29.
9. Gotoh T, Noda H, Ito S. Cardinium symbionts cause cytoplasmic incompatibility in spider mites. Heredity 2007; 98:13-20.
10. Marzorati M, Alma A, Sacchi L et al. A novel Bacteroidetes symbiont is localized in Scaphoideus titanus, the insect vector of Flavescence doree in Vitis vinifera. Appl Environ Microbiol 2006; 72:1467-75.
11. Aksoy S. Control of tsetse flies and trypanosomes using molecular genetics. Vet Parasitol 2003; 115:125-45.
12. Durvasula RV, Sundaram RK, Cordon-Rosales C et al. Rhodnius prolixus and its symbiont, Rhodococcus rhodnii: A model for paratrangenic control of disease transmission. In: Bourtzis K, Miller TA, eds. Insect Symbiosis. Boca Raton: Crc Press Llc, 2003:33487.
13. Chang TL, Chang CH, Simpson DA et al. Inhibition of HIV infectivity by a natural human isolate of Lactobacillus jensenii engineered to express functional two-domain CD4. Proc Natl Acad Sci USA 2003; 100:11672-77.
14. Bextine B, Lauzon C, Potter S et al. Delivery of a genetically marked Alcaligenes sp. to the glassy-winged sharpshooter for use in a paratransgenic control strategy. Curr Microbiol 2004; 48:327-31.
15. Zabalou M, Riegler M, Theodorakopoulou M et al. Wolbachia-induced cytoplasmic incompatibility as a means for insect pest population control. Proc Natl Acad Sci USA 2004; 101:15042-45.
16. Schnepf E, Crickmore N, Van Rie J et al. Bacillus thuringiensis and its pesticidal crystal proteins. Microbiol Mol Biol Rev 1998; 62:775-806.
17. Zchori-Fein E, Gottlieb Y, Kelly SE et al. A newly discovered bacterium associated with parthenogenesis and a change in host selection behavior in parasitoid wasps. Proc Natl Acad Sci USA 2001; 98:12555-60.
18. Beard CB, Dotson EM, Pennington PM et al. Bacterial symbiosis and paratransgenic control of vector-borne Chagas disease. Int J Parasitol 2001; 31:621-27.
19. Baldridge GD, Burkhardt NY, Simser JA et al. Sequence and expression analysis of the ompA gene of Rickettsia peacockii, an endosymbiont of the Rocky Mountain wood tick, Dermacentor andersoni. Appl Environ Microbiol 2004; 70:6628-36.
20. World Health Organization. World malaria report 2005. Geneva, Switzerland: World Health Organization, Online http://www.rbm.who.int/wmr2005.
21. Atkinson PW, Michel K. What's buzzing? Mosquito genomics and transgenic mosquitoes. Genesis 2002; 32:42-48.
22. Grossman GL, Rafferty CS, Clayton JR et al. Germline transformation of the malaria vector, Anopheles gambiae, with the piggyBac transposable element. Insect Mol Biol 2001; 10:597-604.

23. Catteruccia F, Nolan T, Loukeris TG et al. Stable germline transformation of the malaria mosquito Anopheles stephensi. Nature 2000; 405:959-62.
24. Ito J, Ghosh A, Moreira AL et al. Transgenic anopheline mosquitoes impaired in transmission of a malaria parasite. Nature 2002; 417:452-55.
25. Catteruccia F, Godfray HC, Crisanti A. Impact of genetic manipulation on the fitness of Anopheles stephensi mosquitoes. Science 2003; 299:1225-27.
26. Marrelli MT, Li C, Rasgon JL et al. Transgenic malaria-resistant mosquitoes have a fitness advantage when feeding on Plasmodium-infected blood. Proc Natl Acad Sci USA 2007; 104:5580-83.
27. Riehle MA, Moreira CK, Lampe D et al. Using bacteria to express and display anti-Plasmodium molecules in the mosquito midgut. Int J Parasitol 2007; 37:595-603.
28. Khampang P, Chungjatupornchai W, Luxananil P et al. Efficient expression of mosquito-larvicidal proteins in a gram-negative bacterium capable of recolonization in the guts of Anopheles dirus larva. Appl Microbiol Biotechnol 1999; 51:79-84.
29. Lindh JM, Terenius O, Faye I. 16S rRNA gene-based identification of midgut bacteria from field-caught Anopheles gambiae sensu lato and A. funestus mosquitoes reveals new species related to known insect symbionts. Appl Environ Microbiol 2005; 71:7217-23.
30. Hoy MA. Transgenic insects for pest management programs: Status and prospects. Environ Biosafety Res 2003; 2:61-64.
31. Favia G, Ricci I, Damiani C et al. Bacteria of the genus Asaia stably associate with Anopheles stephensi, an Asian malarial mosquito vector. Proc Natl Acad Sci USA 2007; 104:9047-51.
32. Yamada Y, Katsura K, Kawasaki H et al. Asaia bogorensis gen. nov., sp. nov., an unusual acetic acid bacterium in the alpha-Proteobacteria. Int J Syst Evol Microbiol 2000; 50:823-29.
33. Katsura K, Kawasaki H, Potacharoen W et al. Asaia siamensis sp. nov., an acetic acid bacterium in the alpha-proteobacteria. Int J Syst Evol Microbiol 2001; 51:559-63.
34. Dong Y, Taylor HE, Dimopoulos G. AgDscam, a hypervariable immunoglobulin domain-containing receptor of the Anopheles gambiae innate immune system. PLoS Biol 2006; 4:e229.
35. Cheng Q, Aksoy S. Tissue tropism, transmission and expression of foreign genes in vivo in midgut symbionts of tsetse flies. Insect Mol Biol 1999; 8:125-32.
36. Tamas I, Andersson SGE. Comparative genomics in insect endosymbionts. In: Bourtzis K, Miller TA, eds. Insect Symbiosis. Boca Raton: Crc Press Llc, 2003:33487.
37. Ochman H, Moran NA. Genes lost and genes found: Evolution of bacterial pathogenesis and symbiosis. Science 2001; 292:1096-99.
38. Mostafa HE, Heller KJ, Geis A. Cloning of Escherichia coli lacZ and lacY genes and their expression in Gluconobacter oxydans and Acetobacter liquefaciens. Appl Environ Microbiol 2002; 68:2619-23.
39. Moran NA, Dunbar HE. Sexual acquisition of beneficial symbionts in aphids. Proc Natl Acad Sci USA 2006; 103:12803-06.
40. Nalepa CA, Bignell DE, Bandi C. Detritivory, coprophagy and the evolution of digestive mutualisms in Dictyoptera. Insect Socieaux 2001; 48:194-201.
41. Ricci I, Cancrini G, Gabrielli S et al. Searching for Wolbachia (Rickettsiales: Rickettsiaceae) in mosquitoes (Diptera: Culicidae): Large polymerase chain reaction survey and new identifications. J Med Entomol 2002; 39:562-67.
42. Yoshida S, Ioka D, Matsuoka H et al. Bacteria expressing single-chain immunotoxin inhibit malaria parasite development in mosquitoes. Mol Biochem Parasitol 2001; 113:89-96.
43. Ghosh AK, Ribolla PE, Jacobs-Lorena M. Targeting Plasmodium ligands on mosquito salivary glands and midgut with a phage display peptide library. Proc Natl Acad Sci USA 2001; 98:13278-81.
44. Zieler H, Keister DB, Dvorak JA et al. A snake venom phospholipase A(2) blocks malaria parasite development in the mosquito midgut by inhibiting ookinete association with the midgut surface. J Exp Biol 2001; 204:4157-67.

Chapter 5

Proposed Uses of Transposons in Insect and Medical Biotechnology

Peter W. Atkinson*

Abstract

Transposons are small pieces of DNA that can transpose through either RNA or DNA intermediates. They have been found in almost all organisms and are important components of the evolutionary process at the chromosomal level. They have provided the raw genetic material that has produced domesticated genes that now provide important cellular functions and are now being explored as genetic tools in both humans and insects that vector human pathogens. Here I compare the requirements for both insect and human gene therapy and discuss the similarities between them in terms of transposon performance. Recent progress in understanding transposon function in terms of transposase structure is described as is the rapidly emerging role of RNAi in generic transposon regulation. These developments reinforce the view that, autonomous, transposon behavior in host organisms is, in part, determined by the nuclear and cellular environment of the cell and these factors need to be considered when developing transposons as therapeutic agents either in humans or in insects that vector human disease.

Introduction

Transposons are fundamental agents of genome evolution. They are present in almost all organisms and in many, including humans, occupy significant proportions of the entire genome. The degree to which these small pieces of DNA shape genomes and in doing so generate new genes that become domesticated or exapted by their hosts for purposes other than transposition has, through computational biology approaches, recently become measurable.[1,2] For example Britten (2006) used six frame translations of repeated sequences within the human genome to determine how many protein coding human genes contained closely related sequences and found, as a minimal estimate, 1,905 different genes. Volff (2006) identified 24 clear examples of domesticated transposase-derived genes nine of which had derived from *hAT* transposons. Other examples of transposon domestication also exist. For example the RAG1 recombinase of the vertebrate V(D)J recombination system responsible for immunoglobulin gene rearrangement shows structural and functional relationships with the *hAT* transposon superfamily and structural relationships with the *Transib* family of transposons.[3,4]

In addition to their role in evolution, several transposons have been successfully developed for use as genetic tools. The application of these have had significant effects on biotechnology and on basic science in which the ability to rescue genotypes has enabled scientists to directly address gene function. For example the *P* and *piggyBac* transposons now provide *Drosophila* researchers

*Peter W. Atkinson—Department of Entomology, Institute for Integrative Genome Biology, University of California, 900 University Avenue, Riverside, CA 92521, USA.
Email: peter.atkinson@ucr.edu

Transgenesis and the Management of Vector-Borne Disease, edited by Serap Aksoy.

with insertion mutagenesis tools that cover virtually all of the *Drosophila* genome enabling simple access to mutants of most protein-coding genes. The *piggyBac* transposon is an efficient enhancer trap tool in *Tribolium castaneum* while recent work with the resurrected *Sleeping Beauty* (*SB*) transposon from fish shows that it moves at high transposition rates in both transgenic mice and in human cell culture thereby opening up the possibility of its use in human gene therapy.[5-7] The *Tol2 hAT* transposon of medaka fish is also active in human cells and in mice[8] while *piggyBac* has recently been used as a gene transfer agent in human cells with high efficiency.[6,9] The widespread use of transposons as genetic tools in vertebrates and invertebrates is thus increasing. It should be noted that a detailed understanding of their interactions and behavior in their new hosts at the molecular level is not necessarily required for the effective use in some applications in these organisms. For example the *P* and *piggyBac* transposons are used in *Drosophila* transformation and mutagenesis despite the fact that much remains to be discovered about their regulation and their interaction in this host. *PiggyBac* enjoys wide use as a gene vector in non-drosophilid insects in which the purpose is simply to introduce genes which themselves are the target of study. Similarly the rapid progress made with *SB* as a genetic tool in mammals has been made without understanding all of the details of *SB* structure and function. The success of these transposons as genetic tools combined with the confinement of their use within the laboratory allows for the successful operation of these systems in the absence of complete information about the molecular basis of their behavior in new cellular environments. The strength of the artificial selection imposed by the researcher means that only phenotypes that are predicted or deemed useful or interesting are preserved as new genetic strains in the laboratory. The proposed use of transposons as therapeutic agents in humans as fundamental components of human gene therapy programs and as gene vectors in other organisms, such as insect vectors of human pathogens, which may then be released into the field requires however that our knowledge of their behavior in these species rise above the superficial.

This remains a considerable challenge. In human gene therapy the need is for gene vectors that are efficient in targeting cell type, targeting the nucleus once inside the cell and then, once in the nucleus, targeting an insertion point in the genome that enables the therapeutic genes contained within the gene vector to be expressed at levels sufficient to ameliorate the genetic disease being treated. A key point is that this genomic integration cannot itself lead to the progression of disease through, for example, the activation of oncogenes or the inactivation of tumor suppressor genes. The gene vector should be stable following integration, should not at a subsequent stage produce an antigenic response and ideally, the gene therapy treatment should not need to be repeated again during the life of the patient. At present no single transposon fulfills all of these requirements however there are examples of transposons that do possess at least one of these desired properties. For example the Tn*7* transposon of *E. coli* can integrate specifically at the *attTn7* site when integration is mediated by the TnsD protein produced by the tranposase.[10,11] In the absence of TnsD, Tn*7* integrates throughout the *E. coli* genome.[11] Both the *piggyBac* and *SB* transposons move efficiently in human cell culture yet lack the absolute target site specificity demanded for use in human gene therapy.[6,12] By their wide distribution throughout arthropods it is clear that members of the *mariner* transposon superfamily can invade genomes to extraordinary levels, leading to tens of thousands of copies of the transposon. Yet only one naturally occurring active *mariner* transposon, *Mos1*, has been found, and its mobility properties are not extraordinary.

Current attempts in the application of Class II transposons in human gene therapy are beginning to tackle these hurdles. Recently the mobility properties of *SB* and *piggyBac* have been compared in different types of human cell culture although a strict side by side comparison with identical experimental parameters still remains to be performed.[12-16] Targeting the *piggyBac* and *mariner* transposons to integrate at a greater frequency at UAS sites of the GAL4 transcriptional enhancer have been achieved by fusing the Gal4 DNA binding domain upstream of the transposase.[14] The experimental design for these experiments follows on from similar work in integrases performed over a decade ago in which the target site specificities of integrases had been altered by also constructing chimeric enzymes which contain both the GAL4 DNA binding domain and the integrase domain.[17] While the outcomes of these experiment suggest that the targeting properties of transposase can be altered by the construction of fusion proteins they still do not address the fundamental question of the molecular basis of target site selection by these transposases and the mechanism by which the target DNA is brought into the catalytic domain of the enzyme to join with the transposon ends located within the transpososome. While this remains unknown these fusion proteins will never enjoy the absolute target site specificity demanded by human gene therapy since integrations into alternate genomic sites will still occur and some of these may result in progression to disease. It should also be pointed out that the analysis of *piggyBac* and *SB* directed integration into pGDV1 target DNA presented overstates the integration frequencies since no allowance is made for the generation of clonal events within each experiment.[14]

In contrast, in insect biotechnology the properties of naturally occurring transposons are often seen as being too problematic in using them as gene vectors for the genetic control of pest insects even though some of goals are shared with human gene therapy applications. This is in part caused by the very different requirements of these transposons and by the fact that, unlike human gene therapy, genetic control requires germ-line transformation of the pest insect with the subsequent generation and release of millions of insects containing the transposon. The type of insect genetic control program dictates whether genomic stability or high levels of transposition are required and, as a consequence, the requirements of the transposon and its transposase differ. However concerns about unintended consequences following from the release of millions of genetically engineering insects are equivalent in principle to the concerns arising from the infusion of millions of genetically engineered cells into patients for which gene therapy is intended. Rare events arising from the unintended and perhaps unpredicted behavior of transposons can lead to undesired consequences, be they progression to a different disease or the generation of unintended genotypes in the target pest species, or even other species should the transposon be horizontally transferred. In gene therapy patients an unintended progression to diseases such cancer through the selection of cells made highly proliferative and malignant is equivalent in a sense to a rapid spread of an undesired new and unexpected genotype in pest insects in the field, or perhaps even its horizontal transfer into nontarget species.

It must be stressed that these challenges are solvable provided one is prepared to fully understand the molecular basis of transposon behavior and the genetic and biochemical interactions between the transposon and the new host genome. Indeed the same can be said for the use of any genetic vehicle designed to change a genotype for the better, be these viral vectors or, in the case of insects, endosymbionts. Treating these gene vectors as entities that are somehow insulated from the genome and proteome of the new genetic hosts and will simply and quickly achieve the desired effect ignores the complexity of biological systems, particularly when human imposed artificial selection is removed following release (in the case of some forms of insect genetic control) or infusion (in the case of human gene therapy). The behavior, or phenotype, of a given transposon in a new host can be represented as a simple extension of the formulae which governs an organisms phenotype:

$$P_{TE} = G_{TE} + E_H$$

Where P_{TE} is the phenotype or behavior of the transposon, G_{TE} is the genotype of the transposon and can be thought of as its in vitro performance characteristics as an autonomous transposon and E_H is the nuclear and cellular environment of the host organism. G_{TE} is determined by the ability of the transposase to bind transposon DNA, to form a transpososome structure and to excise and transpose the transposon to a new target site. E_H can be influenced by host factors which may be specific for a particular transposon or transposon family and may include the presence in a genome of a transposon of the same family as the transposon being introduced into the organism. Clearly E_H in vitro is zero since host factors are not required for transposition which is, by definition, autonomous. However in vivo E_H can influence transposition. Recent results on the role of piRNA in transposon regulation in *Drosophila* and zebrafish show that components of this pathway may also contribute to E_H and these are discussed below.[18,19] Indeed in genetic control strategies in which transposons are driving genes through a population, the rate at which down regulation of piRNA occurs may set a time frame within which transposon fixation (or near fixation) must be achieved.

Genetic Control Strategies and Transposon Immobility

For genetic control strategies such as the sterile insect technique (SIT) the performance requirements of the transposon are mostly the same as they are human gene therapy. High frequency integration is required in order to make the process of transformation routinely efficient and so achieve the desired effect within a therapeutically or economically desired time frame. Stability post-integration is also required and the same site of integration should be maintained through all subsequent generations with no alteration at all to the internal structure of the transposon carrying the beneficial transgene. As for human gene therapy the site of integration of the transposon should not be disadvantageous to the host. The concern with humans is primarily to avoid secondary disease over what might be a very long life-span while with insects used in genetic control strategies the concern is integration into genomic sites that decrease the ability of the genetically modified insect to disperse and mate with equal efficiency to the field insects.

How might this be achieved? The three principal concerns are transformation efficiency, target site preference and post-integration stability, all in the presence of endogenous transposon systems that may frustrate each of these requirements in as yet unknown ways. Furthermore the recent identification of master regulatory loci controlling transposition in *Drosophila* introduces an added complication to the maintenance of these strains however the immobility of these transposons following initial integration should evade this type of regulation provided integration is not at one of these loci.[19] Progress to date is impressive but not rigorous since, as stated above, the molecular basis of transposon regulation in target pest insects remains unknown. In the face of this ignorance our efforts are understandably reduced to rudimentary approaches.

Transformation Efficiency

The issue of initial transformation frequency can be addressed by either improving the means by which the transposon is introduced or by modifying the transposon or its transposase. For the *P* transposon in *Drosophila* the use of a recipient strain containing the *P* transposase specifically expressed in germ-line tissues increased the frequency of transformation.[20] No such strains exist in non-drosophild insects. The use of *Minos* transposase mRNA instead of helper plasmid increased *Minos*-mediated transformation frequency of both *Ceratitis capitata* and *Drosophila* several-fold most likely through circumventing the need for transcription of the transposase gene located on a helper plasmid.[21] This same strategy was employed in the recent demonstration of the success of the ΦC31 integrase system in *Ae. aegypti*.[22] While this system achieved a higher transformation frequency of *Ae. aegypti* than had been obtained using the *piggyBac* transposon system, the *piggyBac* experiments used helper *piggyBac* plasmid rather than *piggyBac* transposase mRNA so a direct comparison of transformation frequencies at least in these experiments is not possible. Limited attempts at circumventing translation and processing in which transposase protein was coinjected with the transposon-containing plasmid have not been successful in increasing insect transformation frequency. The most illustrative comparative example comes from work performed on *mariner* using the *Mos1* transposase in which hardly any gains were achieved over conventional two-plasmid injections into *Ae.aegypti* embryos.[23] The difficulty in obtaining sufficient levels of pure, active and correctly folded transposase no doubt contribute to the challenge of directly using transposase protein in injections. Success at obtaining active *Hermes* transposase has depended on a strategy that does not ask for maximum production in *E. coli* of a transposase which may result in toxicity to the cell. Cells are grown at low temperature (16°C) following induction with arabinose and with subsequent purification using Ni-column chromatography.[24] Yields are moderate however the transposase is active and provided a crystal structure that could be resolved at 2.1Å. This structure revealed that the transposase catalytic domain consisted of an RNaseH-like fold with the three carboxylates required to position the divalent cation located on two of the five β sheets of the fold and the final glutamate located on an α helix towards the carboxy end of the transposase.[3] The location of the catalytic domain of the *Hermes* transposase combined with the identification of amino acids that protrude in into it enable directed mutagenesis of these residues to be performed with the aim to produce a more active transposase. Furthermore the association of a conserved RW motif downstream of the second aspartate of the catalytic triad most likely is responsible for the stabilization of a flipped out base during the excision step which produces a hairpin intermediate on DNA flanking the *Hermes* terminal inverted repeats.[3] This RW conservation is also seen in the Tn*5* transposase in which the cocrystal structure of both a hyperactive form of the transpoase and the outside ends of the Tn*5* transposase has been determined and remains the paradigm for understanding transposase function from its crystal structure.[25,26] Unlike *Hermes*, hairpin loops are formed not on flanking DNA but on the Tn*5* transposon ends during excision. Assays for *Hermes* activity are performed in yeast in which *Hermes* is active, enabling a high throughput genetic screening strategy to be undertaken (N. L. Craig, personal communication).

The crystal structure of the catalytic domain of the *Mos1* transposase has been obtained and, like *Hermes*, also found to contain the RNaseH domain which is found in many transposases and integrases.[27] This structure, combined with the proposed structure of the *Mos1* DNA binding domain based on the crystal structure of the DNA binding domain of the related Tc*3* transposase was used to develop a model for the paired end complex of the *Mos1* transposome. Unlike both *Hermes* and Tn*5* no evidence of hairpin formation during excision has been recorded for the *Mos1* transposon or related transposons. This has mechanistic consequences since hairpin formation during excision enables a single catalytic site to perform two coupled DNA cleavage reactions while the absence of such an intermediate structure means that these two DNA cleavage reactions must be performed sequentially. The model of the pair end complex proposed by Richardson et al (2006) postulate that after first strand cleavage of the nontransferred strand of the *Mos1* transposon a conformational change occurs which then releases this end from the catalytic site and replaces it with the transferred strand which is then cleaved.[27] The formation of a paired end complex consisting of dimer of transposase at the transposon end facilitates this process.

The point is simply that the crystal structures of these transposases, once they become available, enable links to be made between transposase structure and transposition mechanism. Key amino acid residues can be identified, modified and tested for their effects on transposition frequency enabling transposases with improved mobility properties to be developed and tested in the target organism.

The DNA binding domains of these transposases are also targets for modification. Of significance to this transposase and other transposases which, like the *Hermes* transposase, contain a BED DNA binding domain is the recent elucidation of the crystal structure of the BED domain.[28] This is particularly useful for understanding the function of the *Hermes* transposase since the amino end of the *Hermes* transposase containing its BED domain is absent from the amino-truncated protein used to establish its crystal structure. Analysis of both structures provides insights into the possible function of the *Hermes* transposase and so identifies residues around and within the DNA binding domain that, through directed mutagenesis, may produce mutants with increased transpositional activity. Such hyperactive mutants have already been identified in the Tn*5* system in which E54K change was proposed to increase the binding affinity of the transposase for the Tn5 transposon.[25]

Random mutagenesis has been extensively used for the *Himar1 mariner* synthetic transposon to generate hyperactive mutants in papillation assays performed in *E. coli*.[29] These have been relatively easy to obtain given the strong selection imposed by the papillation assay, the high throughput of mutant screening it allows and the likelihood that naturally occurring transposons are successful since they have sub-optimal activity. None however have resulted in hyperactivity in insects perhaps because the strong selective pressure exerted by these assays simply selects for hyperactive mutants specific for *E. coli*. For example these may result from changes in the folding of the transposase in *E. coli* that are not carried into eukaryotes. More recently targeted mutagenesis of the DNA binding domain at the amino end of the *Himar1* transposase generated 16 alanine scanning mutants deficient in either DNA binding or strand cleavage.[30] Two were hyperactive, once again in *E. coli*, and both were changes of a carboxylate to alanaine which would be predicted to increase the binding affinity of the transposase for the transposon.[30]

The three mutations originally found to cause hyperactivity in *Himar1* in *E. coli* (Q131R, E137K, H267R) were used as a guide to generate the two corresponding changes in the *Mos1* transposase (R131R - not mutated, E137R, E264K).[31] When tested in *Ae. aegypti* these were found to lead to a three-fold increase in transpositional activity however these data may exaggerate the level of hyperactivity since variation between experiments was not accounted for and it is not clear that the most appropriate negative controls were used to generate baseline data.[31] Nevertheless if one accepts that some increase in transpositional efficiency was seen it is notable that both amino acid changes once again serve to increase the binding affinity for the transposon. It may well be that these mutants will be the primary class of hyperactives obtained regardless of the basis of the genetic screen.

Target Site Preference

Little is known about the mechanism that Class II transposons locate target DNA prior to strand transfer. A related key question is whether a single active site of these transposases is simultaneously filled with both donor and target DNA or whether these exist sequentially within it. As described above the model proposed by Richardson et al (2005) for *Mos1* requires a conformational change at the catalytic site to replace the non transferred strand of the transposon with the transferred strand, after which it is assumed the target DNA sequence is brought close to the catalytic site. For the *Rag1* transposase this is thought to be chronologically separate from donor strand excision. Matthews et al (2004) demonstrated that for *Rag1* and its substrates the active site is filled sequentially.[32] Given the relatedness between *Rag1* and the *Hermes* transposase at both the structural and functional level it is possible that the same may be true for the *Hermes* transposase. A consequence of this is that conformational changes at the active site are probably needed to move donor DNA in and out and then target DNA in (which then needs to be cleaved in a manner characteristic of each transposase) and then out of the catalytic site. *SB* as a member of the *mariner/Tc1* superfamily inserts at TA dinucleotides and recently a relationship between insertion site preference and the physical structure of the DNA surrounding these insertion sites was determined.[33] As stated above the tethering of *SB* and *piggyBac* transposases to known DNA binding domains has successfully skewered the target site selection of these chimeric transposases to the DNA binding site recognized by the DNA binding domain however these experiments do not address the mechanism by which the native transpososomes recognize, find and cleave target DNA.

Post Integration Stability

Transposon stability following integration is a key requirement for the use of transposons in human gene therapy and in the generation of insect genetic control strategies such as the SIT. Two different approaches have been successfully used to eliminate transposon mobility in insects post-integration. Neither requires a detailed understanding of the molecular basis of transposition since both seek merely to inactivate the transposon through its subsequent exposure to recombinases which have their target sites preengineered into the transposon. Both require the generation of a recombinase-producing strain which is crossed with the strain containing the transposon to be inactivated. In one strategy the terminal sequences of the transposon are removed through exposure to the FLP recombinase which recognizes FRT sites imbedded in one transposon end, thereby eliminating the transposase binding sites and thereby rendering the transposon immobile and stable.[34] In the second the entire transposon is removed using a similar recombinase-based strategy however the transgene within it remains in the genome.[35] The rationale given for choosing to eliminate the entire transposon is that it may be a target of related transposases within the genome which may lead to instability however in practicality there is little to distinguish between the outcomes of either approach.

Regulation of Transposition by piRNA

Recent work indicates a role for RNAi in the regulation of transposition of both Class I and Class II transposons and may have significant implications for the use of transposons in insect genetic control, especially those applications which rely on the transposon to, via its mobility, drive genes through insect populations. Brennecke et al (2007) used antibodies specific to the piwi, ago3 and aubergine to purify these proteins, which are part of a germ-line specific RISC complex, from the testes of adult flies. Sequence analysis of the small RNAs associated with these proteins revealed that there was an abundance of transposase associated miRNAs suggesting that these RISC proteins may preferentially target and silence transposons. Comparing the distribution of transposons throughout the *Drosophila* genome with the locations of the piRNAs obtained from the three RISC argonaute proteins revealed a strong correspondence between them with particular concurrence at three loci, one being the *flamenco* locus which in previous and unrelated studies had been shown to regulate *gypsy* transposition in *Drosophila.*[36] Other transposons present at the *flamenco* locus were *ZAM* and *Idefix.*[19] The hypothesis developed by Brennecke et al (2007) is that upon invasion of a naïve genome, an active transposon continues to transposase until at least one copy inserts into a piRNA locus at which time piRNA specific to its transposase is made. This then targets the transposase for destruction through the piRNA pathway and the result is the silencing of these transposons. There are several attractive features of this model. First it offers a generic mechanism for the silencing of transposons. Second, the localization of *piwi*, *ago3* and *aubergine* expression to the germ-line targets this process to the germ-line which presumably increases the genetic fitness of the host species. Third, it explains the feature of dysgenesis which results in a genetic memory in which the egg, if it comes from a mother containing, for example once active *P* elements, can recognize incoming active *P* elements in the sperm DNA and quickly inactivate them. Fourth, the colocalization of the *flamenco* locus with one of the major piRNA clusters identified by Brennecke et al (2007) offers an mechanism by which the *gyspy* transposon is regulated in *Drosophila.* Fifth, it builds upon and extends previous work implicating small RNA in transposon regulation. *Piwi* had been shown to affect the mobility of the *mdg1* retrotransposon in the *Drosophila* male germ line through the observation that *piwi* mutants displayed increased *copia* and *mdg1* transcription in the apical cells of testes.[37] The progeny arising from homozygous *piwi* mutant crosses were examined for the remobilization of endogenous *mdg1* transposons and in 19 out of 101 flies examined, *mdg1* appeared at a new genomic location corresponding to a transposition rate of 0.19 per generation. No *mdg1* remobilizations were observed in progeny arising from crosses in which the father was heterozygous for the *piwi* mutation.[37] siRNA was implicated in the regulation of the *Penelope* retrotransposon in *Drosophila virilis*, this time in the maternal germline.[38] Northern blot analysis of RNA prepared from whole adults indicated the presence of small RNA approximately 23nt in size corresponding to the *Penelope* open reading frame was found in a strain containing these transposons but not in a wild type strain.[38] Other transposons also mobilized in dysgenic crosses in *D. virilis* were not detected. The recent identification of the piRNA pathway opens up the possibility that the siRNAs identified by Blumenstiel and Hartl (2005) may well be in fact piRNAs. These data built upon initial observations from both *Drosophila* and *C. elegans* that showed that small RNAs corresponding to different classes of transposons were present during development of both these arthropods.[39-41]

The piRNA amplification-based model proposed by Brennecke et al (2007) has profound implications for non-drosophilid insects in which we can assume the same piRNA mechanism exists. This is because many of these species have more transposons than *Drosophila* and several of the ones that are used as biotechnological tools have members of the same family or subfamily already incumbent in the target insect species. The yellow fever mosquito, *Ae. aegypti*, in which transposons constitute a major part of the genome, is one example but a very pertinent one for vector biology in which it is used as something of a model mosquito even for malaria. Indeed the few transposons used to genetically transform *Ae. aegypti* become almost immobile once integrated.[42] This is in contrast to the use of *piggyBac* in both *Tribolium* and *Drosophila* in which it is a potent genetic tool used for gene tagging and enhancer trapping. This raises the interesting possibility that a piRNA system in *Aedes* may work exceedingly rapidly to inactivate new transposons. Many questions, most of them answerable in insects, are raised by the proposal that piRNA may be a potent regulator of transposition. What degree of relatedness between transposases is needed in order for one incumbent family member to silence another? Does the number of master regulatory loci vary between species and does this number show any correlation with transposon load?

The possibility that piRNA regulates transposition offers at the same time a challenge and an opportunity for those who seek to use transposons as a gene drive engine to introgress a beneficial genotype into a pest species such as mosquitoes. The challenge is simply that the transposon may become inactivated (perhaps along with the genetic cargo it carries) before it has spread through sufficient numbers of the target population to have any beneficial effect. The opportunity is that characterization of the piRNA mechanism in the target species may well offer quantitative parameters that will allow the investigator the ability to predict the rate of spread of the transposon through the population before such inactivation could occur.

For example the frequency of transposition of a given transposon can be measured in vitro, in vivo in a model insect such as *Drosophila*, and in vivo in the target species using both transposition assays and simple experiments with caged populations of transgenics containing an autonomous element in which the transposase is placed under the control of an inducible or tissue regulated promoter. If the genome of the target pest has been sequenced a rudimentary estimate of transposon abundance and distribution, at least along supercontigs, can be made.

Genetic Control Strategies and Transposon Drive

For genetic control strategies which require the transposon to drive a desired transgene through a field population, the requirements of the transposon are somewhat more problematic. Efficient transformation and maintenance of transposon and transgene integrity are still required properties. In addition, rather than being stable at one genomic location, the transposon must have a transposition rate that enables it to spread through a population in what one assumes to be a therapeutically beneficial time frame. This has always remained problematic since active transposons, by their very nature, are mutagens and so can act to reduce the genetic fitness of their host. Indeed a successful transposon can be seen as one that simply propagates itself without compromising the viability of its host and so selecting for hyperactive forms of transposons would, in the long run, serve neither the host nor the transposon. Clearly any attempts to develop transposons as successful gene drive agents in insects will need to find a delicate balance between the selective forces allowing spread and those forces selecting against individuals in which transposition rate becomes detrimental. This remains to be achieved but several recent results offer some illumination and guidance. In the case of developing transposon-based drive systems in *Anopheles* to prevent the spread of *Plasmodium* the recent demonstration that transgenic *An. stephensi* infected with *P. berghei* are more fit than nontransgenic infected females demonstrates that the any fitness burden resulting from transposon acquistion through engineering may be less than the burden of *Plasmodium* infection.[43] Of course this situation is experimentally different from an autonomous transposon which is moving through a mosquito population however it does reveal that the alleviating infection can improve genetic fitness.

The second recent advance is the discovery of what may be a possible generic mechanism of transposon regulation through piRNA. Clearly one consequence of the amplification model proposed by Brennecke et al (2007) is that ultimately an active transposon will be silenced by the piRNA machinery once one copy of it has inserted at a piRNA locus. In contemplating transposon based gene drive strategies the piRNA pathway now introduces some important parameters that previously were cryptic. In principle these can measured and so produce an estimate of whether a given autonomous transposon will be able to introgress quickly enough through a target population to produce the desired therapeutic outcome within the desired time. For example the rate of transposition of a modified or unmodified transposon can be determined and, as described above, the piRNA profile of the target host can be determined, particularly if its genome sequence is known. Whether related transposons incumbent in the host genome can recognize and silence the new transposon can be most likely determined experimentally. Ultimately an estimate may be able to be made as to how many generations would elapse before silencing of the new transposon would occur. If this remains greater than the length of time required to achieve a therapeutic benefit the transposon based gene drive strategy may be viable. Inevitably the transposon will be silenced and this may also be advantageous since it would prevent any reasonable chance of transfer to nontarget organisms and so may be seen as a self-regulating control. It is also possible, however, that the beneficial transgene carried by the transposon would also be silenced and any therapeutic effect lost in the longer term.

Conclusions

Transposons are enjoying something of a resurgence in modern biology. The ability to generate and analyze transposase crystal structures, the revelation that the small RNA pathways play a fundamental role in their regulation in host organisms, the ability to analyze their distribution and abundance in whole genomes both as transposons and as expated genes, and their emerging use as potential gene therapy agents in both humans and insects have converged to generate further interest in these fascinating nuclear parasites. While their development into precise genetic tools that can always be targeted to specific genomic sites or into gene drive agents in insects that can evade, at least temporarily, the RNAi response of the host remains in the future, it is clear that as our understanding of their function and regulation increases, so does the likelihood that they will fulfill their roles in modern medicine and disease control increase. These may well become the tissue-specific or species-specific genetic tools that can be safely used to achieve desired medical outcomes in both economic hemispheres of the world.

References

1. Volff JN. Turning junk into gold: Domestication of transposable elements and the creation of new genes in eukaryotes. Bioessays 2006; 28:913-22.
2. Britten R. Transposable elements have contributed to thousands of human proteins. Proc Natl Acad Sci USA 2005; 103:1798-803.
3. Zhou L, Mitra R, Atkinson PW et al. Transposition of hAT elements links transposable elements and V(D)J recombination. Nature 2004; 432:995-1001.
4. Kapitonov VV, Jurka J. RAG1 core and V(D)J recombination signal sequnces were derived from Transib transposons. PLoS Biol 2005; 3:e181.
5. Lorenzen MD, Berghammer AJ, Brown SJ et al. piggyBac-mediated germline transformation in the beetle Tribolium castaneum. Insect Mol Biol 2003; 12:433-40.
6. Ding S, Wu X, Li G et al. Efficient transposition of the piggyBac (PB) transposon in mammalian cells and mice. Cell 2005; 122(473-483).
7. Collier LS, Carlson CM, Ravimohan S et al. Cancer gene discovery in solid tumors using transposon-based somatic mutagenesis in the mouse. Nature 2005; 436:272-6.
8. Koga A, Iida A, Kamiya M et al. The medaka fish Tol2 transposable element can undergo excision in human and mouse cells. J Hum Genet 2003; 48:231-5.

9. Wilson MH, Coates CJ, George ALJ. PiggyBac transposon-mediated gene transfer in human cells. Molecular Therapy 2007; 15:139-45.
10. Bainton RJ, Kubo KM, Feng JN et al. Tn7 transposition: Target DNA recognition is mediated by multiple Tn7-encoded proteins in a purified in vitro system. Cell 1993; 72:931-43.
11. Craig NL. Tn7: A target site-specific transposon. Mol Microbiol 1991; 5:2569-73.
12. Hackett PB, Ekker SC, Largaespada DA et al. Sleeping Beauty transposon-mediated gene therapy for prolonged expression. Adv Genet 2005; 55:189-232.
13. Geurts AM, Yang Y, Clark KJ et al. Gene transfer into genomes of human cells by the Sleeping Beauty transposon system. Mol Ther 2003; 8:108-17.
14. Maragathavally KJ, Kaminski JM, Coates CJ. Chimeric Mos1 and piggyBac transposases result in site-specific integration. FASEB J 2006; 20:fj.05-05485fje.
15. Huang X, Wilber AC, Bao L et al. Stable gene transfer and expression in human primary T cells by the Sleeping Beauty transposon system. Blood 2006; 107:483-91.
16. Belur LR, Frandsen JL, Dupuy AJ et al. Gene insertion and long-term expression in lung mediated by the Sleeping Beauty transposon system. Mol Ther 2003; 8:501-7.
17. Bushman FD, Miller MD. Tethering human immunodeficiency virus type I preintegration complexes to target DNA promotes integration at nearby sites. J Virol 1997; 71:458-64.
18. Houwing S, Kamminga LM, Berezikov E et al. A role for Piwi and piRNAs in germ cell maintenance and transposon silencing in zebrafish. Cell 2007; 129:69-82.
19. Brennecke JB, Aravin AA, Stark A et al. Discrete small RNA-generating loci as master regulators of transposon activity in Drosophila. Cell 2007; 128:1089-103.
20. Robertson HM, Preston CR, Phillis RW et al. A stable genomic source of P element transposase in Drosophila melanogaster. Genetics 1988; 118:461-70.
21. Kapetanaki MG, Loukeris TG, Livadaris I et al. High frequencies of Minos transposon mobilization are obtained in insects by using in vitro synthesized mRNA as a source of transposase. Nucleic Acids Res 2002; 30:3333-40.
22. Nimmo DD, Alphey L, Meredith JM et al. High efficiency site-specific engineering of the mosquito genome. Insect Mol Biol 2006; 15:129-36.
23. Coates CJ, Jasinskiene N, Morgan D et al. Purified mariner (Mos1) transposase catalyzes the integration of marked elements into the herm line of the yellow fever mosquito, Aedes aegypti. Insect Biochem Mol Biol 2000; 30(11):1003-8.
24. Perez ZN, Musingarmi P, Craig NL et al. Purification, crystallization and preliminary crystallographic analysis of the Hermes transposase. Acta Crystallograph Sect F Struct Biol Cryst Commun 2005; 61(pt 6):587-90.
25. Davies DR, Goryshin IY, Reznikoff WS et al. Three-dimensional structure of the Tn5 synaptic complex transposition intermediate. Science 2000; 289:77-85.
26. Reznikoff WS. Tn5 transpositon: A molecular tool for studying protein structure-function. Biochemical Society Transactions 2006; 34:320-3.
27. Richardson JM, Dawson A, O'Hagan N et al. Mechanism of Mos1 transposition: Insights from structural analysis. EMBO J 2006; 25:1324-34.
28. Sato S, Sasagawa A, Tochio N et al. Solution structures of the C2H2 type zinc finger domain of human zinc finger BED domain containing protein 2. Protein Data Bank 2005; 2DJR:DOI 10.2210/pdb2djr/pdb.
29. Lampe DJ, Akerley BJ, Rubin EJ et al. Hyperactive transposase mutants of the Himar1 mariner transposon. Proc Natl Acad Sci USA 1999; 96(20):11428-33.
30. Butler MG, Chakraborty SA, JLD. The N-terminus of Himar1 mariner transposase mediates multiple activities during transposition. Genetica 2006; 127:351-66.
31. Pledger DW, Coates CJ. Mutant Mos1 mariner transposons are hyperactive in Aedes aegypti.. Insect Biochem Mol Biol 2005; 35:1199-207.
32. Matthews AG, Elkin SK, Oettinger MA. Ordered DNA release and target capture in RAG transposition. EMBO J 2004; 23:1198-206.
33. Liu G, Geurts AM, Yae K et al. Target-site preferences of Sleeping Beauty transposons. J Mol Biol 2005; 346:161-73.
34. Handler AM, Zimowska GJ, Horn C. Post-inetgration stabilization of a transposon vector by terminal sequence deleton in Drosophila melanogaster. Nat Biotechnol 2004; 22:1150-4.
35. Dafa'alla TH, Condon GC, Condon KC et al. Transposon-free insertions for insect genetic engineering. Nat Biotech 2006; 24:820-1.
36. Pelisson A, Song SU, Prud'homme N et al. Gypsy transposition correlates with the production of a retroviral envelope-like protein under the tissue-specific control of the Drosophila flamenco gene. EMBO J 1994; 13:4401-11.

37. Kalmykova AI, Klenov MS, Gvozdev V. Argonaute protein PIWI controls the mobilization of retrotransposons in the Drosophila male germline. Nucleic Acids Res 2005; 33:2052-9.
38. Blumenstiel JP, Hartl DL. Evidence for maternally transmitted small interfering RNA in the repression of transposition in Drosophila virilis. Proc Natl Acad Sci USA 2005; 102:15965-70.
39. Aravin AA, Lagos-Quintana M, Yalcin A et al. The small RNA profile during Drosophila melanogaster development. Dev Cell 2003; 5:337-50.
40. Sijen T, Plasterk RHA. Transposon silencing in the Caenorhabditis elegans germ line by natural RNAi. Nature 2003; 426:310-4.
41. Vastenhouw NL, Fischer SEJ, Robert VJP et al. A genome-wide screen identifies 27 genes involved in transposon silencing in C. elegans. Curr Biol 2003;13:1311-6.
42. O'Brochta DA, Sethuramuran N, Wilson R et al. Gene vector and transposable element behavior in mosquitoes. J Exp Biol 2003:3823-34.
43. Marrelli MT, Li C, Rasgon JL et al. Transgenic malaria-resistant mosquitoes have a fitness advantage when feeding on Plasmodium-infected blood. Proc Natl Acad Sci USA 2007; 104:5580-3.

Chapter 6

The Yin and Yang of Linkage Disequilibrium: Mapping of Genes and Nucleotides Conferring Insecticide Resistance in Insect Disease Vectors

William C. Black IV,* Norma Gorrochetegui-Escalante, Nadine P. Randle and Martin J. Donnelly

Abstract

Genetic technologies developed in the last 20 years have lead to novel and exciting methods to identify genes and specific nucleotides within genes that control phenotypes in field collected organisms. In this review we define and explain two of these methods: linkage disequilibrium (LD) mapping and quantitative trait nucleotide (QTN) mapping. The power to detect valid genotype-phenotype associations with LD or QTN mapping depends critically on the extent to which segregating sites in a genome assort independently. LD mapping depends on markers being in disequilibrium with the genes that condition expression of the phenotype. In contrast, QTN mapping depends critically upon most proximal loci being at equilibrium. We show that both patterns actually exist in the genome of *Anopheles gambiae*, the most important malaria vector in sub-Saharan Africa while segregating sites appear to be largely in equilibrium throughout the genome of *Aedes aegypti*, the vector of Dengue and Yellow fever flaviviruses. We discuss additional approaches that will be needed to identify genes and nucleotides that control phenotypes in field collected organisms, focusing specifically on ongoing studies of genes conferring resistance to insecticides.

Genetic technologies developed in the last 20 years have enabled thousands of genes, of known function and position in a genome, to be analyzed in single organisms.[1] This has in turn translated into new capabilities in population genetics and genetic epidemiology. One of the most novel and exciting of these is the ability to identify genes and specific nucleotides within genes that control the expression of phenotypes in field collected organisms.[2] Genome wide testing for the association between phenotypes and genotypes is called *linkage disequilibrium mapping*[3] while, at a finer scale, analyzing phenotype-nucleotide associations in candidate genes is called *association mapping*[4] or, more recently, *Quantitative Trait Nucleotide (QTN) mapping*.[5] Neither mapping method involves traditional genetic crosses. Instead individuals or their eggs are usually collected from the field, used to raise an F_1 generation in the laboratory and the phenotype of interest is determined in a large number of F_1 individuals. In linkage disequilibrium mapping the genotypes of markers from throughout the genome are examined in these F_1 individuals and tests for statistical independence are performed between marker genotypes and the phenotype under study. Genome regions statistically associated with the

*Corresponding Author: William C. Black IV—Department of Microbiology, Immunology and Pathology, Colorado State University, Fort Collins, Colorado 80523, USA. Email: william.black@colostate.edu

Transgenesis and the Management of Vector-Borne Disease, edited by Serap Aksoy.

phenotype may contain a gene or genes that condition that phenotype. In QTN mapping, the sequences of a candidate gene are examined in the F_1 individuals and the frequencies of alleles and genotypes at each polymorphic site are compared among phenotypes. A putative QTN is detected when phenotypes are not equally distributed among genotype classes.

These processes have been rapidly embraced by genetic epidemiologists as a powerful tool for identifying heritable genetic predisposition to disease in humans.[6-8] *Drosophila* researchers have used these approaches to identify single nucleotide polymorphisms (SNPs) in genes affecting wing morphology[9] and bristle number.[10] Domestic animal and plant breeders use QTN mapping to identify SNPs associated with increased yield or other desirable characters as selectable markers for more rapid crop[5] and animal[11] improvement. The association between polymorphisms in the *Early Trypsin* gene and susceptibility to dengue virus was assessed in the mosquito *Aedes aegypti*[12] using a QTN mapping approach.

Mapping of Genome Regions and SNPs Conferring Insecticide Resistance

Insecticide resistance is widespread in arthropod vectors of disease. Resistance has and will continue to directly impact the incidence of existing arthropod borne diseases, reemergence of previously controlled diseases and the emergence of new diseases.[13-17] Although alternatives to insecticidal control are available, inadequate health service infrastructure, drug resistance and vaccine cost and availability make vector control an important option. The concept of *insecticide resistance management (IRM)*[18-20] treats insecticide susceptibility as a finite resource. As pesticides are applied and the target population becomes resistant, the susceptibility resource is depleted. A key assumption of IRM is that resistance alleles confer lower fitness in the absence of insecticides. Thus when a specific insecticide is discontinued, resistance will decline, susceptibility will be renewed and it may therefore be possible to reintroduce the insecticide. An analogous situation has been observed in the malaria parasite *Plasmodium falciparum* in Malawi, where withdrawal of the first line treatment chloroquine was followed by a decrease in resistance allele frequency (*pfcrt* K76T) from 85% in 1992 to 13% 8 years later.[21] *Resistance surveillance* is an essential step in IRM in providing baseline data for program planning and pesticide selection, for detecting of resistance at an early stage and monitoring its frequency so that alternatives can be used.[22] Bioassays and biochemical tests for monitoring resistance have been developed but these assays do not identify genetic mechanisms of target site insensitivity nor do they provide information on the numbers and types of genes involved in various forms of resistance.[23-26] Furthermore in cases where resistance alleles are partially or fully recessive these tests will only detect resistance when alleles reach a sufficiently high frequency that resistant homozygotes occur. It is clear that molecular genetic information on resistance mechanisms is essential to improve resistance detection and diagnostics.

Linkage disequilibrium and QTN mapping provide a novel and much needed interface between population genetics and the molecular basis of insecticide resistance. A priori hypotheses concerning the role of polymorphisms in resistance genes require knowledge of the basic biochemistry and physiology of target-site resistance or detoxification enzymes. Population geneticists study variation in individual genes and can identify gene regions that are apparently subject to selection. However without an understanding of molecular structure and gene function in target-site resistance or detoxification, the adaptive significance of SNPs will remain obscure. Conversely, while the molecular biologist may understand the structure, function and physiology of a target-site resistance or detoxification, only insights gained through population genetics can indicate whether variation found in a gene actually affects resistance, which parts of a gene or protein control resistance, and the mode of selection acting on that gene. Ultimately, this interface will be used to predict mechanisms of target-site resistance or detoxification. This type of information is also essential for designing new insecticides.

Terminology in Linkage Disequilibrium and QTN Mapping

In this review we will use the term *locus* to refer to a specific location in a genome. This traditionally referred to the location of a gene along a linkage group but now also refers to the location of a specific nucleotide in a gene. A polymorphic nucleotide locus is referred to as a *segregating site*. Similarly an *allele*, traditionally considered as a novel form of a gene that conferred a novel phenotype, now also refers to a unique nucleotide sequence of a gene. Among individuals of a species a gene can exhibit many nucleotide substitutions. Each novel nucleotide sequence is considered a novel allele. A *genotype* is the combination of alleles at a locus in a diploid individual and a *haplotype* is the nucleotide sequence of an allele. *Linkage disequilibrium* (LD) is a quantitative measure used in population genetics to indicate the extent to which markers or segregating sites in a genome assort independently. Segregating sites on different chromosomes or more than 50 centiMorgans (cM) apart on the same chromosome are expected to assort independently (LD = 0). Conversely, we expect that some sites will cosegregate (LD ≠ 0) because they are closely linked on the same chromosome, interact epistatically, reside together within an inversion or occur together near a telomere or centromere.

Measuring Linkage Disequilibrium

As an example, consider a gene that has two segregating sites at nucleotide positions 147 and 186. Three alternate nucleotides C, T, or G are found in the field at segregating site 147 while at position 186 either C or T are found. We define *Pxy* as the observed frequency of gametes containing nucleotide allele *x* at locus 147 and nucleotide allele *y* at locus 186. The expected frequency is estimated as the frequency of nucleotide *x* at locus 147 multiplied by the frequency of nucleotide *y* at locus 186. For example, if *PGT* is the observed frequency of 147_G 186_T gametes then the linkage disequilibrium coefficient *DGT* is calculated as

$$DGT = PGT - p(147_G)\, p(186_T) \quad (1)$$

If the probability of sampling a 147_G186_T gamete equals the product of their independent frequencies ($p(147_G)\, p(186_T)$) then $DGT = 0$ and 147_G and 186_T alleles are said to be in *linkage equilibrium*. Otherwise, alleles 147_G and 186_T are said to be in *linkage disequilibrium* if they occur together more ($DGT > 0$) or less ($DGT < 0$) often than predicted by their independent frequencies.

Assume that we initiate a population with a male that is a 147_G186_T double homozygote and a female that is a 147_C186_C double homozygote. The frequency of 147_G186_T sperm = 1.00 even though the frequency of 147_G and 186_T alleles in the overall population are both 0.50. Thus $DGT = 1.00 - (0.5 \times 0.5) = 0.75$. The same is true of *DCC* in eggs. Sperm fertilize eggs to produce $147_C186_C/147_G186_T$ zygotes that mature to adults. The frequency that nucleotides at the two segregating sites occur together in the next generation of gametes will be a function of the linkage relationship between the loci. There are four potential gamete types 147_G186_T, 147_C186_C, 147_C186_T, or 147_G186_C. 147_C186_C and 147_G186_T gametes are maintained when there is no recombination during meiosis in the parents. 147_C186_T or 147_G186_C gametes can only arise through recombination during meiosis in the parents. The expected frequency of these gametes is predicted by "r," the recombination frequency/meiosis (cM) between the segregating sites. Assuming no mutation, the predicted frequency of 147_G186_T gametes after one generation is:

$$PGT(1) = (1\text{-r})\, PGT(0) + \text{r}\, p(147_G)\, p(186_T) \quad (2)$$

The frequency of 147_G186_T gametes in generation 1 ($PGT(1)$) equals the frequency of 147_G186_T gametes in generation 0 ($PGT(0)$) that did not recombine (1-r) added to the frequency that 147_G and 186_T occur independently in generation 0 ($p(147_G)\, p(186_T)$) and come to occur on a common chromosome through recombination (r). Substituting equation (1) into equation (2):

$$DGT(1) = (1\text{-r})\, DGT(0) \quad (3)$$

which states that the disequilibrium between nucleotides at the 147 and 186 loci will decrease by (1-r) each generation. Equation 3 can be generalized over many generations to:

$$DGT(t) = (1\text{-}r)^t \, DGT(0) \qquad (4)$$

After t generations the initial disequilibrium between alleles will have declined by $(1\text{-}r)^t$. Disequilibrium declines but the rate of decline depends largely on r. Disequilibrium declines more slowly among tightly linked loci (small values of r). In our example (4) can be rearranged to estimate the time in generations for *DGT* to decline to *DGT(t)*.

$$t = logDGT(t) \,/\, log(1\text{-}r) \qquad (5)$$

Putting this into the perspective of a disease vector, the resolution of base pairs/ cM varies from 1.0 - 3.4 x 10^6 base pairs (bp)/ cM along different arms of the *Ae. aegypti* genome,[27] the reciprocal translates to 0.294 - 1.000 x 10^{-6} cM/bp. This is also known as the per-generation recombination rate between sites, designated as r in the literature.[28] Thus, the nucleotides at sites 147 and 186 are 39 bp apart and we would expect r to range from 1.2-3.9 x 10^{-5} cM/ generation = recombination events / meiosis / generation. Substituting r for r in equation (4) the time required for *DGT(t)* to decline to 0.01 would range from ~118,100 - 384,000 generations depending on where it occurs in the genome. Given the conditions of our restricted model, a very long time would be required in *Aedes aegypti* to eliminate the disequilibrium between these two polymorphic sites. But recall that this population was initiated with only two parents with alternate alleles at the two loci. In reality populations contain large numbers of breeding individuals and thus a more realistic population genetic measure of recombination rate is the effective recombination rate ρ[29] where:

$$\rho = 4N_e r \qquad (6)$$

N_e is the effective population size, the number of breeding individuals. In a population with Ne = 100 breeding individuals, ρ = 4 x 100 x 1.2-3.9 x 10^{-5} = 4.8-15.6 x 10^{-3} recombinations/ generation or one recombination event between nucleotides at sites 147 and 186 every 64-208 generations. If Ne = 1000, there would be a recombination every 6-21 generations and if Ne = 10,000, recombination would occur every 1-2 generations.

There are three things to note in this example. First, if recombination events are equally probable in a genome region, then ρ is proportional to physical distance in that region. However, recombination is expected to be reduced near centromeres, telomeres and in regions of heterochromatin and depending upon whether they reside within an inversion. Second, ρ is dependent upon effective population size. Recombination per generation will be slow in species with small *Ne* but rapid in species with large *Ne*. Third, this model predicts that ultimately no linkage disequilibrium should be detected even among tightly linked loci within long established populations or in populations with large *Ne*.

Patterns of Linkage Disequilibrium in Vectors

Having explained linkage disequilibrium, its expectations and the factors that can influence it, what do we in fact find in natural populations of *Ae. aegypti* and *An. gambiae*? Figure 1 presents the results of linkage disequilibrium analyses among 57 segregating sites in the *Abundant Trypsin* gene among 1,661 female *Ae. aegypti* from throughout Mexico. Each half matrix in this figure represents the outcome of all possible pairwise linkage disequilibrium analyses among all 57 segregating sites. Squares in black indicate sites in disequilibrium. In 4 different geographic regions of Mexico, the majority of segregating sites are in linkage equilibrium over a distance of only 777 nucleotides. Furthermore, r was estimated as 88 x 10^{-3} recombinations/ generation.[30] This same pattern was observed in the *Early Trypsin*[12] and r was estimated as 5.2 x 10^{-3} recombinations/generation. *Ne* was then estimated using the coalescent approach.[31,32] Estimates of *Ne* in Mexico in the cities of Merida, Monterrey and Tapachula were respectively 1,076, 2,090, 3,642 breeding individuals.[33] Thus in the *Abundant Trypsin* gene there are

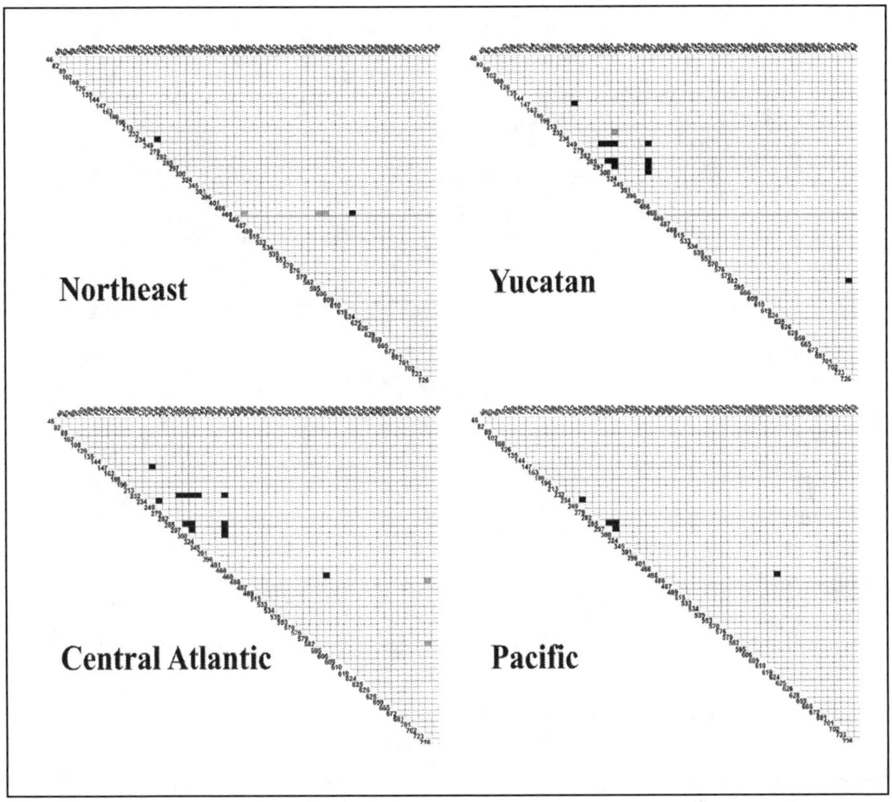

Figure 1. Ohta's linkage disequilibrium D-statistics[36] analyses among 57 segregating sites in the *Abundant Trypsin* gene among 1,661 female *Aedes aegypti* from throughout Mexico. Each half matrix in this figure represents the outcome of all possible pairwise linkage disequilibrium analyses among all 57 segregating sites in four different regions of Mexico. A black square corresponds to a pattern of systematic disequilibrium consistent with epistasis ($D^2_{IS} > D^2_{ST}$ and $D^{2\prime}_{IS} < D^{2\prime}_{ST}$). A gray square corresponds to a pattern of nonsystematic disequilibrium consistent with epistasis in only some collections ($D^2_{IS} > D^2_{ST}$ and $D^{2\prime}_{IS} > D^{2\prime}_{ST}$). A clear square indicates disequilibrium due to genetic drift ($D^2_{IS} < D^2_{ST}$ and $D^{2\prime}_{IS} > D^{2\prime}_{ST}$).

~380-3,640 recombination events/generation while in *Early Trypsin* there are ~20-80 recombination events/generation. Thus in *Ae. aegypti* populations in Mexico ρ is sufficiently large to maintain most segregating sites in linkage equilibrium.

Contrast this with (Fig. 2) which is an analysis of disequilibrium among SNPs in six genes located over 31 Mbp of DNA from division 22E to 27A in or around the 2La inversion in *Anopheles gambiae.* From this analysis it is evident that in contrast to *Ae. aegypti*, a large proportion of segregating sites in these six genes are in linkage disequilibrium. *r* varied widely among these 6 genes from 3 - 687 x 10^{-3} recombinations/generation. Estimates of *Ne* in Kenyan populations of *An. gambiae* in Asembo and Jego, were respectively 6,359 and 4,258 breeding individuals.[34] Thus ρ ranges from 50-17,500 recombinations/generation in *An. gambiae.*

How can the patterns of disequilibrium be so different in the two species if ρ in both are similarly large? Both species have recombination map sizes of 160-170 cM. However, *Aedes* has a ~5x larger genome size. Thus, the rate of recombination / nucleotide is actually much smaller in *Ae. aegypti* and ρ should also be less rather than more as suggested by the different disequilibrium patterns. It is possible that the 2La inversion is reducing recombination in this

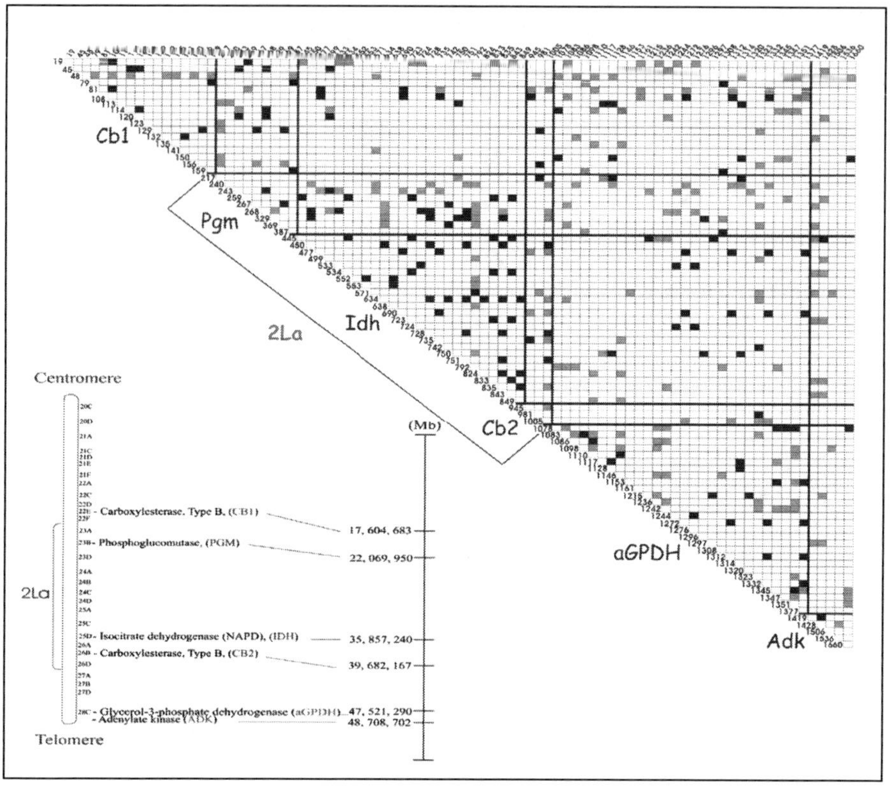

Figure 2. Analyses of disequilibrium among SNPs in six genes located over 31 Mbp of DNA from division 22E to 27A in or around the 2La inversion in *Anopheles gambiae*.

region. A contrasting pattern was observed on the right arm of chromosome 3. Ten microsatellite loci were designed and scored between the published loci Ag3H59 and Ag3ND30E1 which flank a QTL associated with permethrin resistance.[35] The microsatellites were between 5.6kb and 128.7 kb apart (Mean = 46.6kb). In samples of *An. gambiae* from Sao Tome and Principe that were phenotyped for permethrin susceptibility no significant linkage disequilibrium between any pair of loci was detected (66 pairwise tests, $p<0.05$) although the two most closely spaced loci (5.6kb) apart were on the verge of significance. Tests of independence between resistance phenotype and microsatellite genotype revealed that one locus was significantly associated with permethrin phenotype. This locus was 75kb upstream of a cytochrome P450 cluster and in a multivariate analysis genotype information for this locus was sufficient to predict resistance phenotype with 72% accuracy. This association suggests that an upstream *cis* acting regulator factor may be responsible for the phenotype. Figure 3 partitions the variance in linkage disequilibrium between these loci into inter- and intra- phenotype group components.[36] We would predict that if the two phenotype groups are distinct at these loci as a result of hitch-hiking associated with a resistance phenotype that $D^{2'}_{ST}$ should be larger than $D^{2'}_{IS}$ and negatively correlated with physical distance (i.e., $D^{2'}_{ST}$ should be larger among proximal sites). However (Fig. 3) shows the converse and suggests that even with a very recent selective event, such as that imposed by insecticide resistance, that recombination rapidly breaks down the haplotype block that arises by hitch-hiking. Whilst this study used a different marker system to the sequencing studies on 2L they suggest that there could be marked heterogeneity in recombination rates across the genome.

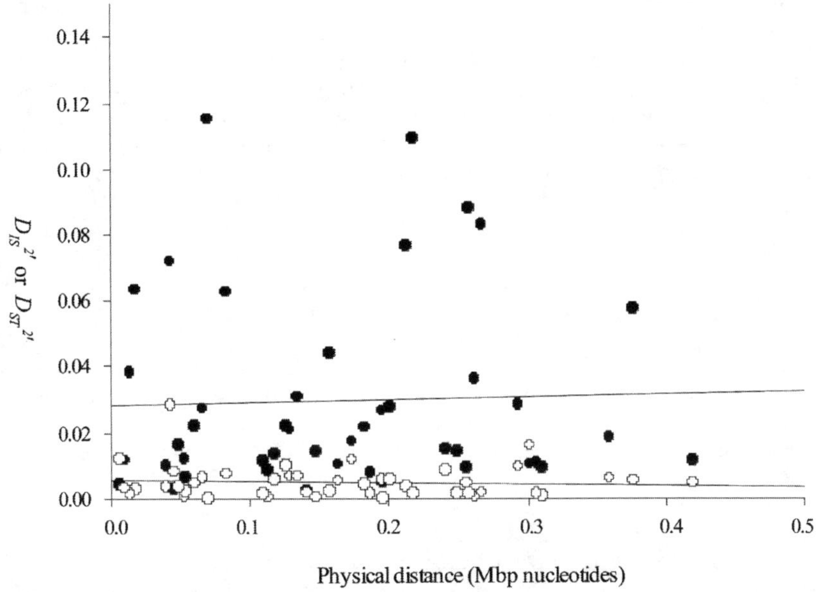

Figure 3. Ohta's linkage disequilibrium D-statistics[36] analyses among 10 microsatellite loci on the right arm of chromosome 3 among 88 female *Anopheles gambiae s.s.* from Sao Tome and Principe. Using World Health Organization definitions 44 mosquitoes were resistant to permethrin and 44 susceptible. For pairwise linkage analyses at all scales the inter-phenotype group component of linkage disequilibrium ($D^{2\prime}_{ST}$) (open circles) was substantially lower than the intra-group component ($D^{2\prime}_{IS}$) (filled circles).

Whether 2L or 3R is more representative of the *An. gambiae* genome requires further investigation and has important implications for the use of genotype/phenotype association studies in these species. The reasons for the potentially contrasting patterns of linkage disequilibrium between *Ae. aegypti* and *An. gambiae* are not immediately apparent. However, all of the models that we have presented in this discussion do not account for mutation, genetic drift, and migration and all of these influence patterns of linkage disequilibrium in natural populations.

The Yin and Yang of Linkage Disequilibrium

The reason for this detailed consideration of linkage disequilibrium in *Aedes aegypti* and *Anopheles gambiae* is that the power to detect valid genotype-phenotype associations with linkage disequilibrium or QTN mapping depends critically on the extent to which segregating sites in a genome assort independently.

Linkage disequilibrium mapping depends on markers being in disequilibrium with the gene(s) that actually condition expression of the phenotype (Figs 2,3). For this reason, if the trypsin and 2L data sets are representative for each organism's genome then the preceding discussion suggests that linkage disequilibrium mapping in *An. gambiae* (at least in genome regions in and around inversions) will probably be successful whereas in *Ae. aegypti* it will not. Linkage disequilibrium mapping was used in *An. gambiae* as a means to identify genome regions that differ between two sympatric, partially isolated subtaxa known as M and S molecular forms.[37] Field studies have shown that these forms mate assortatively but, in the laboratory, no fitness reduction has been detected in heterozygotes, suggesting isolation of M and S forms involves a prezygotic isolation mechanism.[38] Early association studies using microsatellite markers found little genetic differentiation outside the centromeric end of the X chromosome but these studies only genotyped 10-25 loci.[39,40] In contrast, hybridized of genomic DNA from field samples of each form was hybridized to Affymetrix GeneChip microarrays containing 142,065 unique probes and three genome regions were differentiated between M and S forms[37] Two of these regions are adjacent to centromeres, on Chromosomes 2L and X and the third was on Chromosome 2R. Sequenced loci in these regions contain fixed differences between forms and no shared polymorphisms, while no fixed differences were found at nearby control loci. These 3 regions contained 67 predicted genes.

Linkage mapping studies for insecticide resistance in *An. gambiae* are already underway using a targeted SNP approach. We are preferentially targeting SNPs identified *in silico* that are clustered around known detoxification enzymes (e.g., Cytochrome P450s, GSTs and Esterases), insecticide target sites (e.g., Na^+ channel) and genes associated with oxidative stress. Using this clustered SNP approach with markers placed at a frequency of ~ 5kb we expect to be able to identify loci associated with insecticide resistance phenotypes. This approach will be a prelude to whole genome SNP arrays that could be used for any quantifiable phenotype (e.g., refractoriness, morphology, habitat segregation, host preference).

In contrast to LD mapping, QTN mapping depends critically upon most proximal loci being at equilibrium (e.g., Fig. 1). In order to precisely associate individual segregating sites with expression of a phenotype, those sites must segregate independently. This suggests that QTN mapping in *An. gambiae* may be imprecise because polymorphic sites cosegregate. In contrast, QTN mapping will probably be very precise in *Ae. aegypti.*

Thus the Yin and Yang of linkage disequilibrium is that the very same forces that make linkage disequilibrium mapping useful in one species or in specific genome regions of a species may simultaneously limit our ability to perform QTN mapping in that species. Conversely, linkage disequilibrium mapping may not be useful in species such as *Ae. aegypti* which exhibit little disequilibrium among even proximal sites (e.g., Fig. 1) while QTN mapping may be very precise.

As an example, how will we discover which of the 67 predicted genes actually condition prezygotic barriers to mating between M and S forms in *An. gambiae*? Can QTN be individually identified in genome regions with large amounts of disequilibrium in *An. gambiae*? If LD mapping is unlikely to work in *Ae. aegypti,* then how do we go about finding genome regions affecting phenotypes of interest. How will we discover which genome regions confer resistance to a pesticide in natural *Ae. aegypti* populations? One solution may lie in the use of summary statistical approaches that have proved informative in studies of the evolution of resistance to chemotherapeutics in *Plasmodium falciparum*[41-43] and adaptation in *Drosophila.*[44] An allele (either a promoter or expressed gene) conferring insecticide resistance will increase in frequency in populations subject to strong insecticide challenge. The result of this is that neutral loci around the beneficial mutation will, as well as showing increased levels of linkage disequilibrium also show decreased levels of diversity (heterozygosity). It is possible to use the same SNP markers to identify regions with increased linkage disequilibrium and decreased diversity. What is clear is that exploratory studies are required to estimate analytical power and marker densities prior to commencing linkage or association mapping studies[12,45] and that they should be complemented by additional approaches or functional assays.

QTL Mapping

Anopheles gambiae has the advantage of having almost complete genome sequences. This database provided the information necessary to generate high density linkage maps which in turn has lead to the ability to map genome regions containing Quantitative Trait Loci (QTL) that condition phenotypes. QTL mapping differs from LD mapping in starting with only 2 P_1 parents from either a field population but usually from lines selected for opposite phenotypes (refractory vs. susceptible, resistant vs. susceptible to insecticides). The F_1 offspring are intercrossed and the phenotypes of individual F_2 offspring are scored and analyzed alongside marker genotype data. Various computer programs methodically test each locus for a statistical association between the genotypes at marker loci and the phenotypes in the F_2 offspring. QTL mapping has been used in *An. gambiae* to identify genome regions associated with melanotic encapsulation of *Plasmodium* oocysts,[46-48] ookinete killing,[49] insecticide resistance[35,50] and hybrid female sterility between *An. gambiae* and *An. Arabiensis.*[51,52] QTL mapping has been used in *Aedes aegypti* to map genome regions associated with susceptibility to the filarial parasite *Brugia malayi*[53,54] and to malarial parasite *Plasmodium gallinaceum.*[55-57] The approach was used to identify genome regions associated with midgut infection and dissemination barriers to dengue virus.[58-60] However, QTL mapping requires enormous time, labor and expense to map regions in only 4 genomes in the two P_1 parents. By contrast, LD mapping examines many genomes simultaneously in F_1 individuals arising from a field population. Nevertheless, QTL mapping may represent the only means to identify genome regions affecting phenotypes in species such as *Ae. aegypti* with little or no LD in natural populations. Unfortunately, the Ae. aegypti failed to assemble [76] and so there is currently no way of determining the genes underlying a QTL.

Microarray Technology

Microarrays are also a product of genome projects and are a powerful technology for examining the relationship between genome-wide gene expression profiles and insecticide resistance. The use of whole genome microarrays provides an efficient way to screen for new candidate genes associated with a particular trait.[61-69] A 3,840 EST microarray was used to described the complex gene expression responses of *Anopheles gambiae* to microbial and malaria challenges, injury, and oxidative stress.[70] They showed that Gram+ and Gram- bacteria and microbial elicitors up-regulate a diverse set of immunity class genes. These genes only partially correspond to those induced by malaria. Oxidative stress activated a distinctive set of genes, mainly implicated in oxido-reductive processes. More recently, a microarray was constructed that contained unique fragments from 230 *Anopheles gambiae* genes putatively involved in insecticide metabolism (cytochrome P450s, glutathione S-transferases, carboxylesterases) as well as redox genes, partners of the P450 oxidative metabolic complex and various controls.[62] David et al (2005) used this 'detox chip' to monitor the expression of the detoxifying genes in insecticide resistant and susceptible *An. gambiae* laboratory strains. Glutathione S-transferase *GSTe2,* two cytochrome P450s and two peroxidase genes were strongly upregulated in a DDT resistant strain. In addition, cytochrome P450 *CYP325A3,* belonging to a class not previously associated with insecticide resistance was expressed at statistically higher levels in this strain. An *Ae. aegypti* 'detox chip' is presently under construction at the Liverpool School of Tropical Medicine. The *Aedes* detox chip will replicate the design of the *Anopheles* detox chip with an array of *Ae. aegypti* specific, P450s, GSTs and COEs in addition to redox and housekeeping genes.[77] One minor difference will be that the chip will only use oligonucleotide probes rather than the combination of oligonucleotides and cDNA that are presently used on the *Anopheles* detox chip.

Combining the information gathered from QTL mapping and microarray analysis may in the end represent the most promising current approach towards identifying which of the many potential genes in a genome region actually condition the phenotype of interest. But it is important to recognize that microarray analysis will only detect differences in transcription rates between groups of individuals. Furthermore candidate genes are usually those which show the highest expression levels which is an inherent bias in the approach. Microarray analysis will not detect differences due to SNPs in the structural part of a gene, unless these in some way influence the transcriptional profile of the gene. Actually identifying polymorphisms that confer phenotypic differences will require a QTN mapping approach. QTN mapping is a labor intensive process (see ref. 12) that cannot be easily done for large numbers of genes. Furthermore, as indicated in this review it may not prove useful in ultimately identifying QTN in all regions of the *An. gambiae* genome.

None of the technologies discussed in this review will detect post-transcriptional or translational differences in a gene. That will require a comprehensive proteomics approach, something that is only beginning to be used in vector genomes. Furthermore, as candidate genes and QTN are identified, all will require further characterization and validation in the laboratory and ultimately in the field. To date only limited validation has been performed but systems are being developed that will determine whether putative detoxificiation enzymes do render insecticides metabolically inactive and whether regulatory sequence variation mediates differential gene expression. Robust in vitro protein expression systems and functional assays are being developed for GST and P450 enzyme families using technologies from human drug metabolism studies.[71,72] Researchers have also begun to investigate the influence of promoter sequence variation on metabolic resistance gene expression thereby determining both proximal and distal causes of resistance. A luciferase reporter assay was used to investigate the expression of an epsilon class GST that had been implicated in DDT metabolism.[73] They observed that a two base pair indel in the putative promoter region in the DDT resistant strain was associated with a 2.8 fold increase in gene expression. The combination of these functional approaches with linkage disequilibrium and QTN analyses should ultimately enable researchers to overcome problems with inter- or intra- genome variation in recombination rates.

Whilst in this article we have stressed the differences between the two vector species similarities do exist which may aid the identification of insecticide resistance mechanisms using linkage-based approaches. A putative ortholog of GSTe2 that was found to catalyze the dehydrochlorination of DDT in *An. gambiae* has also been implicated as conferring resistance to DDT in *Ae. aegypti.*[74] Given that in theory monogenic resistance is most likely to emerge in natural populations,[75] and that the *GSTe2* data suggest that despite the extensive duplication and diversification within gene families insecticide resistance may be a property of only a limited number of genes interspecific comparative analyses may hold the key to the rapid identification of resistance loci.

Acknowledgements

This research was supported by National Institutes of Health Grants R01-AI49256, U01-AI45430, and R01-040308 and from a Medical Research Council research studentship G78/8109.

References

1. Black WC, Baer CF, Antolin MF et al. Population genomics: Genome-wide sampling of insect populations. Annu Rev Entomol 2001; 46:441-469.
2. Erickson DL, Fenster CB, Stenoien HK et al. Quantitative trait locus analyses and the study of evolutionary process. Mol Ecol 2004; 13:2505-2522.
3. Stephens JC, Briscoe D, O'Brien SJ. Mapping by admixture linkage disequilibrium in human populations: Limits and guidelines. Am J Hum Genet 1994; 55:809-824.
4. Keim P, Diers BW, Olson TC et al. RFLP mapping in soybean: Association between marker loci and variation in quantitative traits. Genetics 1990; 126:735-742.
5. Osterberg MK, Shavorskaya O, Lascoux M et al. Naturally occurring indel variation in the Brassica nigra COL1 gene is associated with variation in flowering time. Genetics 2002; 161:299-306.
6. Maksymowych WP, Rahman P, Reeve JP et al. Association of the IL1 gene cluster with susceptibility to ankylosing spondylitis: An analysis of three Canadian populations. Arthritis Rheum 2006; 54:974-985.
7. Morton NE. Fifty years of genetic epidemiology, with special reference to Japan. J Hum Genet 2006; 51:269-277.
8. Zeng Z, Zhou Y, Zhang W et al. Family-based association analysis validates chromosome 3p21 as a putative nasopharyngeal carcinoma susceptibility locus. Genet Med 2006; 8:156-160.
9. Palsson A, Dodgson J, Dworkin I et al. Tests for the replication of an association between Egfr and natural variation in Drosophila melanogaster wing morphology. BMC Genet 2005; 6:44.

10. Genissel A, Pastinen T, Dowell A et al. No evidence for an association between common nonsynonymous polymorphisms in delta and bristle number variation in natural and laboratory populations of Drosophila melanogaster. Genetics 2004; 166:291-306.
11. Grisart B, Farnir F, Karim L et al. Genetic and functional confirmation of the causality of the DGAT1 K232A quantitative trait nucleotide in affecting milk yield and composition. Proc Natl Acad Sci USA 2004; 101:2398-2403.
12. Gorrochotegui-Escalante N, Lozano-Fuentes S, Bennett KE et al. Association mapping of segregating sites in the early trypsin gene and susceptibility to dengue-2 virus in the mosquito Aedes aegypti. Insect Biochem Mol Biol 2005; 35:771-788.
13. Curtis CF. Should DDT continue to be recommended for malaria vector control? Med Vet Entomol 1994; 8:107-112.
14. Curtis CF, Miller JE, Hodjati MH et al. Can anything be done to maintain the effectiveness of pyrethroid-impregnated bednets against malaria vectors? Philos Trans R Soc Lond B Biol Sci 1998; 353:1769-1775.
15. Donnelly MJ, Simard F, Lehmann T. Evolutionary studies of malaria vectors. Trends Parasitol 2002; 18:75-80.
16. Hemingway J, Hawkes NJ, McCarroll L et al. The molecular basis of insecticide resistance in mosquitoes. Insect Biochem Mol Biol 2004; 34:653-665.
17. Roberts DR, Andre RG. Insecticide resistance issues in vector-borne disease control. Am J Trop Med Hyg 1994; 50:21-34.
18. Kolaczinski JH, Curtis CF. Investigation of negative cross-resistance as a resistance-management tool for insecticide-treated nets. J Med Entomol 2004; 41:930-934.
19. Kurtak D, Meyer R, Ocran M et al. Management of insecticide resistance in control of the Simulium damnosum complex by the Onchocerciasis Control Programme, West Africa: Potential use of negative correlation between organophosphate resistance and pyrethroid susceptibility. Med Vet Entomol 1987; 1:137-146.
20. Tabashnik BE. Implications of gene amplification for evolution and management of insecticide resistance. J Econ Entomol 1990; 83:1170-1176.
21. Kublin JG, Cortese JF, Njunju EM et al. Reemergence of chloroquine-sensitive Plasmodium falciparum malaria after cessation of chloroquine use in Malawi. J Infect Dis 2003; 187:1870-1875.
22. Brogdon WG, McAllister JC. Insecticide resistance and vector control. Emerg Infect Dis 1998; 4:605-613.
23. Brogdon WG, Beach RF, Barber AM et al. A generalized approach to detection of organophosphate resistance in mosquitoes. Med Vet Entomol 1992; 6:110-114.
24. Brogdon WG, McAllister JC. Simplification of adult mosquito bioassays through use of time-mortality determinations in glass bottles. J Am Mosq Control Assoc 1998; 14:159-164.
25. Saelim V, Brogdon WG, Rojanapremsuk J et al. Bottle and biochemical assays on temephos resistance in Aedes aegypti in Thailand. Southeast Asian J Trop Med Public Health 2005; 36:417-425.
26. Vulule JM, Beach RF, Atieli FK et al. Elevated oxidase and esterase levels associated with permethrin tolerance in Anopheles gambiae from Kenyan villages using permethrin-impregnated nets. Med Vet Entomol 1999; 13:239-244.
27. Brown SE, Severson DW, Smith LA et al. Integration of the Aedes aegypti mosquito genetic linkage and physical maps. Genetics 2001; 157:1299-1305.
28. Wall JD. A comparison of estimators of the population recombination rate. Mol Biol Evol 2000; 17:156-163.
29. Carvajal-Rodriguez A, Crandall KA, Posada D. Recombination estimation under complex evolutionary models with the coalescent composite-likelihood method. Mol Biol Evol 2006; 23:817-827.
30. Hudson RR, Kreitman M, Aguade M. A test of neutral molecular evolution based on nucleotide data. Genetics 1987; 116:153-159.
31. Beerli P, Felsenstein J. Maximum-likelihood estimation of migration rates and effective population numbers in two populations using a coalescent approach. Genetics 1999; 152:763-773.
32. Beerli P, Felsenstein J. Maximum likelihood estimation of a migration matrix and effective population sizes in n subpopulations by using a coalescent approach. Proc Natl Acad Sci USA 2001; 98:4563-4568.
33. Gorrochotegui-Escalante N, Gomez-Machorro C, Lozano-Fuentes S et al. Breeding structure of Aedes aegypti populations in Mexico varies by region. Am J Trop Med Hyg 2002; 66:213-222.
34. Lehmann T, Hawley WA, Grebert H et al. The effective population size of Anopheles gambiae in Kenya: Implications for population structure. Mol Biol Evol 1998; 15:264-276.
35. Ranson H, Paton MG, Jensen B et al. Genetic mapping of genes conferring permethrin resistance in the malaria vector, Anopheles gambiae. Insect Mol Biol 2004; 13:379-386.

36. Ohta T. Linkage disequilibrium due to random genetic drift in finite subdivided populations. Proc Natl Acad Sci USA 1982; 79:1940-1944.
37. Turner TL, Hahn MW, Nuzhdin SV. Genomic islands of speciation in Anopheles gambiae. PLoS Biol 2005; 3:e285.
38. Tripet F, Toure YT, Taylor CE et al. DNA analysis of transferred sperm reveals significant levels of gene flow between molecular forms of Anopheles gambiae. Mol Ecol 2001; 10:1725-1732.
39. Stump AD, Fitzpatrick MC, Lobo NF et al. Centromere-proximal differentiation and speciation in Anopheles gambiae. Proc Natl Acad Sci USA 2005; 102:15930-15935.
40. Stump AD, Shoener JA, Costantini C et al. Sex-linked differentiation between incipient species of Anopheles gambiae. Genetics 2005; 169:1509-1519.
41. Nair S, Williams JT, Brockman A et al. A selective sweep driven by pyrimethamine treatment in southeast asian malaria parasites. Mol Biol Evol 2003; 20:1526-1536.
42. Roper C, Pearce R, Nair S et al. Intercontinental spread of pyrimethamine-resistant malaria. Science 2004; 305:1124.
43. Wootton JC, Feng X, Ferdig MT et al. Genetic diversity and chloroquine selective sweeps in Plasmodium falciparum. Nature 2002; 418:320-323.
44. Harr B, Kauer M, Schlotterer C. Hitchhiking mapping: A population-based fine-mapping strategy for adaptive mutations in Drosophila melanogaster. Proc Natl Acad Sci USA 2002; 99:12949-12954.
45. Black W, Severson D. Genetics of vector competence. In: Marquardt W, ed. Biology of Disease Vectors. 2nd ed. Harcourt Academic Press, 2002.
46. Menge DM, Zhong D, Guda T et al. Quantitative trait loci controlling refractoriness to plasmodium falciparum in natural anopheles gambiae from a malaria endemic region in western kenya. Genetics 2006.
47. Zheng L, Wang S, Romans P et al. Quantitative trait loci in Anopheles gambiae controlling the encapsulation response against Plasmodium cynomolgi Ceylon. BMC Genet 2003; 4:16.
48. Gorman MJ, Severson DW, Cornel AJ et al. Mapping a quantitative trait locus involved in melanotic encapsulation of foreign bodies in the malaria vector, Anopheles gambiae. Genetics 1997; 146:965-971.
49. Niare O, Markianos K, Volz J et al. Genetic loci affecting resistance to human malaria parasites in a West African mosquito vector population. Science 2002; 298:213-216.
50. Ranson H, Jensen B, Wang X et al. Genetic mapping of two loci affecting DDT resistance in the malaria vector Anopheles gambiae. Insect Mol Biol 2000; 9:499-507.
51. Slotman M, Della Torre A, Powell JR. Female sterility in hybrids between Anopheles gambiae and A. arabiensis, and the causes of Haldane's rule. Evolution Int J Org Evolution 2005; 59:1016-1026.
52. Slotman M, Della Torre A, Powell JR. The genetics of inviability and male sterility in hybrids between Anopheles gambiae and An. arabiensis. Genetics 2004; 167:275-287.
53. Severson DW, Mori A, Zhang Y et al. Chromosomal mapping of two loci affecting filarial worm susceptibility in Aedes aegypti. Insect Mol Biol 1994; 3:67-72.
54. Beerntsen BT, Severson DW, Klinkhammer JA et al. Aedes aegypti: A quantitative trait locus (QTL) influencing filarial worm intensity is linked to QTL for susceptibility to other mosquito-borne pathogens. Exp Parasitol 1995; 81:355-362.
55. Severson DW, Thathy V, Mori A et al. Restriction fragment length polymorphism mapping of quantitative trait loci for malaria parasite susceptibility in the mosquito Aedes aegypti. Genetics 1995; 139:1711-1717.
56. Severson DW, Zaitlin D, Kassner VA. Targeted identification of markers linked to malaria and filarioid nematode parasite resistance genes in the mosquito Aedes aegypti. Genet Res 1999; 73:217-224.
57. Zhong D, Menge DM, Temu EA et al. AFLP mapping of quantitative trait loci for malaria parasite susceptibility in the yellow fever mosquito, Aedes aegypti. Genetics 2006.
58. Bennett KE, Flick D, Fleming KH et al. Quantitative trait loci that control dengue-2 virus dissemination in the mosquito Aedes aegypti. Genetics 2005; 170:185-194.
59. Gomez-Machorro C, Bennett KE, del Lourdes Munoz M et al. Quantitative trait loci affecting dengue midgut infection barriers in an advanced intercross line of Aedes aegypti. Insect Mol Biol 2004; 13:637-648.
60. Bosio CF, Fulton RE, Salasek ML et al. Quantitative trait loci that control vector competence for dengue-2 virus in the mosquito Aedes aegypti. Genetics 2000; 156:687-698.
61. Dana AN, Hong YS, Kern MK et al. Gene expression patterns associated with blood-feeding in the malaria mosquito Anopheles gambiae. BMC Genomics 2005; 6:5.
62. David JP, Strode C, Vontas J et al. The Anopheles gambiae detoxification chip: A highly specific microarray to study metabolic-based insecticide resistance in malaria vectors. Proc Natl Acad Sci USA 2005; 102:4080-4084.

63. Chen H, Wang J, Liang P et al. Microarray analysis for identification of Plasmodium-refractoriness candidate genes in mosquitoes. Genome 2004; 47:1061-1070.
64. Hall N, Karras M, Raine JD et al. A comprehensive survey of the Plasmodium life cycle by genomic, transcriptomic, and proteomic analyses. Science 2005; 307:82-86.
65. Sanders HR, Evans AM, Ross LS et al. Blood meal induces global changes in midgut gene expression in the disease vector, Aedes aegypti. Insect Biochem Mol Biol 2003; 33:1105-1122.
66. Ranson H, Claudianos C, Ortelli F et al. Evolution of supergene families associated with insecticide resistance. Science 2002; 298:179-181.
67. Christophides GK, Zdobnov E, Barillas-Mury C et al. Immunity-related genes and gene families in Anopheles gambiae. Science 2002; 298:159-165.
68. Morel CM, Toure YT, Dobrokhotov B et al. The mosquito genome—a breakthrough for public health. Science 2002; 298:79.
69. Zdobnov EM, von Mering C, Letunic I et al. Comparative genome and proteome analysis of Anopheles gambiae and Drosophila melanogaster. Science 2002; 298:149-159.
70. Dimopoulos G, Christophides GK, Meister S et al. Genome expression analysis of Anopheles gambiae: Responses to injury, bacterial challenge, and malaria infection. Proc Natl Acad Sci USA 2002; 99:8814-8819.
71. Kemp CA, Flanagan JU, van Eldik AJ et al. Validation of model cytochrome P450 2D6: An in silico tool for predicting metabolism and inhibition. J Med Chem 2004; 47:5340-5346.
72. Lycett GJ, McLaughlin A, Ranson H et al. Anopheles gambiae P450 reductase is highly expressed in oenocytes and in vivo knock down increases permethrin susceptibility. Insect Mol Biol 2006; 15:321-327.
73. Ding Y, Ortelli F, Rossiter LC et al. The Anopheles gambiae glutathione transferase supergene family: Annotation, phylogeny and expression profiles. BMC Genomics 2003; 4:35.
74. Lumjuan N, McCarroll L, Prapanthadara LA et al. Elevated activity of an Epsilon class glutathione transferase confers DDT resistance in the dengue vector, Aedes aegypti. Insect Biochem Mol Biol 2005; 35:861-871.
75. ffrench-Constant RH, Daborn PJ, Le Goff G. The genetics and genomics of insecticide resistance. Trends Genet 2004; 20:163-170.
76. Nene V, Wortman JR, Lawson D et al. Genome sequence of Aedes aegypti, a major arborvirus vector. Science 2007; 316:1718-1723.
77. Strode C, Wondji CS, David J-P et al. Genomic analysis of detoxification genes in the mosquito Aedes aegypti. Insect Biochem Mol Biol 2007; 2008, 38(1):1130123.

Chapter 7

Impact of Technological Improvements on Traditional Control Strategies

Mark Q. Benedict* and Alan S. Robinson

Introduction

Since 1982 when transgenesis of *Drosophila melanogaster* splashed onto the scientific scene,[1,2] members of the vector biology community (e.g., refs. 3,4) and international public health organizations[5] have recognized the potential utility of transgenesis to produce a modern incarnation of a historically puzzling observation: "anophelism without malaria:" The presence of anophelines but without disease. The corresponding concept among arbovirologists does not have a similarly appealing description, but the essence is the same: replacement of mosquito populations that are capable of transmitting disease with modified populations that are not. Visionary proponents argue that such a strategy is not only technically feasible, but that it leverages the power of biotechnology and a modern understanding of the means by which modified phenotypes can be spread through populations using transposable elements, cytoplasmic incompatibility, homing endonucleases, and meiotic drive. If theoretical possibilities are realized, such a vector population transformation would have minimal disruption to an ecosystem due to specific modification of only one vector/pathogen interaction and that without the use of drugs or environmentally harmful insecticides. Advocates cautiously emphasize that such applications will likely never be magic bullet solutions, and that they will be implemented only in integrated vector management programs. Nonetheless, realization of such a goal would be a remarkable demonstration of biological power for the benefit of human health.

Operational realization of such strategies will require years of study during which development of useful—but less obvious—applications of insect transgenesis are possible. These have the potential to change the complexion of existing control strategies, make their conduct more effective, and open experimental manipulations for the development of novel variations on traditional control approaches. Such activities also provide an essential translational basis on which the behavior, safety, and use of transgenesis in public settings can be tested.

In this chapter, we describe several such uses of insect transgenesis, first to improve the sterile insect technique and conceptual control strategies, and second to facilitate entomological research activities. We also describe how transgenic applications are consistent with existing integrated pest management operations. We hope to describe how these applications can make an immediate and significant impact on disease intervention efforts and serve as research tools that enhance the discovery of novel methods.

*Corresponding Author: Mark Q. Benedict—CDC, 4770 Buford Highway, Chamblee, Georgia 30341 USA. Email: MBenedict@cdc.gov

Transgenesis and the Management of Vector-Borne Disease, edited by Serap Aksoy.

Applications

Sterile Insect Technique (SIT) and Potential Genetic Control Methods

Many genetic control techniques have been tested against pest insects, but only one, the sterile insect technique, has reached the stage of an economic, practical and sustainable technology for insect control.[6] Therefore, it is likely that the use of transgenics will first impact this method. In the following section, we describe several applications that are envisioned or that have already been demonstrated.

Sexual Sterilization

Ionizing radiation is now the technology of choice for sterilizing insects for release, however in the early days, the use of chemosterilants was widely evaluated,[7] and a chemosterilant was used in the successful sterile release of *Anopheles albimanus* in El Salvador.[8] Ionizing radiation induces dominant lethal mutations in sperm of treated males which, when used for fertilization by wild females following their mating with sterile males, lead to the death of developing zygotes.[9] Radiation is generally applied to the insect as late in the developmental cycle as possible to reduce somatic effects to a minimum.

There are some disadvantages associated with the use of radiation: (1) high capital cost of irradiators and their housing and the strict regulatory framework for use of radiation sources, (2) narrow developmental window for application, (3) some deleterious somatic effects, (4) an extra handling procedure, (5) strict quality control protocols to prevent release of nonirradiated insects, and (6) restrictions on the stage of insect that can be released. Despite these disadvantages, radiation has provided a safe and very reliable physical method for insect sterilization in large scale field programmes.

Genetic methods for introducing sterility into field populations have also been extensively researched; these include chromosomal translocations, hybrid sterility, cytoplasmic incompatibility and compound chromosomes.[10] Translating the potential of these biological systems into effective field application has proven difficult, and to date none have been extensively used although many field trials have been carried out. Two factors tend to compromise the usefulness of these systems: firstly the response of the wild population following interaction with the released insects cannot always be accurately predicted, and secondly, mass rearing of strains with specific biological characteristics is not straightforward. These two factors are generally of little consequence when essentially wild-type insects are mass reared and sterilized for release as is typical in SIT. The exception would be the use of genetic sexing strains for Mediterranean fruit fly SIT programmes (see below).

Transgenesis offers the possibility of creating new means to accomplish sterility including at least two approaches. In the first, the males would transmit a trait to their progeny that, outside of the laboratory, would result in lethality. Lethality in the field would rely on the released males mating successfully with wild females and that the progeny of such a mating would inherit at least one copy of the transgene. In the absence of the conditional constraint these progeny would die. The system obviously requires that the released transgenic adults are not in themselves "susceptible" to the effector in the absence of the constraint. In some systems presently being tested, lethality is generated by a single effector molecule which produces a toxic product. If the target of the effector were encoded by a single gene, this would seem a risky strategy considering the ease with which insects can develop resistance to many natural and man-made toxic products. A single mutational event in the wild population would rapidly spread and render the lethality obsolete. However, if there were multiple targets, resistance problems might be avoided.

Transgenesis has already opened the door to realization of one form of these approaches for insect sterilization via dominant lethality[11-13] but with the same requirements for the lethal effects as those generated by radiation i.e., they must act dominantly in the field and be controllable in the mass rearing facility. Dominance can be achieved by selecting the appropriate effector molecule and controllability by placing the effector gene under a conditional promoter. For application in the field of vector control where the pestiferous stage is generally the adult, lethality at any time before adulthood is acceptable. This is in contrast to many agricultural applications where early lethality is required to prevent damage to a crop by developing larvae.

In a second approach that has not been experimentally developed in vectors, an inducible transgene effector would be used to cause male sterility not by dominant lethality experienced by progeny, but by incapacitating the ability of males to fertilize females. The details of how this could be accomplished in various species would differ for reasons more subtle than the mere molecular considerations of promoter, inducer, and effector function in that particular species. This is because the effectiveness of SIT depends partly on the ability of released males to minimize subsequent remating by wild females. It is necessary that the mating activity itself cause this response. Therefore the appropriate signal must be received by females to change their behavior appropriately to minimize remating. For example, mosquitoes use at least two different signals to accomplish this. *Aedes aegypti* males transfer a substance called "male accessory gland substance" to females to reduce subsequent mating.[14] In contrast, *Anopheles gambiae* females require sperm transfer.[15] Therefore, a strategy that interfered with spermatogenesis in transgenic *Ae. aegypti* might be effective, but it likely would not be in *An. gambiae.* Given the small number of male-encoded genes that interfere only with a sperm's ability to fertilize eggs in *Drosophila*[16] it is possible that such transgenic sterility could be developed for *Anopheles* spp.

Regardless of which approaches are adopted, conditional expression in the mass rearing facility will be achieved by creating a permissive environment in which the transgenic strain can be efficiently mass reared using either a physical or a chemical constraint. The absence of this constraint in the field will lead to lethality. The physical or chemical manipulation of one or more components of the rearing process is fairly straightforward under normal laboratory conditions but can become a significant problem when expanded production is required. Accurate control of temperature during the rearing process is not always reliable, and if a chemical treatment is used problems related to its distribution in a diet and its disposal can become important. However, for vectors with an aquatic stage, the larval and pupal environment has obvious advantages in terms of conditional regulation.

The stability of the transgene both in terms of expression, mutation and genomic location could become important under large scale mass rearing where many millions or even hundreds of millions of insects will be produced every week. Quality control procedures must be introduced to monitor the stability of the strain and this will likely include the use of a filter rearing system (FRS).[17] The FRS concept was specifically designed to maintain stability in Mediterranean fruit fly genetic sexing strains constructed using translocations and a temperature sensitive lethal mutation. It enables exceptional individuals to be identified and removed from the rearing process. In the case of the Mediterranean fruit fly genetic sexing strains, the exceptional individuals are those in which rare male recombination events have resulted in the generation of females that are resistant to elevated temperature. The use of an FRS is now standard practice in many large facilities rearing Mediterranean fruit flies and the principle could probably be transferred to other species and strains.

Producing Only Males for Release

Release of only males is essential for any mosquito sterile release as even sexually sterile females can contribute to disease transmission. Even in cases in which transmission can result from both sexes, production considerations may make it desirable to retain females in the factory. Two technical and operational approaches have been used to accomplish this, based either on population or individual selection. To be effective in large scale release programmes, sexing systems have to be implemented at the population level as opposed to the individual level. That is to say that a selective treatment is most effective when all individuals can be exposed simultaneously as opposed to treatment or selection of individuals sequentially. The best examples among vectors of the population selection methods are based on dominant insecticide resistance markers pseudo-linked to male sex determination.[18-20] These have been sufficiently successful that operational releases have been resulted in field programme use.[21,22] In a conventional SIT programme the use of transgenesis to effect sex-separation would be restricted to the mass-rearing facility with the transgenic males being irradiated and released. Consequently, there is no interaction between the genome of the wild insects and the transgene in the sterile males.

Potential for eliminating females during factory culture has been demonstrated in *Drosophila melanogaster* using a transgenic system in which a pro-insecticidal drug is converted to the active form.[23] In this system, a bacterial activating enzyme, *cytosine deaminase* (CD), is expressed only in females and converts nontoxic 5-fluorocystine to a toxic product, 5-fluorouracil. Sex specificity is achieved by placing the CD gene under the control of a female-specific yolk protein promoter. Consequently, lethality on selective media is not manifested until the adult stage. While this specific promoter might not be suitable for insects in which lethality must be accomplished during the preadult stages, a conserved larval female-specific promoter has been identified[24] and could be considered for this purpose.

Several individual selection methods have also been developed. The use of sexual dimorphism in the pupal stage of *Aedes aegypti*[25] and *Aedes albopictus*[26] is sufficient for these species and >99% female-elimination can be accomplished. Other individual selection methods have involved the use of pupal colour mutations and pseudo-linkage to males (see ref 27 for references). Sexing can then be accomplished by passing the pupae through modified agricultural seed sorters. A similar approach is now being followed in tsetse flies using differences in near infra-red absorbance between male and female pupae.[28]

Similar dimorphic phenotypic traits are often not available for medically important arthropods. However, a generic transgenic solution for individual sorting has been accomplished by expressing eGFP under the control of the male-specific *beta tubulin* promoter in testes of *Anopheles stephensi*.[29] The authors were able to separate males from females using an automated sorter, however much development and testing remains before the technology is ready for large-scale mosquito production.[30]

The above sexing systems have relied on techniques which either kill females or which enable them to be separated from males. In the latter case the females can be returned for colony maintenance whereas in the former the females are lost. An attractive alternative to killing or separating females before release is to transform them into males, and this has now been demonstrated in the Mediterranean fruit fly[31] using RNAi technology. In this species the gene *transformer* (*tra*), is autoregulating in females and maintains sexual development in the female mode. In males the Y chromosome disrupts the feminizing activity of the maternal *tra* protein. By inserting a transgene carrying a heat shock promoter and an inverted repeat corresponding to the *tra* sequence it was possible to transform XX embryos which would normally develop into females into XX males.[33] This approach is also being extended to mosquitoes (Saccone, pers. comm.). Details of the sex-determination pathway of vectors must be determined before such approaches can be applied more widely, and the sexual competitiveness of such insects would be of critical importance.

Novel Genetic Control Strategies—RIDL, Selfish Genes, and Female Lethals

Thomas et al[12] and Heinrich et al[13] were the first to demonstrate the use of a transgenic repressible dominant system that has the potential to introduce sterility into natural populations. Thomas et al coined the term "release of insects carrying a dominant lethal" (RIDL) to describe the principle and it is discussed in detail elsewhere in this book (Chapter 6). This method, similar to SIT, requires releases of large numbers of insects, but it has been argued that it offers several advantages including avoiding the loss of vigor resulting from irradiation.[32]

As an alternative to this approach, it has been proposed that certain classes of selfish genes can also be used to kill insects in natural populations through the disruption of essential genes.[33]

These selfish genes have the advantage that they can replicate and increase in frequency thus allowing the release of a relatively few individuals. Finally, killing females in the field is equivalent to sterilizing them in terms of population dynamics, and there are several transgenic approaches that have been proposed to achieve this.[34,35] In these models the female killing transgene can be transmitted via the males which can carry the transgene at many loci.

Tsetse SIT and Refractoriness

Tsetse flies are the sole vectors of African trypanosomes in the majority of countries in Sub-saharan Africa[36] and the trypanosomes cause a wasting disease (nagana) in domestic animals and sleeping sickness in humans. The use of sterile tsetse as a component of an integrated approach to control has been shown to be successful.[37] In contrast to mosquitoes, both sexes feed on blood and can therefore transmit the parasite. This could pose problems for the use of the SIT as large numbers of potential vectors would be released. There is also a confounding factor that irradiated female tsetse live longer than their nonirradiated sisters, presumably as a consequence of the termination of larviparous reproduction.[38] A longer lifespan increases the chances that a fly will become infected and be able to transmit the parasite. This requires that only male tsetse be released that have had their vectorial competence compromised through treatment with an anti-trypanocide.[39] Although these procedures are acceptable and effective where only animal trypanosomiasis is present in an area, this might not be the case in areas where there is active human transmission. In these cases the use of para-transgenesis to produce refractory strains[40] (see Chapter 3) would facilitate the use of sterile insects in a release program.

Laboratory and Field Research Reagents

Strain Identity in Laboratory Culture

Strains of insects are cultured in numerous laboratories for uses including physiological, insecticide, genetic, and molecular biology studies. Distinguishing such strains is usually based on a visible or biochemical phenotype. However, "wild-type" strains that are designated only by their location of origin are often identifiable only by the label on the culture containers. Contamination is common and may be difficult or impossible to detect: once they enter a laboratory inhabited by other isolates, mixing is always possible. It is probably impossible to confirm the identity of the majority of the wild-type strains being used in different laboratories. While molecular markers that distinguish such strains (e.g., microsatellites, RFLPs, SNPs) could be developed, they are very labor-intensive to identify and very few, if any, strains have unique markers.

An ideal solution would be to confer a unique heritable trait on all members of the population. A deliberate transposable element insertion could serve such a purpose when followed by a breeding scheme to make the insertion homozygous in the wild-type background. In those species for which transformation technology is available, it is feasible to genetically transform the strain and to purify the insertion after several generations of assortment or back-crossing to the background population. These genetic procedures would ensure that the overall genetic variation of the resulting stock would be as similar to that of the original as possible.

The DNA sequence of the insertion site by inverse or universal PCR and the insertion itself would ultimately define the unique marker. If desired, a visible marker could be left in place, or the marker and inverted terminal repeats could be removed by specific methods that are now available[41,42] in which case PCR would be used to confirm the integrity of the stock. Such an insertion or its remnants containing either a unique visible marker or partial transposable element that would be simple to detect yet would positively distinguish such a stock from others.

Genome sequencing projects are a specific application in which such tagging seems essential. Strains that are sequenced often become standard research strains since the known DNA sequence facilitates cloning and gene expression analyses. During the sequencing of the *Anopheles gambiae* PEST strain,[43] several stocks in various laboratories were lost. In an urgent effort to identify authentic copies, the pink eye mutation which was fixed in this stock proved invaluable for determining which putative PEST strains were authentic. While the existence of pink eye was superior to having no marker, even this example could have failed to solve the problem as the same marker exists in other stocks. The current *Aedes aegypti* genome project standard strain, LVP-IB12 is unmarked by either molecular or visible markers though efforts to develop the former are underway (D. Severson, pers comm.). In spite of the paucity of microsatellites in this species, finding molecular markers should be accomplished easily since the strain has been full-sib inbred for 12 generations. However, an even simpler alternative that would positively distinguish it from all others would be to transform the strain and select it to homozygosity. The unique insertion site would definitively identify the strain and would have no negative effects on its utility as a reference stock for genomic comparisons. An elegant solution would be use of a site specific docking cassette[44] which could also be used for gene expression studies in a controlled genetic background of known sequence.

Vector Bionomics and Genetics in Field Studies

Population Genetics

Studies of population genetics for many species have been sufficiently abundant that spatially and temporally higher resolution methods are now needed. Even in cases where there is little population genetic information, it is likely that releases of genetically unique animals would complement classical population genetics. For example, though population genetic information for *Anopheles gambiae* s.l. is by no means exhaustive, it is abundant, and some have recognized that releases of uniquely marked fertile mosquitoes into wild populations would provide high resolution data and have planned accordingly (G. Lanzaro, pers. comm.). In the example cited above, a rare microsatellite was selected for release in a village in Mali, but production problems prevented implementation. None-the-less, such releases would provide gene flow and dispersal information that conventional population genetic studies cannot.

As described above, any "wild-type" strain can be modified with one or more insertions whose fate could be followed subsequent to release into natural populations. Insertions with no detectable phenotype affecting natural behavior and life history would be ideal. If located within naturally polymorphic inversions, these could also provide a readily assayed marker for determining the behavior and significance of these aberrations. Similarly, release of genetically marked "wild-type" strains would allow definitive studies of oviposition behavior that have previously been studied using genetic relatedness.[45,46] Unlike using natural markers whose polymorphism reduces the resolution, transgenic markers would provide absolute identification of related progeny.

Marking for Mating Studies

For any genetic control technique, knowing the proportion of wild females that mated with released males versus wild males provides critical information on the competitiveness of the released males. Currently, females must be collected, and individual female egg-hatching rates determined—a tedious and not always reliable procedure. A very desirable assay would be one in which direct observation of the sperm contained within the spermatheca would indicate whether a released or wild male had been her mate. Only two methods that do not require obtaining eggs are available for mosquitoes: detection of ^{13}C[47] and ^{32}P in females[48] mated to males labeled during larval culture. These methods are limited however, as the former method requires individual mass-spectroscopy analysis and the latter introduces a radioactive isotope into the environment.

An alternative is to label the sperm or accessory gland fluid with a visible marker. This has now been achieved by placing expression of eGFP under the control of a testes-specific *beta-tubulin* promoter in transgenic *An. stephensi*[29] and *An. arabiensis* (H.C. Bossin, pers. comm.). This approach is especially relevant for monitoring genetic control programmes for mosquitoes where methods to trap males are not efficient and the effect of male release can only be determined through inference from more easily captured females.

Marking for Release-Recapture

Releases of adults are performed for various purposes, and distinguishing those released from the wild individuals is essential. Estimation of adult population size and daily survival are critical data to estimate the entomological component of vector-borne disease transmission and these can be determined by capture-mark-release-recapture or by release of marked laboratory cultured insects (e.g., ref. 49). Traditionally, discriminating between released insects and wild insects has been achieved by dusting the released insects with a fluorescent powder and the trapped insects are then distinguished under UV light. This system has many flaws not the least of which are the negative effects on survival on the insect and the costs associated with screening of large numbers of individuals. Obviously the fluorescent markers that are routinely used to identify transgenic individuals could also be used as markers for released insects, provided that they fulfill the necessary requirements associated with the use of transgenics in general.

Are Transgenic Insects Compatible Partners in the IPM Mix?

As indicated above, no individual technology is likely to be successful when used alone: this is true for conventional approaches and will undoubtedly be true for transgenic ones as well. It will therefore be essential that any effective control approach currently in use must remain effective when combined with modern biotechnology. One oft-heard criticism of moving forward with transgenic releases for control is that in order for them to be successful, one would necessarily need to discontinue use of insecticides (e.g., ref. 50).

Considering a specific example of supreme importance to public health of why insecticide use in fact is compatible with transgenic release is illuminating. The two most common forms of malaria mosquito control are insecticide treated bednets and indoor residual spraying, both of which are being actively promoted by numerous international, governmental and nongovernmental organizations. Both techniques target the female mosquito during her attempts to obtain a blood meal or while resting in habitations. By nature most of the new approaches being developed will rely on the act of mating to exert their effect on the population. To the extent that mating occurs in outdoor swarms, mating and blood feeding occur in two different spatial compartments and the two types of control should be fully compatible and may even be synergistic. Similarly, larval control, either by using insecticides or source reduction, could also be integrated with transgenic approaches, especially those which are targeted at population reduction via inundative releases of adults. Again, a combination of environmental and genetic loads on two different developmental stages could be advantageous.

Even considering the most ambitious transgenic replacement scenario in which a gene is being "driven" through a population, insecticide application that is not preferentially directed against the transgenic type will not prevent gene spread. Because both the altered and native forms would be affected to the same degree, the drive mechanism would continue to spread the effector gene or allele at the same rate as if insecticides were not applied. In fact, suppression of a population by the use of insecticides prior to release would make the release numbers more efficient.

Conclusions

The increasing variety, availability, simplicity, and subtle characteristics of transgenic organisms will necessarily make their uses more appealing. To the extent that biotechnology is accompanied by thorough testing, appropriate regulatory control and public acceptance, their potential will be realized. Moreover, the examples of imminent applications of transgenic technology in SIT as described above will provide real-world experience with transgenic organism's behavior in large-scale projects that will provide the essential safety demonstrations and biological knowledge to apply them in more ambitious manners.

References

1. Rubin GM, Spradling AC. Genetic transformation of Drosophila with transposable element vectors. Science 1982; 218:348-353.
2. Spradling AC, Rubin GM. Transposition of cloned P elements into Drosophila germ line chromosomes. Science 1982; 218:341-347.
3. Crampton J, Morris A, Lycett G et al. Transgenic mosquitoes - A future vector control strategy. Parasitol Today 1990; 6:31-36.
4. Moreira LA, Ghosh AK, Abraham EG et al. Genetic transformation of mosquitoes: A quest for malaria control. Int J Parasiolt 2002; 32:1599-1605.
5. Anonymous. Prospects For Malaria Control by Genetic Manipulation of its Vectors. World Health Organization, 1991.
6. Dyck VA, Hendrichs J, Robinson AS. The Sterile Insect Technique: Principles and Practice in Area-Wide Integrated Pest Management. Dordrecht, The Netherlands: Springer, 2005.
7. Borkoveck AB. Insect Chemosterilants. New York: Wiley Interscience, 1966.
8. Lofgren CS, Dame DA, Breeland SG et al. Release of chemosterilized males for the control of Anopheles albimanus in El Salvador. III. Field methods and population control. Am J Trop Med Hyg 1974; 23:288-297.
9. Curtis CF. Induced sterility in insects. Adv Reprod Physiol 1971; 5:119-165.
10. Pal R, Lachance LE. The operational feasibility of genetic methods for control of insects of medical and veterinary importance. Ann Rev Entomol 1974; 19:269-291.
11. Alphey LS, Andreasen M. Dominant lethality and insect population control. Molec Biochem Parasitol 2002; 121:173-178.
12. Thomas DD, Donnelly CA, Wood RJ et al. Insect population control using a dominant, repressible, lethal genetic system. Science 2000; 287:2474-2476.
13. Heinrich JC, Scott MJ. A repressible female-specific lethal genetic system for making transgenic insect strains suitable for a sterile-release program. Proc Natl Acad Sci USA 2000; 97(15):8229-8232.
14. Craig GB. Mosquitoes: Female monogamy induced by male accessory gland substance. Science 1967; 156:1499-1501.
15. Klowden MJ. Sexual receptivity in Anopheles gambiae mosquitoes: Absence of control by male accessory gland substances. J Insect Physiol 2001; 47:661-666.
16. Perotti ME, Cattaneo F, Pasini ME et al. Male sterile mutant casanova gives clues to mechanisms of sperm-egg interactions in Drosophila melanogaster. Mol Reprod Dev 2001; 60:248-259.
17. Fisher K, Caceres C. A filter rearing system for mass reared genetic sexing strains of Mediterranean fruit fly (Diptera: Tephritidae). Penang, Malaysia: Paper presented at: International Conference on Area-wide Control of Insect Pests and of the Fifth International Symposium on Fruit Flies of Economic Importance, 2002, (1998).
18. Reisen WK, Baker RH, Sakai F et al. Anopheles culicifacies Giles: Mating behavior and competitiveness in nature of chemosterilized males carrying a genetic sexing system. Ann Entomol Soc Am 1981; 74:395-401.
19. Kaiser PE, Seawright JA, Dame DA et al. Development of a genetic sexing system for Anopheles albimanus. J Econ Entomol 1978; 71:766-771.
20. Lines JD, Curtis CF. Genetic sexing systems in Anopheles arabiensis Patton (Diptera: Culicidae). J Econ Entomol 1985; 78:848-851.
21. Baker RH, Sakai RK, Raana K. Genetic sexing for a mosquito sterile-male release. J Heredity 1981; 72:216-218.
22. Kaiser PE, Bailey DL, Lowe RE et al. Mating competitiveness of chemo sterilized males of a genetic sexing strain of Anopheles albimanus in laboratory and field tests. Mosquito News 1979; 39:768-775.
23. Markaki M, Craig RK, Savakis C. Insect population control using female specific pro-drug activation. Insect Biochem Mol Biol 2004; 34:131-137.
24. Zakharkin SO, Headley VV, Kumar NK et al. Female-specific expression of a hexamerin gene in larvae of an autogenous mosquito. Eur J Biochem 2001; 268:5713-5722.
25. McCray EM. A mechanical device for the rapid sexing of Aedes aegypti pupae. J Econ Entomol 1961; 54:819.
26. Bellini R, Calvitti A, Carrier M, et al. Use of the sterile insect technque against Aedes albopictus in Italy: first results of a pilot trial. In: Vreysen MJB, Robinson AS, Hendrichs J, eds. Area-wide Control of Insect Pests: From Research to Field Implementation. Dordrecht: Springer, 2007.
27. Robinson AS. Genetic sexing strains in medfly, Ceratitis capitata, sterile insect technique programmes. Genetica 2002; 116:5-13.
28. Dowell FE, Parker AG, Benedict MQ et al. Sex separation of tsetse fly pupae using near-infrared spectroscopy. Bull Entomol Res 2005; 95:249-257.

29. Catteruccia F, Benton JP, Crisanti A. An Anopheles transgenic sexing strain for vector control. Nat Biotechnol 2005; 23:1414-1417.
30. Knols BG, Hood-Nowotny RC, Bossin H, et al. GM sterile mosquitoes—a cautionary note. Nat Biotechnol 2006; 24: 1067-8.
31. Pane A, Salvemini M, Delli Bovi P et al. The transformer gene in Ceratitis capitata provides a genetic basis for selecting and remembering the sexual fate. Development 2002; 129:3715-3725.
32. Alphey LS. Reengineering the sterile insect technique. Insect Biochem Mol Biol 2002; 32:1243-1247.
33. Burt A. Site-specific selfish genes as tools for the control and genetic engineering of natural populations. Proc Biol Sci 2003; 270:921-928.
34. Gould F, Magori K, Huang Y. Genetic strategies for controlling mosquito-borne diseases. Am Scientist 2006; 94:238-246.
35. Schliekelman P, Gould F. Pest control by the release of insects carrying a female-killing allele on multiple loci. J Econ Entomol 2000; 93:1566-1579.
36. Anonymous. Scientific working group on African trypanosomiasis (sleeping sickness): (WHO) Committee on African Trypanosomiasis. WHO/Tropical Disease Research Unit, 2001.
37. Vreysen MJ, Saleh KM, Ali MY et al. Glossina austeni (Diptera: Glossinidae) eradicated on the island of Unguja, Zanzibar, using the sterile insect technique. J Econ Entomol 2000; 93:123-135.
38. Van Der Vloedt AMV, Barnor HF. Effects of ionizing radiation on tsetse biology. Their relevance to entomological monitoring during integrated control programs using the sterile insect technique. Insect Sci Applied 1984; 5:431-437.
39. Van Den Bossche P, Akoda K, Djagmah B et al. Effect of isometamidium chloride treatment on susceptibility of tsetse flies (Diptera: Glossinidae) to trypanosome infections. J Med Entomol 2006; 43:564-7.
40. Aksoy S, Maudlin I, Dale C et al. Prospects for control of African trypanosomiasis by tsetse vector manipulation. Trends Parasitol 2001; 17:29-35.
41. Handler AM, Zimowska GJ, Horn C. Post-integration stabilization of a transposon vector by terminal sequence deletion in Drosophila melanogaster. Nat Biotechnol 2004; 22:1150-1154.
42. Dafa'alla TH, Condon GC, Condon KC et al. Transposon-free insertions for insect genetic engineering. Nat Biotechnol 2006; 24:820-821.
43. Holt RA, Subramanian GM, Halpern A et al. The genome sequence of the malaria mosquito Anopheles gambiae. Science 2002; 298:129-149.
44. Nimmo DD, Alphey L, Meredith JM et al. High efficiency site-specific genetic engineering of the mosquito genome. Insect Mol Biol 2006; 15:129-136.
45. Chen H, Fillinger U, Yan G. Oviposition behavior of female Anopheles gambiae in Western Kenya inferred from microsatellite markers. Am J Trop Med Hyg 2006; 75:246-250.
46. Apostol BL, Black WC, Reiter P et al. Use of randomly amplified polymorphic dna amplified by polymerase chain reaction markers to estimate the number of Aedes aegypti families at oviposition sites in San Juan, Puerto Rico. Am J Trop Med Hyg 1994; 5:89-97.
47. Helinski ME, Hood-Nowotny R, Mayr L, et al. Stable isotope-mass spectrometric determination of semen transfer in malaria mosquitoes. J Exp Biol 2007; 210:1266-74.
48. Dame DA. P-32-labeled semen for mosquito mating studies. J Econ Entomol 1964; 57:669-672.
49. Service MM. Mosquito Ecology: Field Sampling Methods. 1st ed. New York: Chapman and Hall, 1993.
50. Spielman A. Why entomological antimalaria research should not focus on transgenic mosquitoes. Parasitol Today 1994; 10:374-376.

Chapter 8

Insect Population Suppression Using Engineered Insects

Luke Alphey,* Derric Nimmo, Sinead O'Connell and Nina Alphey

Abstract

Suppression or elimination of vector populations is a tried and tested method for reducing vector-borne disease, and a key component of integrated control programs. Genetic methods have the potential to provide new and improved methods for vector control. The required genetic technology is simpler than that required for strategies based on population replacement and is likely to be available earlier. In particular, genetic methods that enhance the Sterile Insect Technique (e.g., RIDL™) are already available for some species.

Introduction

Suppression of vector populations has for many years been at the forefront of efforts to control vector-borne diseases. This was one of the earliest deliberate control measures—the ancient Romans drained swamps to control malaria. More recent successes include the elimination of the malaria vector *Anopheles gambiae* from Brazil in the 1930s by the program led by Soper and the Rockefeller Foundation, the control of mosquito-borne diseases in South America by the Pan-American Health Organization, and the large-scale malaria control and eradication programs of the 1950s and 60s. All of these mosquito control programs combined the use of insecticidal chemicals, breeding site restriction and physical barriers such as screens and bed nets, but with a heavy emphasis on insecticides. Increased prevalence of insecticide resistance in vector populations, combined with greater awareness of the negative environmental impacts of widespread insecticide use, means that equivalent programs are probably impractical today. However, vector control remains a key component of any integrated program intended to control vector-borne disease. It is therefore imperative to develop new and improved methods for vector control, especially ones that do not depend on the use of toxic chemicals.

Several molecular genetic methods have been proposed for the suppression or eradication of pest insect populations. This chapter will focus on genetics as applied to the pest insect itself, but there may also be uses for engineered plants and microbes in this context, for example for enhanced versions of microbial biocontrol.

Suppressing or eradicating a target pest population means reducing its reproductive capacity, in other words the average number of progeny from each adult that themselves survive to reproduce. If the average reproductive capacity can be brought below unity, then the population is not self-replacing and will decline. Most natural populations are regulated by density-dependence; in this case a moderate reduction in reproductive capacity will lead to the

*Corresponding Author: Luke Alphey—Department of Zoology, Oxford University, South Parks Road, Oxford OX1 3PS, UK. and Oxitec Limited, 71 Milton Park, Abingdon, OX14 4RX, UK. Email: luke.alphey@zoo.ox.ac.uk

Transgenesis and the Management of Vector-Borne Disease, edited by Serap Aksoy.

population reaching a new, lower equilibrium level—this is suppression, but not eradication. However, any population is limited in its ability to accommodate excess mortality or sterility. This ability is related to the basic reproductive rate in the absence of density dependent effects; if it can be overcome, then the population can indeed be eliminated. There are numerous precedents for this, both for disease vectors and for agricultural pests. Most methods of population control act independently or even synergistically, and can be combined to give a greater impact on the target population, so an effective program is likely to involve several different approaches. In turn, vector control is only one route to reducing the intensity of transmission, and so vector population control methods are themselves likely to form one part of a larger integrated disease control program.

Two broad strategies for genetics-based population suppression have been proposed. In one of these, released engineered insects would produce nonviable progeny, either immediately or within a few generations, thereby tending to suppress the target population. Implementation of such a program would require the release of large numbers of engineered insects, and would resemble the currently-used sterile insect technique (SIT), to a degree that depends on the precise nature of the genetic system used. This is discussed further below. The aim of the other proposed method is to impose a genetic load on the population by spreading within it one or more genetic elements that will reduce the average fitness of the insects that carry them. This second method has no obvious analog in present pest control methods. One potential way to do this is to introduce genetic elements that will drastically reduce the fitness of the target population. There are a number of different types of genetic element that might have the potential to do this. Several variant schemes have been proposed for using one such class of element, homing endonucleases.[1,2] However, practical use remains a distant prospect as daunting technical and regulatory issues remain to be solved.

In contrast, sterile insect methods have a long history of successful field use.[3] The SIT depends on the rearing, sterilization and release of large numbers of insects. These sterile insects compete for mates with the wild population. Mating to sterile insects reduces the reproductive potential of the wild population; if sufficient sterile insects can be released for a sufficient period of time, the target population will collapse and may even be locally eliminated. A ground-breaking experiment in the 1940s by Vanderplank eliminated a tsetse population through hybrid sterility following mass-release of a sibling species.[4,5] Later, radiation was used to sterilize mass-reared insects of the target species itself. One early program, and the paradigm for this approach, was the complete elimination of the New World screwworm (*Cochliomyia hominivorax*) from North and Central America, and the successful prevention of reinvasion from South America by regular release of sterile insects in a barrier zone in Panama.[6] Large-scale programs based on the SIT have also been successfully conducted against tsetse and against a range of agricultural pests, e.g., tephritid fruit flies.[7-9]

The SIT has a number of significant advantages as a control method. It does not depend on the use of toxic chemicals in the environment, and it uses the insects' own mate-seeking behavior to find the wild females. It can therefore work well against targets which are difficult to find, for example because of low prevalence, or resting or breeding behaviors that make adults or juveniles difficult to reach with less target-seeking methods, e.g., toxic chemicals. Since it depends on mating with released insects, the SIT is exquisitely species-specific, with minimal nontarget impacts.

Though it has been very effectively deployed against some agricultural pests, the SIT has had limited impact on disease vectors, the elimination of tsetse from Zanzibar being one of rather few examples.[10,11] Following the success of the New World screwworm program, a number of trials were conducted in the 1970s, with some success, but there are no large-scale SIT programs in operation today against any mosquito,[12,13] indeed the number of pest species against which SIT programs are currently deployed is quite limited. These early trials, taken together, indicated that the SIT had considerable potential for the control of mosquitoes, but that the technology then available was not quite adequate to provide a cost-effective intervention except in special circumstances.[13] Problems with mass-rearing, sex-separation, sterilization and distribution, and maintaining the competitiveness of the insect through all of this, tend to reduce the cost-effectiveness of the SIT. For some of these key issues, especially sex-separation and sterilization, modern genetic methods could potentially provide dramatic improvements in the cost-effectiveness of the SIT, particularly for some vector species. This is likely to improve the cost-effectiveness of the SIT to the point where it becomes attractive option for vector control in a range of contexts, allowing SIT to fulfill its potential as an effective, environmentally friendly vector control system.

Genetic Sexing and Genetic Sterilization

It is highly desirable that the released sterile insects are all male. There are two distinct reasons for this. Firstly, for insects such as mosquitoes, the adult females are potentially dangerous—even sterile females will bite and might transmit disease—whereas males do not bite. Secondly, the presence of females in the release population may distract the sterile males from seeking out wild females. Since mating to wild females is the basis of SIT, this could have a significant impact on the program. Indeed, for the Mediterranean fruit fly (*Ceratitis capitata*, "Medfly"), eliminating females from the release population ("male-only release") was found to give a 3- to 5-fold improvement in effectiveness per male.[14] Even if females were merely neutral to the program, there would still be a considerable financial benefit in eliminating them from the release population, by eliminating the costs of rearing and distributing these females. For some species there may be sufficient sexual dimorphism, for example in size or time from pupation to eclosion, to allow efficient sex-separation on the basis of this trait. This has been used for tsetse,[15] and may also be possible for some mosquitoes. However, for many insects no such method is available. In these cases a genetics-based sex separation system would be highly desirable.

Several genetic sexing systems have been constructed by classical genetics. These were all based on the translocation of a dominant selectable marker to the Y chromosome.[16-23] Unfortunately, the chromosome aberrations of these systems tend to reduce the performance of the flies that carry them, making them less effective agents for the SIT. Despite much effort to minimize the problem,[24,25] the translocations are also unstable and reversions are therefore a potential problem when large populations of insects are grown for release. This problem has been greatly mitigated by the introduction of "clean filter rearing systems" to maintain strain integrity;[26] indeed such a system would seem advisable for any nonwild type strain, whether made by classical genetics or recombinant DNA methods. The most successful of the classical systems, the *temperature-sensitive-lethal* (*tsl*) based system developed for Medfly, is now widely used; almost all sterile Medfly production facilities have converted to this strain. However, the absence of equivalent strains, even for other species of fruit fly, highlights another problem with the classical approach—genetic tools developed in one species by classical mutagenesis cannot be transferred to another species. So, despite the success of the Medfly *tsl* strains, development of a sexing system remains a high priority for other species, including the New World Screwworm, various fruit flies (e.g., the Mexican fruit fly, *Anastrepha ludens*) and mosquitoes, and has also been advocated for some moth species.[27]

Recombinant DNA methods offer the prospect of simpler systems for genetic sexing, and ones more readily transferred to new species. One potential method would be to arrange for the sex-limited expression of a visible marker, for example a fluorescent protein. Following several earlier studies in *Drosophila melanogaster*, this was recently demonstrated in the laboratory for *Anopheles stephensi*, using a promoter from a testis-specific gene to give testis-specific, and therefore male-specific, expression of DsRed, a red fluorescent protein.[28-30] This allowed males and females to be separated on the basis of their fluorescence, either manually or using an automated fluorescence-based sorter.[28] Testis-specific expression has several potential advantages—testis-specific promoters are relatively easy to identify, and labeled sperm may allow the determination of the mating status of females. However, expression is limited to a small tissue, can be detected only relatively late in development, and may not be practical on the scale required. On the other hand, labeled sperm have other uses, for example to determine the mating status of trapped females, and thereby monitor the mating success of released males.

Differential expression of a marker, such as fluorescence, requires that every insect be individually examined and sorted. Given the large scale of SIT programs—the Medfly facility at El Pino, Guatemala, has a production capacity of 3.7 billion sterile male Medfly per week—individual inspection may not be practical. It would be highly preferable to use a system in which females are eliminated by a sex-specific lethal system without any need for individual sorting.

Unlike a sex-limited visible marker, a female-lethal system has to be conditional. The females must survive under some "permissive" conditions, to allow the strain to be propagated, but die under some other "restrictive" conditions, to allow elimination of females from the population intended for release. One could arrange that "normal" rearing conditions are permissive, with the lethal system being induced by some change to an abnormal condition. This would be an "inducible" system; the trigger might be heat, for example, as is used in the *tsl*-based system. Alternatively, one could arrange that "normal" rearing conditions are restrictive, so that the lethal system needs to be suppressed, for example by the addition of a chemical "antidote" to the diet; this would be a "repressible" system. For a sexing system, one would then simply omit the antidote from the diet of the last generation before release, thereby automatically eliminating the females. The feasibility of this "repressible" approach has been demonstrated in *Drosophila melanogaster.*[31,32] A female-specific fat body enhancer was used to drive expression of a tetracycline repressible transcription factor, which in turn controlled the expression of a lethal gene product. Under "permissive" conditions—when flies were raised on media containing tetracycline—equal numbers of male and female progeny were found (376 males: 342 females). When tetracycline was removed from the rearing media to generate "restrictive" conditions no female progeny were recovered in comparison to >5,000 males.[31,32] We have recently been able to generate equivalent genetic systems and strains for two pest fruit flies, the Mediterranean fruit fly *Ceratitis capitata* and the Mexican fruit fly, *Anastrepha ludens* (unpublished). Several female-specific promoters are now available from vector species, for example, Actin4 from *Aedes aegypti*,[33] or the carboxypeptidase genes from *Anopheles gambiae*[34] and *Aedes aegypti.*[35] The range of female-specific promoters now available should enable the construction of repressible female-specific lethal systems in a wide range of insect species.

A repressible system has several potential advantages. A suitable repressible lethal system could potentially remove the need for a physical sterilization step, in other words it could remove the need for irradiation. This would work as follows: an insect strain homozygous for a repressible female-specific lethal would be reared to large numbers in the presence of the repressor. Then, for the last generation, before release, the strain would be reared in the absence of the repressor; all the females would therefore die, giving a male-only population for release (Fig. 1). These males would be released into the environment without irradiation. They would seek out and mate wild females. The progeny of such matings would each inherit one copy of the female-specific lethal. Lacking the repressor, the female progeny would die. This is the key to this method of population control—eliminating these females reduces the reproductive capacity of the target population. The male progeny would inherit the female-lethal system, so that half of their daughters would also die (Fig. 2). This gives a modest additional benefit, however it should be obvious that the female-lethal gene will disappear from the population extremely rapidly, by natural selection, if is not maintained by repeated release of large numbers of engineered insects. A repressible lethal system would therefore provide a genetic containment system as a back-up to conventional physical containment (Fig. 1C). In this respect, and operationally, this system—known as RIDL[32,36]—still closely resembles conventional SIT, but with both genetic sexing and genetic sterilization being provided by a single genetic system. Mathematical modeling predicts that this system can be more effective for population control than is conventional SIT, especially if the female-killing alleles are present at several loci (Fig. 3).[32,37,38]

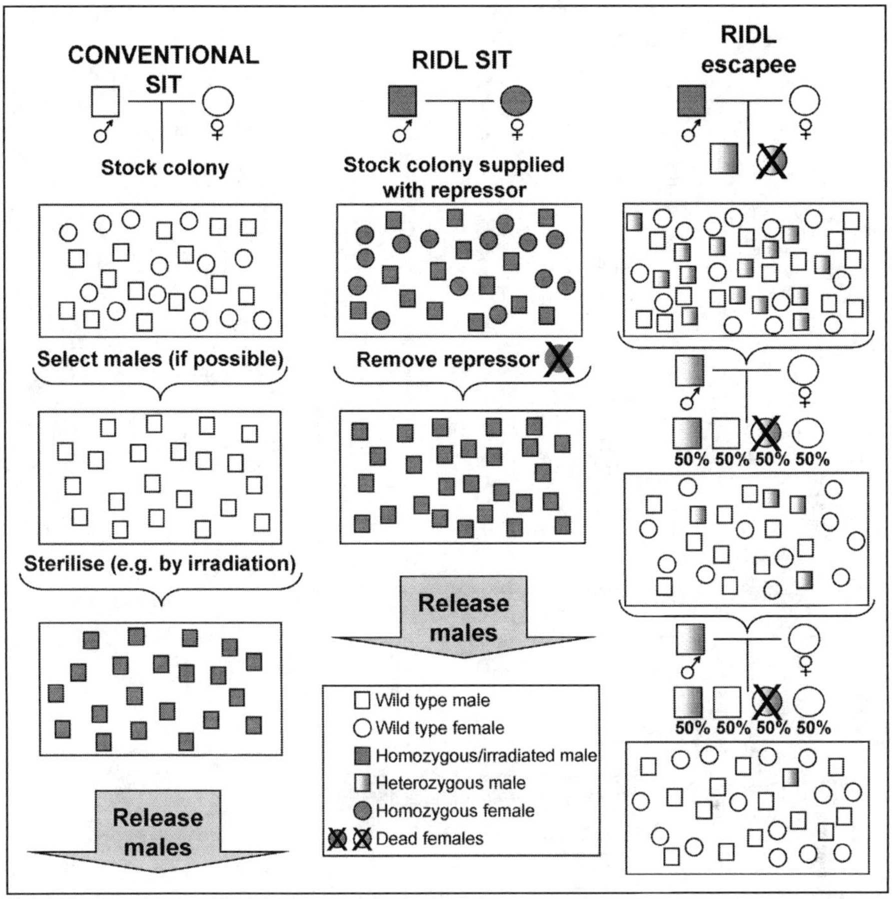

Figure 1. Comparison of conventional SIT and RIDL-SIT (using a female-specific repressible lethal system) in a mass rearing facility. In a conventional SIT program males are isolated from a stock colony, where possible—this is technically challenging in all but a few species—otherwise males and females have to be released together. The males are then sterilised by ionising radiation, which tends to reduce their competitiveness. In RIDL-SIT a stock colony is maintained using a repressor of female-specific lethality, females are eliminated by simply not providing this repressor during the rearing of the release generation. These RIDL males are not irradiated and therefore do not suffer the negative effects associated with irradiation. The female-specific lethal gene also acts as an automatic containment mechanism. This acts at two levels: (i) no viable female progeny are produced, so the escapees would not increase the pest population (unlike nonirradiated escapees from a conventional SIT program), (ii) as all RIDL females die, the RIDL construct itself will be rapidly lost from the population.

The system outlined above combines both genetic sexing and genetic sterilization, but of course this is not esssential. Fryxell and Miller proposed that a dominant conditional (cold-sensitive) lethal mutation could be used in place of radiation, in a system that they called autocidal biological control (ABC).[41] This system would not provide genetic sexing, but would kill all of the affected progeny, rather than just the daughters. This may be desirable under some circumstances, particularly where both males and females are released ("bisex" release). A similar system was demonstrated more recently, using an engineered lethal and

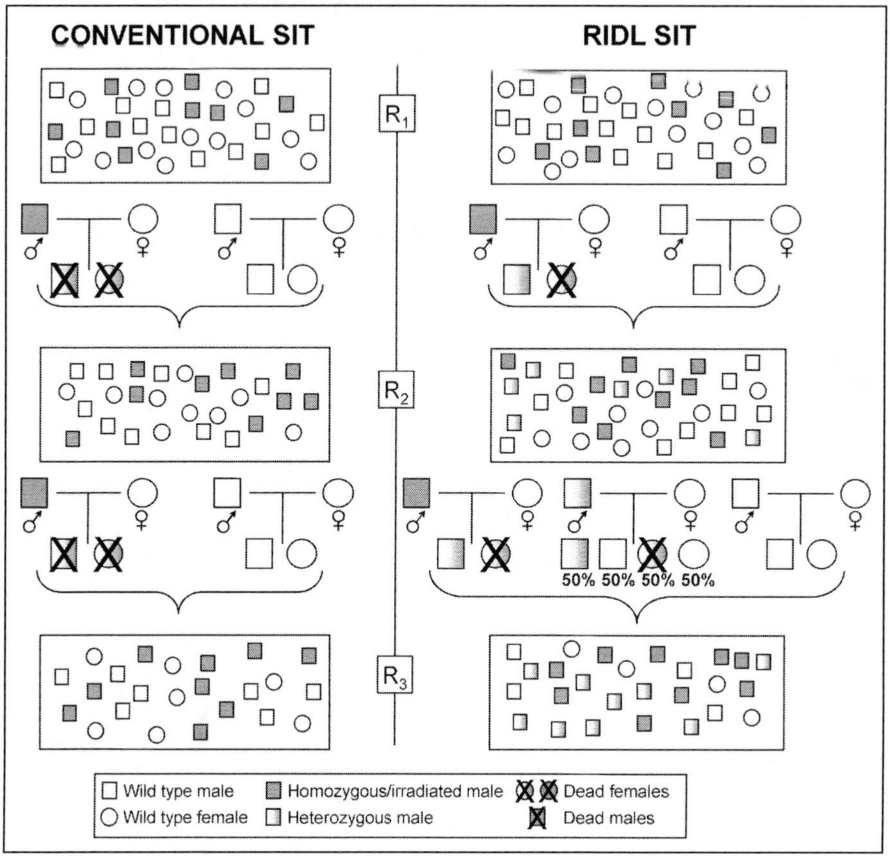

Figure 2. RIDL SIT (using a female-specific repressible lethal system) compared to conventional SIT in controlling insect populations. As part of a control program using conventional SIT or RIDL SIT there would be a series of releases of males (R1-R3). Conventional SIT relies on releasing irradiated sterile males which then compete with wild type males for females; any wild female mating a sterile male produces no offspring. However, mass-reared, irradiated males tend to be less competitive than wild type males. In female-specific RIDL, males are not irradiated and should be able to compete more effectively with wild type males. In the subsequent generations there is some added benefit from heterozygous male progeny (R2-R3); these are produced in the wild, are targeted to the rearing and breeding sites of the insect, and have a more wild type genetic background.

with tetracycline rather than temperature as the repressor (a version of RIDL affecting both males and females).[32] This was further refined by Horn and Wimmer to give embryonic lethality.[42] Radiation sterilization also typically kills the affected individuals as early embryos, before they hatch into larvae and start eating the crop; this is highly desirable for many agricultural pests but unimportant for most disease vectors, where only adults can transmit disease.

A cold-sensitive system, such as that proposed by Fryxell and Miller, could be used in a different manner. If summer temperatures do not drop below the critical point, the gene may spread through the target population for several generations before being activated by lower winter temperatures.[41] The population dynamics of this sort of control are rather different

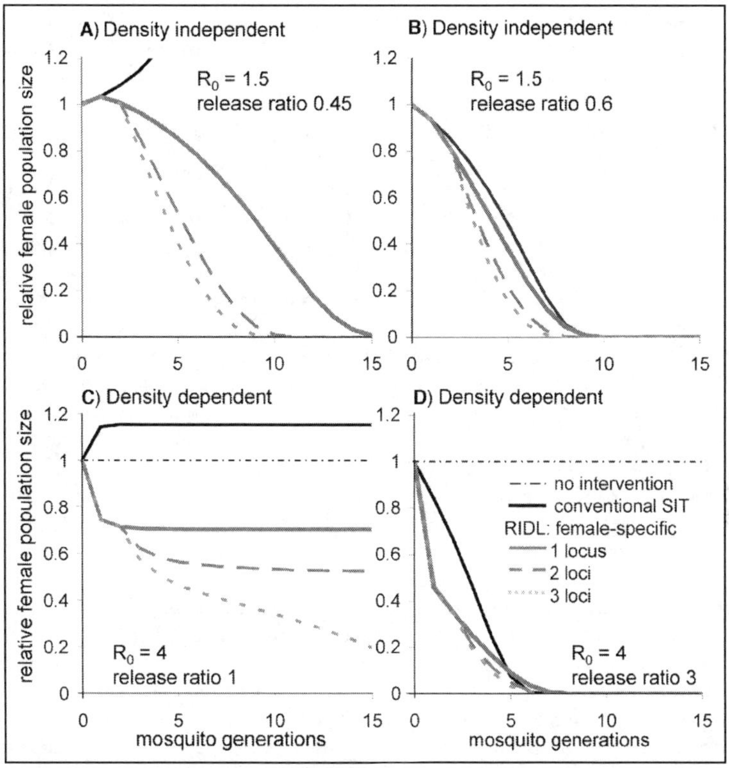

Figure 3. Female-specific lethality as "genetic sterilization". The reproductive potential of an insect population depends primarily on the number of females—increasing the number of females will typically increase the number of progeny, but increasing the number of males normally will not. Killing both sexes, or specifically killing females, might therefore be expected to have equivalent effects on the size of the target population. We used a deterministic discrete-generation population genetics model (based on that of ref. 32) to predict the effects of releasing into a target population nonirradiated males homozygous at one or more loci for a female-specific lethal gene and to compare this with the effectiveness of conventional SIT. Assuming that the lethal gene acts on the developing insect before they reach reproductive maturity, for example at an embryonic or larval stage, a non-sex-specific version of the RIDL system that kills both males and females is equivalent in this model to conventional SIT; negative effects of irradiation or genetic engineering are ignored. The population is assumed to be closed and homogeneous, with a natural sex ratio of 1:1 and females selecting mates in proportion to their relative abundance. The panels show the number of females in the population in each generation, relative to the initial number. The key in panel d applies to all panels. A constant number of adult males are released per mosquito generation; this "release ratio" is relative to the initial wild male population. R_0 denotes the number of female offspring produced by each female that would survive to adulthood if there were no intervention and no density dependent mortality. Panels a and b assume no effects of pest density and $R_0 = 1.5$, and compare release ratios of 0.45 (A) and 0.6 (B). Without any intervention, the population would experience exponential growth (not shown). Panels c and d assume density dependent mortality of larvae (using formulae from refs 39 and 40) and $R_0 = 4$, and compare release ratios of 1 (A) and 3 (B). The population is assumed to start at carrying capacity and will therefore remain at the initial level if there is no intervention. Panels a and c illustrate marginal situations where SIT does not control the population but the RIDL release strategies do. In the density dependent case (C) SIT can actually increase the equilibrium size of the adult female population.[40] Panels b and d show that, with a sufficiently high release ratio, SIT and the RIDL release strategies all control the population.

from SIT—the target population is allowed to grow unchecked for several generations. On the whole this method is likely to be less effective than RIDL or conventional SIT, but may have applications in specific circumstances.[38,43]

Molecular Biology of Repressible Lethal Systems

In principle, the condition could be any controllable feature of the environment—temperature, photoperiod, diet components, etc. In practice, few suitable regulated gene expression systems are available around which such a system could be built. Two well-known ones are heat-shock (temperature regulation) and tetracycline-regulated systems (diet component). The feasibility of using temperature as the regulated condition is shown by the Medfly *tsl*-based system, in which heat treatment of embryos is used to kill females. However, these are inducible systems, in which a specific treatment needs to be applied in order to produce the desired effect, i.e., kill females. As discussed above, there are distinct advantages to using a repressible system rather than an inducible one. A repressible system confers a "fail-safe" property—females can then only survive in the presence of an antidote; if any escape, either they or their progeny will die for lack of the antidote. This system has the additional significant advantage that it can be adapted to provide genetic sterilization as well (see above).

The tetracycline regulated gene expression systems harness the properties of a bacterial protein, TetR. This protein binds to a specific DNA sequence (tetO) in the absence of tetracycline, but not in its presence. TetR does not confer tetracycline resistance; its function in the bacteria is simply to regulate the expression of another gene. Gossen and Bujard developed a eukaryotic gene expression system by fusing TetR to a eukaryotic transcriptional activator.[44] This system has several desirable properties—it has a very good on/off ratio (ratio of expression under induced and repressed conditions), works in a very wide range of species, and responds to low concentrations of tetracycline, a very well-characterized, safe and inexpensive chemical.[45-48] Other eukaryotic expression systems regulated by specific chemicals have been developed, but these features of the tetracycline regulated systems have made them the most widely used.[49]

The tet system is conventionally used as a two-component system. One component expresses the tetracycline-repressible transactivator (tTA); the other has the effector molecule under the control of a tTA-responsive element containing multiple copies of tetO. This is similar to the GAL4/UAS bipartite expression system that has been widely used in *Drosophila melanogaster*.[50,51] Recently, we demonstrated that this system could be condensed into a single component system by placing the tTA coding region under the transcriptional control of the tTA-responsive element (Fig. 4).[48] In the absence of tetracycline, a positive feedback loop is established in which tTA drives the expression of more tTA. We showed that this could provide a repressible lethal system with various desirable characteristics in a major pest insect (Medfly)[48] and have also developed a similar system for the mosquito *Aedes* (*Stegomyia*) *aegypti* (unpublished).

One advantage of the "positive feedback" system is that it does not require any species-specific promoters to be characterized. Koukidou and colleagues exploited this to make a genetic marker, by using a nonlethal positive feedback system to drive expression of a fluorescent protein.[52] They showed that this system could give detectable expression of the fluorescent protein in insects (*Drosophila melanogaster* and *Ceratitis capitata*), a nematode (*Caenorhabditis elegans*), a plant (*Nicotiana tabacum*) and a human cell line, thus illustrating the very broad applicability of this approach.

Regulatory Issues and Concluding Remarks

Some applications of insect genetics to disease control require the long-term persistence of novel DNA sequences in wild vector populations, for example sequences that prevent the insect transmitting a particular pathogen. There are daunting technical difficulties involved in arranging this, particularly the "gene driver" technology required to spread the pathogen-resistance gene through a wild population. There are likely also to be significant regulatory and political issues, particularly where the proposed modification would be irrevers-

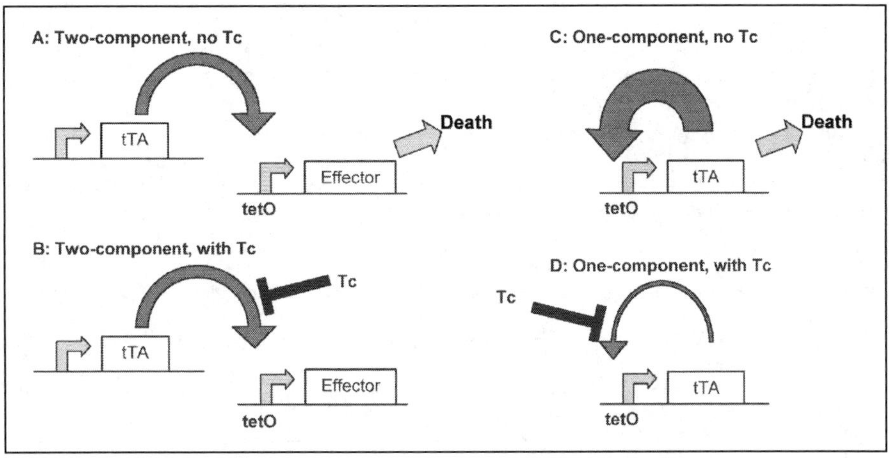

Figure 4. Tetracycline-repressible lethal systems. A,B) Two-component system. tTA is placed under the control of a suitable promoter, e.g., constitutive, female-specific or embryo-specific. A) In the absence of tetracycline (Tc) tTA binds tetO, drives expression of an effector molecule leading, in the case of a lethal effector, to death. B) In the presence of Tc, tTA binds Tc; the Tc-bound form of tTA does not bind DNA, it therefore does not activate expression of the effector and the system is inactivated. C,D) Simplified one-component system. C) In the absence of Tc, basal expression of tTA leads to the synthesis of more tTA, which accumulates to high level. This level can be regulated by modifying the stability and translational efficiency of the tTA mRNA. At the highest levels, expression is lethal, so tTA is both the driver and the effector. D) In the presence of Tc, tTA is inactivated by Tc and is therefore expressed only at basal levels. (Adapted from ref. 48).

ible, or would tend to spread across national boundaries. In contrast, the use of recombinant DNA methods to improve the SIT appears to be relatively straightforward from a technical perspective; the first potentially suitable strains are already available, both for agricultural pests and for at least one disease vector. The regulatory and political hurdles may also be less severe for this approach. Such use would be readily reversible, in that the introduced DNA sequences are not designed to persist, rather they tend to eliminate their hosts and so would rapidly disappear from the wild unless maintained by frequent, large-scale reintroduction in an SIT-like program. Some applications, for example the use of a genetic marker and/or genetic sexing with radiation-sterilization, would not have any significant introgression of recombinant DNA sequences into the wild population at all. Repressible lethal systems would have the additional advantage of a "fail-safe" action not found in present SIT strains.

Despite the above comments, it is important not to focus exclusively on the genetic modification technology; genetic control strategies should be assessed as one would any other novel intervention (such as a new drug or vaccine, or use of treated bed nets). This means focusing on questions of cost-effectiveness, affordability, accessibility, equity and sustainability. In particular, it is important to compare the proposed genetic strategies with current technologies, and to consider what role genetic methods may play as part of an integrated vector control and disease control program.

Recombinant DNA methods clearly have the potential to make substantial improvements to the SIT as currently practiced, and thereby to provide new and effective tools for vector control. This is not a panacea; these new methods will not be cost-effective in all circumstances or against all species. However, the threat from exotic vector-borne diseases is increasing, and the range of other control options is narrowing as insecticide resistance and environmental concerns reduce the range of available treatments. The use of SIT-like genetic control strategies has considerable potential in the control of major human and livestock diseases and this potential needs to be tested and explored. Unlike other proposed genetic strategies, the necessary technology, and in some cases actual strains, already exist to allow experimental trials in the very near future.

Acknowledgements

We thank the many colleagues, especially at Oxford University and at Oxitec Ltd, who have provided advice and encouragement over the years.

References

1. Burt A. Site-specific selfish genes as tools for the control and genetic engineering of natural populations. Proc Biol Sci 2003; 270:921-928.
2. Burt A, Trivers R. Genes in conflict: The biology of selfish genetic elements. Cambridge, MA: Harvard University Press, 2006.
3. In: Dyck V, Hendrichs J, Robinson A, eds. Sterile Insect Technique: Principles and practice in area-wide Integrated Pest Management. Dordrecht: Springer, 2005.
4. Klassen W, Curtis CF. In: Dyck VA, Hendrichs J, Robinson AS, eds. Sterile Insect Technique. Principles and practice in area-wide integrated pest management. The Netherlands: Springer, 2005:3-36.
5. Vanderplank FL. Hybridization between Glossina species and suggested new method for control of certain species of Tsetse. Nature 1944; 154:607-608.
6. Wyss JH. In: Tan KH, ed. Area-Wide Control of Fruit Flies and Other Insect Pests. Penang: Penerbit Universiti Sains Malaysia, 2000:79-86.
7. In: Keng-Hong T, ed. Area-wide control of fruit flies and other insect pests. Penang: Penerbit Universiti Sains Malaysia, 2000.
8. Krafsur E. Sterile insect technique for suppressing and eradicating insect populations: 55 years and counting. J Agric Entomol 1998; 15:303-317.
9. Koyama J, Kakinohana H, Miyatake T. Eradication of the Melon Fly Bactrocera cucurbitae in Japan: Importance of behaviour, ecology, genetics and evolution. Ann Rev Entomol 2004; 49:331-349.
10. Msangi AR et al. Current tsetse control operations in Botswana and prospects for the future. In: Tan KH, ed. Area-Wide Control of Fruit Flies and Other Insect Pests. Penang: Penerbit Universiti Sains Malaysia, 2000:57-66.
11. Vreysen MJ, Saleh KM, Ali MY et al. Glossina austeni (Diptera: Glossinidae) eradicated on the island of Unguja, Zanzibar, using the sterile insect technique. J Econ Entomol 2000; 93:123-135.
12. Asman S, McDonald P, Prout T. Field studies of genetic control systems for mosquitoes. Ann Rev Entomol 1981; 26:289-343.
13. Benedict M, Robinson A. The first releases of transgenic mosquitoes: An argument for the sterile insect technique. Trends Parasitol 2003; 19:349-355.
14. Rendón P, McInnis D, Lance D et al. Medfly (Diptera:Tephritidae) genetic sexing: Large-scale field comparison of males-only and bisexual sterile fly releases in Guatemala. J Econ Entomol 2004; 97:1547-1553.
15. Opiyo E, Luger D, Robinson AS. In: Tan K, ed. Proceedings: Area-wide control of fruitflies and other insect pests. Pulau Pinang, Malaysia: Penerbit Universiti Sains Malaysia, 2000:337-344, (International Conference on area-wide control of insect pests and the 5th International Symposium on fruit flies of economic importance, 28 May-5 June 1998, Penang, Malaysia).
16. Franz G, Willhoeft U, Kerremans P et al. In: IAEA, ed. Evaluation of genetically altered medflies for use in SIT programmes. Vienna: IAEA, 1997:85-95.
17. Hendrichs J, Franz G, Rendón P. Increased effectiveness and applicability of the sterile insect technique through male-only release for control of Mediterranean fruit-flies during fruiting seasons. J Appl Entomol 1995; 119:371-377.
18. Robinson A. In: Robinson A, Hooper G, eds. Fruit Flies. Their Biology, Natural Enemies and Control. Vol. 3A. Amsterdam: Elsevier, 1989:57-65.
19. Robinson A. Genetic sexing strains in medfly, Ceratitis capitata, sterile insect technique programmes. Genetica 2002; 116:5-13.
20. Robinson A, Franz G, Fisher K. Genetic sexing strains in the medfly, Ceratitis capitata: Development, Mass Rearing and Field Application. Trends in Entomology 1999; 2:81-104.
21. Seawright J, Kaiser P, Dame D et al. Genetic method for the preferential elimination of females of Anopheles albimanus. Science 1978; 200:1303-1304.

22. Whitten M. Automated sexing of pupae and its usefulness in control by sterile insects. J Econ Entomol 1969; 62:272-273.
23. Whitten M, Foster G. Genetical methods of pest control. Annu Rev Entomol 1975; 20:461-476.
24. Franz G, Gencheva E, Kerremans P. Improved stability of genetic sex-separation strains for the Mediterranean fruit-fly, Ceratitis capitata. Genome 1994; 37:72-82.
25. Kerremans P, Franz G. Isolation and cytogenetic analyses of genetic sexing strains for the Medfly, Ceratitis capitata. Theor Appl Gen 1995; 91:255-261.
26. Fisher K, Caceres C. In: Hong TK, ed. Area-wide management of fruit flies and other major insect pests. Penang, Malaysia: Universiti Sains Malaysia Press, 2000:543-550.
27. Marec F, Neven LG, Robinson AS et al. Development of genetic sexing strains in Lepidoptera: From traditional to transgenic approaches. J Econ Entomol 2005; 98:248-259.
28. Catteruccia F, Benton J, Crisanti A. An Anopheles transgenic sexing strain for vector control. Nature Biotechnology 2005; 23:1414-1417.
29. Lukyanov KA, Fradkov AF, Gurskaya NG et al. Natural animal coloration can be determined by a nonfluorescent green fluorescent protein homolog. J Biol Chem 2000; 275:25879-25882.
30. Matz MV, Fradkov AF, Labas YA et al. Fluorescent proteins from nonbioluminescent Anthozoa species. Nat Biotechnol 1999; 17:969-973.
31. Heinrich J, Scott M. A repressible female-specific lethal genetic system for making transgenic insect strains suitable for a sterile-release program. Proc Nat'l Acad Sci (USA) 2000; 97:8229-8232.
32. Thomas DD, Donnelly CA, Wood RJ et al. Insect population control using a dominant, repressible, lethal genetic system. Science 2000; 287:2474-2476.
33. Muñoz D, Jimenez A, Marinotti O et al. The AeAct-4 gene is expressed in the developing flight muscles of females Aedes aegypti. Insect Molecular Biology 2004; 13:563-568.
34. Edwards M, Lemos F, Donelly-Doman M et al. Rapid induction by a blood meal of a carboxypeptidase gene in the gut of the mosquito Anopheles gambiae. Insect Biochem Mol Biol 1997; 27:1063-1072.
35. Edwards MJ, Moskalyk LA, Donelly-Doman M et al. Characterization of a carboxypeptidase A gene from the mosquito, Aedes aegypti. Insect Molecular Biology 2000; 9:33-38.
36. Alphey L, Andreasen MH. Dominant lethality and insect population control. Mol Biochem Parasitol 2002; 121:173-178.
37. Schliekelman P, Gould F. Pest control by the release of insects carrying a female-killing allele on multiple loci. J Econ Entomol 2000; 93:1566-1579.
38. Gould F, Schliekelman P. Population genetics of autocidal control and strain replacement. Annu Rev Entomol 2004; 49:193-217.
39. Maynard Smith J, Slatkin M. The stability of predator-prey systems. Ecology 1973; 54:384-391.
40. Rogers D, Randolph S. From a case study to a theoretical basis for tsetse control. Insect Sci Applic 1984; 5:419-423.
41. Fryxell K, Miller T. Autocidal biological control: A general strategy for insect control based on genetic transformation with a highly conserved gene. J Econ Entomol 1995; 88:1221-1232.
42. Horn C, Wimmer E. A transgene-based, embryo-specific lethality system for insect pest management. Nat Biotech 2003; 21:64-70.
43. Schliekelman P, Gould F. Pest control by the introduction of a conditional lethal trait on multiple loci: Potential, limitations, and optimal strategies. J Econ Entomol 2000; 93:1543-1565.
44. Gossen M, Bujard H. Tight control of gene expression in mammalian cells by tetracycline- responsive promoters. Proc Natl Acad Sci USA 1992; 89:5547-5551.
45. Gossen M, Bujard H. Studying gene function in eukaryotes by conditional gene inactivation. Annu Rev Genet 2002; 36:153-173.
46. Bello B, Resendez-Perez D, Gehring W. Spatial and temporal targeting of gene expression in Drosophila by means of a tetracycline-dependent transactivator system. Development 1998; 125:2193-2202.
47. Lycett G, Kafatos F, Loukeris T. Conditional expression in the malaria mosquito Anopheles stephensi with Tet-on and Tet-off systems. Genetics 2004; 167:1781-1790.
48. Gong P, Epton MJ, Fu G et al. A dominant lethal genetic system for autocidal control of the Mediterranean fruitfly. Nat Biotech 2005; 23:453-456.
49. Fussenegger M. The impact of mammalian gene regulation concepts on functional genomic research, metabolic engineering, and advanced gene therapies. Biotechnol Prog 2001; 17:1-51.
50. Brand A, Manoukian A, Perrimon N. Ectopic expression in Drosophila. Meth Cell Biol 1994; 44:635-654.
51. Brand A, Perrimon N. Targeted gene expression as a means of altering cell fates and generating dominant phenotypes. Development 1993; 118:401-415.
52. Koukidou M, Klinakis A, Reboulakis C et al. Germ line transformation of the olive fly Bactrocera oleae using a versatile transgenesis marker. Insect Molecular Biology 2006; 15:95-103.

Chapter 9

Wolbachia-Based Technologies for Insect Pest Population Control

Kostas Bourtzis*

Abstract

Wolbachia are a group of obligatory intracellular and maternally inherited bacteria found in many arthropod species, including insects, mites, spiders, springtails, crustaceans, as well as in certain nematodes. Several PCR-based surveys suggest that over 20% of the arthropod species may be *Wolbachia*-infected, rendering this bacterium the most ubiquitous intracellular symbiont yet described. *Wolbachia* have recently attracted attention for their potential as novel and environmentally friendly bio-control agents. *Wolbachia* are able to invade and maintain themselves in the arthropod species through manipulation of the host's reproduction. Several strategies can be distinguished, one of which is cytoplasmic incompatibility (CI). *Wolbachia*-induced cytoplasmic incompatibility can be used beneficially in the following ways: (a) as a tool for insect pest population control in a way analogous to the "Sterile Insect technique" (SIT) and (b) as a drive system to spread desirable genotypes in field arthropod populations. In addition, virulent *Wolbachia* strains offer the potential to control vector species by modifying their population age structure. In the present chapter, I summarize the recent developments in *Wolbachia* research with an emphasis on the applied biology of *Wolbachia* and conclude with the challenges that *Wolbachia* researchers will face if they want to use and/or introduce *Wolbachia* into pest and vector species of economic, environmental and public health relevance and, through *Wolbachia*-based technologies, to suppress or modify natural populations.

Introduction

Wolbachia pipientis (*Wolbachia* for the purpose of this chapter) is a widespread group of obligatory intracellular maternally transmitted bacteria belonging to the α-Proteobacteria.[1] *Wolbachia* bacteria were initially discovered in the ovaries of the mosquito *Culex pipiens* complex—hence the name *Wolbachia pipientis*—and were formally described in 1936.[2] Since then, *Wolbachia* has been found in several hundreds of arthropod species including all major insect groups, mites, isopods, spiders and filarial nematodes.[3-10] The results of PCR surveys suggest that 20-70% of insect species may be infected, rendering *Wolbachia* the most ubiquitous endosymbiont on earth. *Wolbachia* mainly reside in the reproductive tissues of arthropods and cause a number of reproductive alterations including cytoplasmic incompatibility (CI), thelytokous parthenogenesis, feminization of genetic males, and male-killing.[5,8,11]

*Kostas Bourtzis—Department of Environmental and Natural Resources Management, University of Ioannina, 2 Seferi St, 30100 Agrinio, Greece.
Email: kbourtz@uoi.gr

Transgenesis and the Management of Vector-Borne Disease, edited by Serap Aksoy.

Based on 16 rDNA gene analysis, *Wolbachia* belong to the alpha-2 subdivision of the Proteobacteria, forming a monophyletic group closely related to intracellular bacteria of the genera *Ehrlichia*, *Anaplasma*, *Rickettsia* and *Cowdria*.[1,12-13] Many members of these genera are arthropod-borne pathogens of mammals. The phylogenies of *Wolbachia* isolates so far generated suggest the existence of eight major clades (A-H), which have been named 'supergroups'.[14-16] Most of the supergroups occur in arthropods (A, B, E, F, G and H), while supergroups C and D are so far only found in filarial nematodes. The majority of insect Wolbachia belong to supergroups A and B.[3,14,16-18] Relationships among the various supergroups are not well understood because of the absence of a suitable outgroup for rooting the inferred trees.[14] In order to resolve the branching order of the supergroups, sequence analysis of a large number of genes is necessary. Several such sequencing projects are currently underway. In addition, two "Multi Locus Sequencing Typing" (MLST) systems were developed, which allow genotyping of any given Wolbachia strain on the basis of a combination of alleles of a sample of housekeeping genes.[19-20]

A number of studies have conclusively shown that the unique interaction of *Wolbachia* with the reproductive tissues and organs is responsible for the induction of several reproductive abnormalities such as: (a) feminization, which is the conversion of genetic males into functional females (mainly observed in isopods); (b) parthenogenesis, which is the exclusive production of female offspring by infected females (observed in many parasitoids); (c) male killing, which is the death of male embryos during early embryonic development (observed in insects) and (d) cytoplasmic incompatibility, which is a kind of male sterility observed in all major orders of insects. Each of these reproductive alterations favors the transmission of the bacterium at the expense of the uninfected arthropod population. Treatment of the infected hosts with antibiotics results in restoration of the normal reproductive phenotypes.[5,8,11]

Wolbachia Induced Cytoplasmic Incompatibility

Reproductive isolation phenomena between different populations of the mosquito *C. pipiens* were reported as early as in the 1930s and 1950s, but it was only in the early 1970s that the presence of *Wolbachia* in the gonads of *Culex* mosquitoes was uncovered by Yen and Barr, who demonstrated that these bacteria are responsible for the expression of cytoplasmic incompatibility.[9,21-24]

Cytoplasmic incompatibility is the most common and widespread reproductive abnormality *Wolbachia* infections induce in their arthropod hosts.[25-26] In most insects, the expression of cytoplasmic incompatibility manifests itself as lethality in the developing embryo. In insects with haplodiploid sex determination (Hymenoptera), the result of cytoplasmic incompatibility is, however, a sex ratio shift to the haploid sex, which is usually male. Cytoplasmic incompatibility can be either unidirectional or bi-directional. Unidirectional cytoplasmic incompatibility is typically expressed when an infected male is crossed with an uninfected female. The reciprocal cross (infected female and uninfected male) is fully compatible, as are crosses between infected individuals. Bidirectional cytoplasmic incompatibility usually occurs in crosses between infected individuals harbouring different strains of *Wolbachia*. The penetrance of CI depends on the respective host-*Wolbachia* strain combination and can range from a few to 100%. As a consequence of cytoplasmic incompatibility, *Wolbachia* infections can spread and persist in nature by replacing uninfected populations.[27]

Wolbachia-induced cytoplasmic incompatibility has been reported for almost all major insect orders (Coleoptera, Diptera, Hemiptera, Hymenoptera, Orthoptera and Lepidoptera) as well as in some terrestrial isopod species.[6,26] The mechanism of *Wolbachia*-induced cytoplasmic incompatibility has not yet been elucidated at the molecular level. Indeed, little is known about the mechanistic basis of cytoplasmic incompatibility, except that the bacterial density and the percentage of infected sperm cysts are positively correlated with the penetrance of the incompatibility phenotype. However, a number of genetic, cytogenetic and cellular studies indicate that *Wolbachia* somehow modifies the paternal chromosomes during spermatogenesis (mature sperm do not contain the bacteria), thus influencing their behaviour during the first mitotic divisions and resulting in loss of mitotic synchrony. The presence of the same *Wolbachia* strain in the egg cells can rescue the modification, restoring normal chromosomal behavior.[28-34]

Based on the genetic and cytogenetic data, Werren proposed the so-called modification/rescue model, (mod/resc), which assumes the presence of two distinct bacterial functions.[5] According to this model, the mod function is a kind of 'imprinting' effect which takes place in the male germline during spermatogenesis, while the resc function acts exclusively in the egg. The sperm imprinting effect may be either due to (a) secreted *Wolbachia* factor(s) that modify the paternal chromosomes or to the removal of (a) host factor(s) that is (are) necessary for proper condensation/decondensation of the paternal chromosomal set before and/or during zygote formation. On the other hand, the presence of the same *Wolbachia* strain in the egg may result in the production and secretion of (a) resc factor(s) or alternatively the recruitment of host molecules which are capable of rescuing the sperm imprinting effect in a *Wolb*achia strain-specific manner. In mechanistic terms, it has been suggested that *mod* and *resc* interact in a lock-and-key manner, with a direct inhibition of the *mod* factor (the lock) by the *resc* factor (the key), but other models are as likely.[35]

Recent studies suggest that the temperate bacteriophage WO-B of *Wolbachia*, which contains ankyrin encoding genes and virulence factors, may be implicated in the induction of CI.[36-40] In an elegant study, Bordenstein and colleagues (2006) considered the separate roles that lytic and lysogenic phage might have on bacterial fitness and phenotype through the characterization of phage densities and their association with bacterial densities and cytoplasmic incompatibility.[41] Their results indicate a low to moderate frequency of lytic phage development in *Wolbachia* and an overall negative density relationship between bacteriophage and *Wolbachia*. The above findings motivated a novel phage density model of cytoplasmic incompatibility in which the lytic phage represses *Wolbachia* densities and therefore reproductive parasitism.[41]

A number of factors have been identified which may affect the expression of cytoplasmic incompatibility in the host. These include: (a) the host nuclear genome: for example, the nuclear genome of *D. simulans* is more permissive than the nuclear genome of *D. melanogaster* for the expression of cytoplasmic incompatibility;[42-43] (b) male age: in some species, older males do not express the incompatibility phenotype strongly, while in others the age does not influence the incompatibility;[44-51] (c) repeated copulation of males: it has been shown in *Drosophila simulans* and in *Culex pipiens* that the repeated copulation of males results in less infected sperm cysts and consequently in reduced levels of cytoplasmic incompatibility;[44,52-53] (d) temperature: maximum levels of CI expression in *Drosophila* species are observed when crosses are performed at about 25°C. Crosses performed at higher (i.e., 28°C) or lower temperatures (i.e., 19°C) result in reduced levels of CI;[45,47,49,51] (e) other factors: the presence of antibiotics as well as of low levels of nutrition reduce CI expression levels.[54-55]

Wolbachia-Based Applications

There is a great interest in the use of *Wolbachia* infections as a potential tool for the development of novel and environmentally friendly strategies for the management of arthropod species that are major pests or beneficial species.[9,56-57] *Wolbachia* might be used as a tool to improve beneficial insect species, to suppress natural population of pests through cytoplasmic incompatibility, as a para-transformation tool for insects, as a driving mechanism to replace natural insect populations and as an organism that can alter the senescence of disease vectors.[9,56,58-61]

Wolbachia *Transfers*

Despite the widespread distribution of *Wolbachia*, many important agricultural pests (e.g., *Bactrocera oleae* Gmelin) and disease vectors (e.g., *Aedes aegypti* L. and *Anopheles gambiae* Giles) are not naturally *Wolbachia*-infected. In these cases, it should be possible to transfer *Wolbachia* into the naive species and establish a stable host-*Wolbachia* association. The mechanism(s) through which *Wolbachia* infects a new species in nature is not yet known. Hovever, several studies have shown that a new uni-directional, bi-directional and/or superinfected host strain can be obtained through embryonic cytoplasmic microinjection, even when the respective host-*Wolbachia* combination does not exist in nature.[42-43,62-72] Interspecific and intergeneric transfers have been performed between both closely and distantly related species. The majority of the newly established host-*Wolbachia* associations were stable and expressed cytoplasmic incompatibility.[42-43,67,70-72] These transfer experiments suggest that it is feasible to transfer a *Wolbachia* strain to a naive host, to achieve establishment and expression of the expected reproductive phenotype.

Incompatible Insect Technique

Even before the etiology of CI was determined in mosquitoes,[23-24] Laven had recognised the potential of CI to control mosquitoes and was the first to assess its feasibility in a field trial.[22] This was followed by other field trials with other species.[73-76] In recent years, a renewed interest developed in using *Wolbachia* as a tool to apply the "Incompatible Insect Technique" (IIT) for pest population control. IIT can be defined as the use of the mechanism of *Wolbachia*-induced cytoplasmic incompatibility (or other symbiont-based induced reproductive incompatibility) for the control of populations of pest insects.[77] The term IIT was initially proposed by Boller and colleagues in a study on the incompatible races of European cherry fruit fly, *Rhagoletis cerasi*.[73,78]

The Mediterranean fruit fly, *Ceratitis capitata* is a very important agricultural pest.[79] (Robinson and Hopper, 1989). There are no reports of the presence of Wolbachia in field and/ or laboratory populations with the exception of a recent case from natural and laboratory populations of Mediterranean fruit fly in Brazil.[80-81] Given the potential of cytoplasmic incompatibility, the availability of mass rearing technology and genetic sexing strains, it would seem logical to assess whether *Wolbachia* can be established in medfly, resulting in the expression of CI.[79,82-83]

Wolbachia strains from the host species *R. cerasi* have been used to stably infect the Mediterranean fruit fly through embryonic injection.[71] At the time of this writing, 50 generations (about 4 years) post injection, two transinfected lines, namely WolMed 88.6 (singly infected with *w*Cer2) and WolMed S10.3 (singly infected with *w*Cer4), are stably infected with transmission rates of 100%.

Test crosses were performed in different generations post injection between transinfected lines and the parental uninfected medfly strains. The results of these crosses were the following: (a) crosses between uninfected females and *Wolbachia*-infected males resulted in 100% egg mortality, while the reciprocal crosses resulted in between 16 to 32% egg mortality; (b) crosses between the two transinfected Mediterranean fruit fly lines WolMed 88.6 (*w*Cer2) and WolMed S10.3 (*w*Cer4), each infected with a different *Wolbachia* strain, were 100% bidirectionally incompatible (100% egg mortality).[71] Similar results have been obtained in test crosses performed three years post injection (Bourtzis and Zabalou, unpublished observations). This was the first report of a newly transinfected host species with such high transmission rates and exhibiting 100% CI (uni-directional and bi-directional).[71]

Laboratory cage populations of medfly containing different ratios of transinfected males: uninfected males: uninfected females were set up to determine whether cytoplasmic incompatibility expressed by the *Wolbachia*-infected medfly lines could be used for population suppression. The caged medfly populations were suppressed by these single "releases" of incompatible males in a ratio-dependent manner. Population suppression was extremely efficient reaching levels of >99% at "release" ratios of 50:1.[71] Although these laboratory experiments are encouraging, they need to be extended to field cage systems, where wild flies are used as the targeted population.

Dobson and colleagues recently reported the establishment of *Wolbachia* strain *w*AlbB, naturally occurring in *Aedes albopictus*, in *A. aegypti* (a naive host).[72] Experimental crosses indicated strong CI (100%): no egg hatch was observed from >3800 eggs from crosses of uninfected females and *Wolbachia*-infected males.[72]

The above studies show that *Wolbachia* endosymbionts can be experimentally transferred over genus barriers into a novel host, forming associations which express complete CI (unidirectional and, importantly, bi-directional), thus supporting the concept of introduction of *Wolbachia* into pest and vector species of environmental, economic and health relevance in order to suppress natural populations.[71-72]

The advantage of the *Wolbachia*-based Incompatible Insect Technique over the presently used Sterile Insect Technique system is mainly that the insects do not have to be irradiated before release. This does require however a very effective sexing strain of the insect pest, so that only infected males are released. With the release of only a few infected females, an infected population could establish itself in the field. This problem can be overcome by: (a) developing a highly effective sexing strain; (b) creating two mutually incompatible infected lines of the insect pest sexing strain that can be released in alteration, thus even if the first release includes a few infected females, they will be suppressed by the following release of incompatible males. Once the field population is knocked down to a sufficiently low level, the final step in the eradication of the insect pest would be the release of irradiated males that are completely sterile, thus rendering crosses with infected females of either CI type unproductive; (c) an alternative solution could lie in combining radiation with incompatibility, where females can be sterilized with lower doses of radiation than males. In this scenario, a conventional genetic sexing strain is used to produce almost exclusively males and these males are then irradiated with a dose of radiation sufficient to guarantee sterility in any contaminating females.[77] This scheme was first tested experimentally in *Culex pipiens.*[84]

Cytoplasmic Incompatibility as a Drive Mechanism

A number of well known natural gene drive systems have been described.[85] One of the most widely studied is the infectious mobility pattern of transposons, which have been examined as potential artificial drive systems.[86] Transposon based approaches are promising, but also have a number of potential drawbacks, among which are the propensity of transposons to cross species boundaries at unpredicted rates and their tendency to accumulate internal mutations. Meiotic drive or other forms of asymmetric chromosomal inheritance present alternative population replacement mechanisms,[87-89] however, the most promising genetic control drive system is based on Wolbachia. Indeed, these intracellular bacteria have the ability to spread rapidly through populations by means of cytoplasmic incompatibility.[27,90-91] In addition, it has been suggested that any maternally inherited factor (mitochondria, symbiont or virus) or genes which are expressed by these factors will also be spread along with *Wolbachia* since all maternally transmitted entities are strictly coinherited as maternal lineages.[56-57] It has also been suggested that *Wolbachia* can be used to spread genes that are chromosomally located, once the *Wolbachia* genes responsible for cytoplasmic incompatibility are isolated and introduced onto an insect chromosome.[57-58] Theoretical modeling performed by Sinkins and colleagues suggests that if these genes are expressed appropriately, they can spread into an arthropod population along with any other linked chromosomal gene.[92-93] Bidirectional cytoplasmic incompatibility and/or the establishment of multiple infections of mutually incompatible strains provide the tools for repeated *Wolbachia* sweeps in order to replace a target insect population.[57]

Laboratory cage tests demonstrated that *Wolbachia* can be spread into a targeted unifected *A. aegypti* population, reaching infection fixation within seven generations.[72] These data clearly indicate that *Wolbachia* can be used as a vehicle to drive desirable genotypes into mosquito populations and into populations of other medically important disease vectors.

Virulent Wolbachia *Strains*

Wolbachia mainly reside within the reproductive tissues and organs of their hosts.[5-6,8,11] However, recent studies showed the presence of *Wolbachia* also in other tissues and organs.[94-95] Min and Benzer further reported that in some cases the presence of a certain *Wolbachia* strain, namely popcorn, in the nervous system of adult *Drosophila* flies results in a virulent effect which significantly reduces the life span of the host due to the bacterial over-replication.[96] The authors showed that this phenomenon is due to a genetic property of the particular *Wolbachia* strain. The virulent property of the *Wolbachia* popcorn strain was further confirmed by MacGraw and colleagues when they transferred the strain from *D. melanogaster* to *D. simulans.*[68] The transinfected flies also exhibited a significantly reduced life span. Similarly virulent *Wolbachia* strains have also been described in isopod species, suggesting that this genetic property may be widespread among *Wolbachia* strains infecting a great variety of arthropod species.

Sinkins and O'Neill suggested that virulent *Wolbachia* strains could be used to control disease transmission by insect vector species.[60] The idea is the following: It is known, for example, that *Anopheles* mosquitoes first receive the malaria pathogen through a blood meal. The pathogen undergoes its developmental cycle within the vector for a certain amount of time and is then transferred to a novel host. This means that only older individuals are able to transmit the malaria pathogen to human populations. Thus, if virulent *Wolbachia* strains were established in *Anopheles* mosquitoes, their lifespan would be reduced, which would also mean a reduction of the transmission rate of the malaria parasite. It should be noted, thatthese virulent strains may not significantly influence the reproductive fitness of their moquito hosts. The reason is that these virulent *Wolbachia* will only not be pathogenic to the youngest part of the population, which is expected to contribute the most to reproduction.[60]

Conclusions and Future Challenges

The specific effects of the endosymbiotic bacterium *Wolbachia* on the reproduction of its hosts can suggest its beneficial use in a variety of ways: (a) *Wolbachia*-induced cytoplasmic incompatibility, the basis of the Incompatible Insect Technique, could be used for insect pest population control in a way analogous to the "Sterile Insect Technique"; (b) *Wolbachia*-induced cytoplasmic incompatibility could also be used as a drive system to spread desirable genotypes in field arthropod populations; (c) virulent *Wolbachia* strains offer the potential to control vector species by modifying their population age structure and (d) *Wolbachia* could also be used to transform a sexually reproducing biocontrol agent, a parasitoid species for example, to asexual. This potential *Wolbachia*-based application was not discussed above, since it was beyond the scope of this chapter. However, it is worth noting that thelytoky is much more preferable than arrhenotoky in wasps used in biological control, because only female wasps kill insect pests. The advantages of a parthenogenetic population are obvious: reduction of cost since there will be no male-production, faster population growth because of the increased number of females and easier establishment in the field since no mating is necessary.[97] Despite these obvious effects, there have been no efforts to take advantage of *Wolbachia*-induced parthenogenesis (or parthenogenesis caused by any other symbiont). In addition, *Wolbachia* itself could be used in paratransgenesis approaches, once a genetic transfomation system for this bacterium is developed. The genetic transformation of *Wolbachia* as well as that of other insect symbionts, with the exception of *Sodalis glossinidius*, remains the major challenge for researchers in the field of insect symbiosis.[98] Another major challenge is the unraveling of the mechanism(s) of the *Wolbachia*-induced reproductive phenotypes, A wealth of *Wolbachia* genomic information is currently available. The genome of two *Wolbachia* strains have recently been completed; that of the *w*Mel strain, which is found in *Drosophila melanogaster* and induces cytoplasmic incompatibility and that of the *w*Bm strain, which has established a mutualistic association with the filarial nematode *Brugia malayi.*[99-100] Moreover, the genome sequences of several other *Wolbachia* strains are at the gap closure stage (*w*Ri which induces cytoplasmic incompatibility in *D. simulans*; *w*Pip which induces cytoplasmic incompatibility in *Culex pipiens*; *w*Vul which induces feminization in the isopod *Armadillidium* *vulgare*), in progress (*w*Uni which induces parthenogenesis in the parasitoid wasp *Muscidifurax uniraptor*; *w*Ovo which forms a mutualistic association with the nematode *Oncocherca volvulus*) or have been partially completed (*w*Sim, *w*Ana and *w*Wil).[101-104] Comparative genomics and post genomics analysis will certainly help both the development of a genetic transformation system as well as the elucidation of the molecular mechanism of the *Wolbachia*-induced reproductive alterations. These will be essential breakthroughs which will not only advance basic biological knowledge but will also allow the development of paratransgenic approaches towards the safe, biological and environmentally friendly management of invertebrate species of economic, environmental and medical importance.

Acknowledgments

Kostas Bourtzis is grateful to European Union, the International Atomic Energy Agency, the Greek Secretariat for Research and Technology, the Greek Ministry of Education, the Empirikeion Foundation and the University of Ioannina which have supported the research from his laboratory. The author also thanks Stefan Oehler for his comments on an earlier version of this chapter.

References

1. O'Neill SL, Giordano R, Colbert AME et al. 16S rRNA phylogenetic analysis of the bacterial endosymbionts associated with cytoplasmic incompatibility in insects. Proc Natl Acad Sci USA 1992; 89:2699-2702.
2. Hertig M. The rickettsia, Wolbachia pipientis and associated inclusions of the mosquito, Culex pipiens. Parasitology 1936; 28:453-490.
3. Werren JH, Zhang W, Guo LR. Evolution and phylogeny of Wolbachia-reproductive parasites of arthropods. Proc R Soc Lond B Biol Sci 1995; 261:55-63.
4. Breeuwer JAJ, Jacobs G. Wolbachia: Intracellular manipulators of mite reproduction. Exp Appl Acarol 1996; 20:421-434.
5. Werren JH. Biology of Wolbachia. Annu Rev Entomol 1997; 42:587-609.
6. Werren JH, O'Neill SL. The evolution of heritable symbionts. In: O'Neill SL, Hoffmann AA, Werren JH, eds. Influential passengers: Inherited microorganisms and arthropod reproduction. Oxford: Oxford University Press, 1997:1-41.
7. Bandi C, Anderson TJC, Genchi C et al. Phylogeny of Wolbachia in filarial nematodes. Proc R Soc Lond B Biol Sci 1998; 265:2407-2413.
8. Stouthamer R, Breeuwer JAJ, Hurst GDD. Wolbachia pipientis: Microbial manipulator of arthropod reproduction. Annu Rev Microbiol 1999; 53:71-102.
9. Bourtzis K, Braig HR. The many faces of Wolbachia. In: Raoult D, Brouqui P, eds. Rickettsiae and Rickettsial Diseases at the Turn of the Third Millennium. Paris: Elsevier, 1999:199-219.
10. Jeyaprakash A, Hoy MA. Long PCR improves Wolbachia DNA amplification: wsp sequences found in 76% of 63 arthropod species. Insect Mol Biol 2000; 9:393-405.
11. In: Bo5urtzis K, Miller T, eds. Insect Symbiosis. Florida: CRC Press, 2003:1-347.
12. Breeuwer JAJ, Stouthamer R, Barns SM et al. Phylogeny of cytoplasmic incompatibility microorganisms in the parasitoid wasp genus Nasonia (Hymenoptera: Pteromalidae) based on 16S ribosomal DNA sequences. Insect Mol Biol 1992; 1:25-36.
13. Rousset F, Bouchon D, Pintureau B et al. Wolbachia endosymbionts responsible for various alterations of sexuality in arthropods. Proc R Soc Lond B Biol Sci 1992; 250:91-98.
14. Lo N, Casiraghi M, Salati E et al. How many Wolbachia supergroups exist? Mol Biol Evol 2002; 19:341-346.
15. Rowley SM, Raven RJ, McGraw EA. Wolbachia pipientis in Australian spiders. Curr Microbiol 2004; 49:208-214.
16. Bordenstein S, Rosengaus RB. Discovery of a novel Wolbachia supergroup in isoptera. Curr Microbiol 2005; 51:393-398.
17. Werren JH, Winsdor DM. Wolbachia infection frequencies in insects: Evidence of a global equilibrium? Proc R Soc Lond B Biol Sci 2000; 267:1277-1285.
18. Casiraghi M, Bordenstein SR, Baldo L et al. Phylogeny of Wolbachia pipientis based on gltA, groEL and ftsZ gene sequences: Clustering of arthropod and nematode symbionts in the F supergroup, and evidence for further diversity in the Wolbachia tree. Microbiology 2005; 151:4015-4022.
19. Parask-evopoulos C, Bordenstein SR, Wernegreen J et al. Towards a Wolbachia multilocus sequence typing system: Discrimination of Wolbachia strains present in Drosophila species. Curr Microbiol 2006, 388-395.

20. Baldo L, Dunning Hotopp JC, Jolley KA et al. A multilocus sequence typing system for the endosymbiont Wolbachia. Appl Environ Microbiol 2006, (in press).
21. Ghelelovitch S. Sur le diterminisme ginitique de la stiriliti dans les croisements entre diffirentes souches de Culex autogenicus Roubaud. C R Acad Sci III Vie 1952; 234:2386-2388.
22. Laven H. Speciation and evolution in Culex pipiens. In: Wright J, Pal R, eds. Genetics of insect vectors of disease. Amsterdam: Elsevier, 1967:251-275.
23. Yen JH, Barr AR. New hypothesis of the cause of cytoplasmic incompatibility in Culex pipiens L. Nature 1971; 232:657-658.
24. Yen JH, Barr AR. The etiological agent of cytoplasmic incompatibility in Culex pipiens. J Invertebr Pathol 1973; 22:242-250.
25. Hoffman AA, Turelli M. Cytoplasmic incompatibility in insects. In: O'Neill SL, Hoffmann AA, Werren JH, eds. Influential Passengers: Inherited Microorganisms and Arthropod Reproduction. Oxford: Oxford University Press, 1997:42-80.
26. Bourtzis K, Braig HR, Karr TL. Cytoplasmic Incompatibility. In: Bourtzis K, Miller T, eds. Insect Symbiosis. Florida: CRC Press, 2003:217-246.
27. Turelli M, Hoffmann AA. Rapid spread of an inherited incompatibility factor in California Drosophila. Nature 1991; 353:440-442.
28. Breeuwer JA, Werren JH. Microorganisms associated with chromosome destruction and reproductive isolation between two insect species. Nature 1990; 346:558-560.
29. O'Neill SL, Karr TL. Bidirectional incompatibility between conspecific populations of Drosophila simulans. Nature 1990; 348:178-180.
30. Reed KM, Werren JH. Induction of paternal genome loss by the paternal-sex-ratio chromosome and cytoplasmic incompatibility bacteria (Wolbachia): A comparative study of early embryonic events. Mol Reprod Dev 1995; 40:408-418.
31. Callaini GM, Riparbelli G, Giordano R et al. Mitotic defects associated with cytoplasmic incompatibility in Drosophila simulans. J Invert Pathol 1996; 67:55-64.
32. Callaini G, Dallai R, Riparbelli MG. Wolbachia-induced delay chromatin condensation does not prevent maternal chromosomes from entering anaphase in incompatible crosses of Drosophila simulans. J Cell Sci 1997; 110:271-280.
33. Tram U, Sullivan W. Role of delayed nuclear envelope breakdown and mitosis in Wolbachia-induced cytoplasmic incompatibility. Science 2002; 296:1124-1126.
34. Tram U, Ferree PA, Sullivan W. Identification of Wolbachia-host interacting factors through cytological analysis. Microbes Infect 2003; 11:999-1011.
35. Poinsot D, Charlat S, Mercot H. On the mechanism of Wolbachia-induced cytoplasmic incompatibility: Confronting the models with the facts. Bioessays 2003; 25:259-265.
36. Masui S, Kamoda S, Sasaki T et al. Distribution and evolution of bacteriophage WO in Wolbachia, the endosymbiont causing sexual alterations in arthropods. J Mol Evol 2000; 51:491-497.
37. Masui S, Kuroiwa H, Sasaki T et al. Bacteriophage WO and virus-like particles in Wolbachia, an endosymbiont of arthropods. Biochem Biophys Res Commun 2001; 283:1099-1104.
38. Fujii Y, Kubo T, Ishikawa H et al. Isolation and characterization of the bacteriophage WO from Wolbachia, an arthropod endosymbiont. Biochem Biophys Res Commun 2004; 317:1183-1188.
39. Sinkins SP, Walker T, Lynd AR et al. Wolbachia variability and host effects associated with crossing type in Culex mosquitoes. Nature 2005; 436:257-260.
40. Duron O, Fort P, Weill M. Hypervariable prophage WO sequences describe an unexpected high number of Wolbachia variants in the mosquito Culex pipiens. Proc R Soc Lond B Biol Sci 2006; 273:495-502.
41. Bordenstein SR, Marshall ML, Fry AJ et al. The tripartite associations between bacteriophage, Wolbachia, and arthropods. PLoS Pathogens 2:e43.
42. Boyle L, O'Neill SL, Robertson HM et al. Inter- and intraspecific horizontal transfer of Wolbachia in Drosophila. Science 1993; 260:1796-1799.
43. Poinsot D, Bourtzis K, Markakis G et al. Wolbachia transfer from Drosophila melanogaster to D. simulans: Host effect and cytoplasmic incompatibility relationships. Genetics 1998; 150:227-237.
44. Singh KRP, Curtis CF, Krishnamurthy BS. Partial loss of cytoplasmic incompatibility with age in males of Culex fatigans Wied. Ann Trop Med Parasit 1976; 70:463-466.
45. Hoffmann AA, Turelli M, Simmons GM. Unidirectional incompatibility between populations of Drosophila simulans. Evolution 1986; 40:692-701.
46. Hoffmann AA. Partial cytoplasmic incompatibility between two Australian populations of Drosophila melanogaster. Entomol Exp Appl 1988; 48:61-67.
47. Clancy DJ, Hoffmann AA. Environmental effects on cytoplasmic incompatibility and bacterial load in Wolbachia-infected Drosophila simulans. Entomol Exp Appl 1998; 86:13-24.
48. Jamnongluk W, Kittayapong P, Baisley KJ et al. Wolbachia infection and expression of cytoplasmic incompatibility in Armigeres subalbatus (Diptera: Culicidae). J Med Entomol 2000; 37:53-57.

49. Reynolds KT, Hoffmann AA. Male age and the weak expression or nonexpression of cytoplasmic incompatibility in Drosophila strains infected by maternally-transmitted Wolbachia. Genet Res 2002; 80:79-87.
50. Kittayapong P, Mongkalangoon P, Baimai V et al. Host age effect and expression of cytoplasmic incompatibility in field populations of Wolbachia-superinfected Aedes albopictus. Heredity 2002; 88:270-274.
51. Reynolds KT, Thomson LJ, Hoffmann AA. The effects of host age, host nuclear background and temperature on phenotypic effects of the virulent Wolbachia strain popcorn in Drosophila melanogaster. Genetics 2003; 164:1027-1034.
52. Karr TL, Yang W, Feder ME. Overcoming cytoplasmic incompatibility in Drosophila. Proc R Soc Lond B Biol Sci 1998; 265:391-395.
53. Champion de Crespigny FE, Wedell N. Wolbachia infection reduces sperm competitive ability in an insect. Proc R Soc Lond B Biol Sci 2006; 273:1455-1458.
54. Stevens L. Environmental factors affecting reproductive incompatibility in flour beetles, genus Tribolium. J Invert Pathol 1989; 53:78-84.
55. Sinkins SP, Braig HR, O'Neill SL. Wolbachia pipientis: Bacterial density and unidirectional incompatibility between infected populations of Aedes albopictus. Exp Parasitol 1995; 81:284-291.
56. Beard CB, O'Neill SL, Tesh RB et al. Modification of arthropod vector competence via symbiotic bacteria. Parasitol Today 1993; 9:179-183.
57. Bourtzis K, O'Neill SL. Wolbachia infections and their influence on arthropod reproduction. Bioscience 1998; 48:287-293.
58. Sinkins SP, Curtis CF, O'Neill SL. The potential application of inherited symbiont systems to pest control. In: O'Neill SL, Hoffmann AA, Werren JH, eds. Influential Passengers: Inherited Microorganisms and Arthropod Reproduction. Oxford: Oxford University Press, 1997:155-175.
59. Ashburner M, Hoy MA, Peloquin JJ. Prospects for the genetic transformation of arthropods. Insect Mol Biol 1998; 7:201-213.
60. Sinkins SP, O'Neill SL. Wolbachia as a vehicle to modify insect populations. In: Handler A, James A, eds. Insect Transgenesis. Boca Raton: CRC Press, 2000:271-287.
61. Aksoy S, Maudlin I, Dale C et al. Prospects for control of African trypanosomiasis by tsetse vector manipulation. Trends Parasitol 2001; 17:29-35.
62. Braig HR, Guzman H, Tesh RB et al. Replacement of the natural Wolbachia symbiont of Drosophila simulans with a mosquito counterpart. Nature 1994; 367:453-455.
63. Giordano R, O'Neill SL, Robertson HM. Wolbachia infections and the expression of cytoplasmic incompatibility in Drosophila sechelllia and D. mauritiana. Genetics 1995; 140:1307-1317.
64. Clancy DJ, Hoffmann AA. Behavior of Wolbachia endosymbionts from Drosophila simulans in Drosophila serrata, a novel host. Am Nat 1997; 149:975-988.
65. Rousset F, Braig HR, O'Neill SL. A stable triple Wolbachia infection in Drosophila with nearly additive incompatibility effects. Heredity 1999; 82:620-627.
66. Sasaki T, Ishikawa, H. Transfection of Wolbachia in the Meditteranean flour moth, Ephestia kuehniella, by embryonic microinjection. Heredity 2000; 85:130-135.
67. Charlat S, Nirgianaki A, Bourtzis K et al. Evolution of Wolbachia-induced cytoplasmic incompatibility in Drosophila simulans and D. sechellia. Evolution 2002; 56:1735-1742.
68. McGraw EA, Merritt DJ, Droller JN et al. Wolbachia density and virulence attenuation after transfer into a novel host. Proc Natl Acad Sci USA 2002; 99:2918-2923.
69. Riegler M, Charlat S, Stauffer C et al. Wolbachia transfer from Rhagoletis cerasi to Drosophila simulans: Investigating the outcomes of host-symbiont coevolution. Appl Environ Microbiol 2004; 70:273-279.
70. Zabalou S, Charlat S, Nirgianaki A et al. Natural Wolbachia infections in the Drosophila yakuba species complex do not induce cytoplasmic incompatibility but fully rescue the wRi modification. Genetics 2004a; 167:827-834.
71. Zabalou S, Riegler M, Theodorakopoulou M et al. Wolbachia-induced cytoplasmic incompatibility as a means for insect pest population control. Proc Natl Acad Sci USA 2004b; 101:15042-15045.
72. Xi Z, Khoo CCH, Dobson SL. Wolbachia establishment and invasion in an Aedes aegypti laboratory population. Science 2005; 310:326-328.
73. Boller EF, Russ K, Vallo V et al. Incompatible races of European cherry fruit fly Rhagoletis cerasi (Diptera: Tephritidae) their origin and potential use in biological control. Entomol Exp Appl 1976; 20:237-247.
74. Brower JH. Suppression of laboratory populations of Ephestia cautella (Walker) (Lepidoptera: Pyralidae) by release of males with cytoplasmic incompatibility. J Stored Prod Res 1979; 15:1-4.
75. Brower JH. Reduction of almond moth populations in simulated storages by the release of genetically incompatible males. J Econ Entomol 1980; 73:415-418.

76. Boller EF. Cytoplasmic incompatibility in Rhagoletis cerasi. In: Robinson AS, Hooper G, eds. Fruit Flies, Their Biology, Natural Enemies and Control. World Crop Pests 3B. Amsterdam: Elsevier, 1989:69-74.
77. Bourtzis K, Robinson AS. Insect pest control using Wolbachia and/or radiation. In: Bourtzis K, Miller T, eds. Insect Symbiosis 2. Florida: Taylor and Francis Group, LLC, 2006, (in press).
78. Boller EF, Bush GL. Evidence for genetic variation in populations of the European cherry fruit fly, Rhagoletis cerasi (Diptera: Tephritidae) based on physiological parameters and hybridization experiments. Entomol Exp Appl 1974; 17:279-293.
79. In: Robinson AS, Hooper G, eds. Fruit Flies, Their Biology, Natural Enemies and Control. World Crop Pests 3B. Amsterdam: Elsevier, 1989.
80. Bourtzis K, Nirgianaki A, Onyango P et al. A prokaryotic dnaA sequence in Drosophila melanogaster: Wolbachia infection and cytoplasmic incompatibility among laboratory strains. Insect Mol Biol 1994; 3:131-142.
81. Rocha LS, Mascarenias RO, Perondini ALP et al. Occurrence of Wolbachia in Brazilian samples of Ceratitis capitata (Wiedemann) (Diptera: Tephritidae). Neotropical Entomol 2005; 34:1013-1015.
82. Robinson AS, Franz G, Fisher K. Genetic sexing strains in the medfly, Ceratitis capitata: Development, mass rearing and field application. Trends Entomol 1999; 2:81-104.
83. Franz GF. Genetic sexing strains in Mediterannean fruit fly, an example for other species amenable to large-scale rearing for the sterile insect technique. In: Dyck VA, Hendrichs J, Robinson AS, eds. Sterile Insect Technique. Principles and Practice in Area-Wide Integrated Pest Management. Dordrecht: Springer, 2005:427-451.
84. Arunachalam N, Curtis CF. Integration of radiation with cytoplasmic incompatibility for genetic control in the Culex pipiens complex (Diptera: Culicidae). J Med Entomol 1985; 22:648-653.
85. Sinkins SP, Gould F. Gene drive systems for insect disease vectors. Nat Rev Genetics 2006; 7:427-435.
86. Ribeiro JM, Kidwell MG. Transposable elements as population drive mechanisms: Specification of critical parameter values. J Med Entomol 1994; 31:10-16.
87. Braig HR, Yan G. The spread of genetic constructs in natural insect populations. In: Letourneau DK, Burrows BE, eds. Genetically Engineered Organisms: Assessing Environmental and Human Health Effects. Boca Raton: CRC Press, 2001:251-314.
88. Lyttle TW. Cheaters sometimes prosper: Distortion of mendelian segregation by meiotic drive. Trends Genet 1993; 9:205-210.
89. Taylor DR, Ingvarsson PK. Common features of segregation distortion in plants and animals. Genetics 2003; 117:27-35.
90. Turelli M, Hoffman AA. Cytoplasmic incompatibility in Drosophila simulans: Dynamics and parameter estimates from natural populations. Genetics 1995; 140:1319-1338.
91. Hoshizaki S, Shimada T. PCR-based detection of Wolbachia, cytoplasmic incompatibility microorganisms, infected in natural populations of Laodelphax striatellus (Homoptera: Delphacidae) in central Japan: Has the distribution of Wolbachia spread recently? Insect Mol Biol 1995; 4:237-243.
92. Sinkins SP. Wolbachia and cytoplasmic incompatibility in mosquitoes. Insect Biochem Mol Biol 2004; 34:723-729.
93. Sinkins SP, Godfray HCJ. Use of Wolbachia to drive nuclear transgenes through insect populations. Proc R Soc Lond B Biol Sci 2004; 271:1421-1426.
94. Dobson SL, Bourtzis K, Braig HR et al. Wolbachia infections are distributed throughout insect somatic and germ line tissues. Insect Biochem Mol Biol 1997; 29:153-160.
95. Cheng Q, Ruel TD, Zhou W et al. Tissue distribution and prevalence of Wolbachia infections in tsetse flies, Glossina spp. Med Vet Entomol 2000; 14:44-50.
96. Min KT, Benzer S. Wolbachia, normally symbiont of Drosophila, can be virulent, causing degeneration and death. Proc Natl Acad Sci USA 1997; 94:10792-10796.
97. Stouthamer R. The use of sexual versus asexual wasps in biological control. Entomophaga 1993; 38:3-6.
98. Cheng Q, Aksoy S. Tissue tropism, transmission and expression of foreign genes in vivo in midgut symbionts of tsetse flies. Insect Mol Biol 1999; 8:125-132.
99. Wu M, Sun LV, Vamathevan J et al. Phylogenomics of the reproductive parasite Wolbachia pipientis wMel: A streamlined genome overrun by mobile genetic elements. PLoS Biology 2004; 2:e69.
100. Foster J, Ganatra M, Kamal I et al. The Wolbachia genome of Brugia malayi: Endosymbiont evolution within a human pathogenic nematode. PLoS Biology 2005; 3:e121.
101. Oehler S, Bourtzis K. First International Wolbachia Conference: Wolbachia 2000. Symbiosis 2000; 29:151-161.
102. Salzberg SL, Dunning Hotopp JC, Delcher AL et al. Serendipitous discovery of Wolbachia genomes in multiple Drosophila species. Genome Biol 2005a; 6:R23.
103. Salzberg SL, Dunning Hotopp JC, Delcher AL et al. Correction: Serendipitous discovery of Wolbachia genomes in multiple Drosophila species. Genome Biol 2005b; 6:402.
104. Fenn K, Blaxter M. Wolbachia genomes: Revealing the biology of parasitism and mutualism. Trends Parasitol 2006; 22:60-65.

Chapter 10

Using Predictive Models to Optimize *Wolbachia*-Based Strategies for Vector-Borne Disease Control

Jason L. Rasgon*

Abstract

The development of resistance to insecticides by vector arthropods, the evolution of resistance to chemotherapeutic agents by parasites and the lack of clinical cures or vaccines for many diseases has stimulated a high-profile effort to develop vector-borne disease control strategies based on release of genetically-modified mosquitoes. Because transgenic insects are likely to be less fit than their wild-type counterparts, transgenic traits must be actively driven into the population in spite of fitness costs (population replacement). *Wolbachia* are maternally-inherited symbionts that are associated with numerous alterations in host reproductive biology. By a variety of mechanisms, *Wolbachia*-infected females have a reproductive advantage relative to uninfected females, allowing infection to spread rapidly through host populations to high frequency in spite of fitness costs. In theory, *Wolbachia* can be exploited to drive costly transgenes into vector populations for disease control. Before conducting an actual release, it is important to be able to predict how released *Wolbachia* infections are expected to behave. While inferences can be made by observing the dynamics of naturally-occurring infections, there is no ideal way to empirically test the efficacy of a *Wolbachia* gene driver under field conditions prior to the first actual release. Mathematical models are a powerful way to predict the outcomes of transgenic insect releases and allow one to identify knowledge gaps, identify parameters that are critical to the success of releases, conduct risk-assessment analysis and investigate worst-case scenarios, and ultimately identify the most effective, most logistically feasible control method or methods. In this chapter, I review current and historical advances in applied models of *Wolbachia* spread, specifically within the context of applied population replacement strategies for vector-borne disease control.

Introduction

The development of resistance to insecticides by vector arthropods, the evolution of resistance to chemotherapeutic agents by parasites and the lack of clinical cures or vaccines for many diseases[1-4] has stimulated a high-profile effort to develop vector-borne disease control strategies based on release of genetically-modified mosquitoes.[5-6] In general, these strategies aim to replace existing pathogen-susceptible vector populations with those unable to transmit

*Jason L. Rasgon—The W. Harry Feinstone Department of Molecular Microbiology and Immunology, Bloomberg School of Public Health, Johns Hopkins University, and The Johns Hopkins Malaria Research Institute, Baltimore, Maryland 21205, USA. Email: jrasgon@jhsph.edu

Transgenesis and the Management of Vector-Borne Disease, edited by Serap Aksoy.

pathogens (i.e., "population replacement").[6] Because transgenic insects are likely to be less fit than their wild-type counterparts, they would be expected to be out-competed and rapidly eliminated from the population. Thus, transgenic traits are unlikely to spread by genetic introgression and must be actively driven into the population in spite of fitness costs.[6-7]

Multiple "transgene drivers" are currently under theoretical consideration, including transposable elements and maternally-transmitted bacterial endosymbionts.[5-7] For some systems, we have some idea how released gene drivers are expected to behave due to observations of naturally-occurring gene drive occurrences, such as the spread of *P* element in *Drosophila melanogaster* or the spread of *Wolbachia* in *Drosophila simulans*,[5-8] but in all cases there is no ideal way to formally test the efficacy of an engineered gene driver until the first actual releases are made. It is critical to be able to assess the potential impact of transgenic insects prior to making an actual release. Mathematical models provide a theoretical framework for accomplishing this task. Models allow one to identify knowledge gaps, identify parameters that are critical to the success of releases, conduct risk-assessment analysis and investigate worst-case scenarios, and ultimately identify the most effective, most logistically feasible control method or methods.[7,9]

Wolbachia Endosymbionts

Wolbachia are maternally-inherited symbionts that are associated with numerous alterations in host reproductive biology, including parthenogenesis, feminization, male-killing and cytoplasmic incompatibility (CI).[10] CI causes reduced egg hatch when infected males mate with uninfected females. Matings of infected females are fertile regardless of the infection status of the male. Infected females contribute more offspring to the next generation relative to uninfected females, allowing infection to spread rapidly through host populations to high frequency even if infection induces fitness costs.[7] If transgenic traits are linked to *Wolbachia* (either by inserting them directly into the *Wolbachia* genome or carried on a separate maternally-inherited cytoplasmic factor) the transgene will "hitch-hike" along with infection into the host population. In essence, the fitness advantage conferred by CI can counteract the fitness disadvantage conferred by the transgene, allowing the transgene to increase in frequency to epidemiologically relevant levels.[7]

There are 3 items that are critical to know in order to use *Wolbachia* in an applied manner to control disease: (1) how many infected mosquitoes must be initially released, (i.e., what is the threshold frequency that infection must surpass before *Wolbachia* can invade the population), (2) What frequency will infection ultimately reach in the population if *Wolbachia* successfully invades, and (3) how long will invasion take? Mathematical models have been developed to predict these values based on three basic empirically measurable parameters.[7] These parameters are (1) μ, the percentage uninfected offspring from an infected female ($\mu = 0$ if transmission is 100%), (2) H, the relative hatch rate of an incompatible compared to a compatible cross ($H = 0$ if an incompatible cross is 100% sterile) and (3) F, the relative fitness of an infected female compared to an uninfected female ($F = 1$ if there is no fitness cost associated with infection). Assuming random mating and discrete generations, the change in infection frequency (p) between generations can be described by the following equation,[7] where $s_f = (1\text{-}F)$ and $s_h = (1\text{-}H)$:

$$p_{t+1} = \frac{p_t(1-\mu)F}{1 - s_f p_t - s_h p_t(1-p_t) - \mu F p_t^2 s_h} \tag{1}$$

Eq. 1 predicts two equilibrium points, p_s and p_u.

$$p_s = \frac{s_f + s_h + \sqrt{(s_f + s_h)^2 - 4(s_f + \mu F)s_h(1-\mu F)}}{2s_h(1-\mu F)} \tag{2}$$

and

$$p_u = \frac{s_f + s_h - \sqrt{(s_f + s_h)^2 - 4(s_f + \mu F)s_h(1-\mu F)}}{2s_h(1-\mu F)} \tag{3}$$

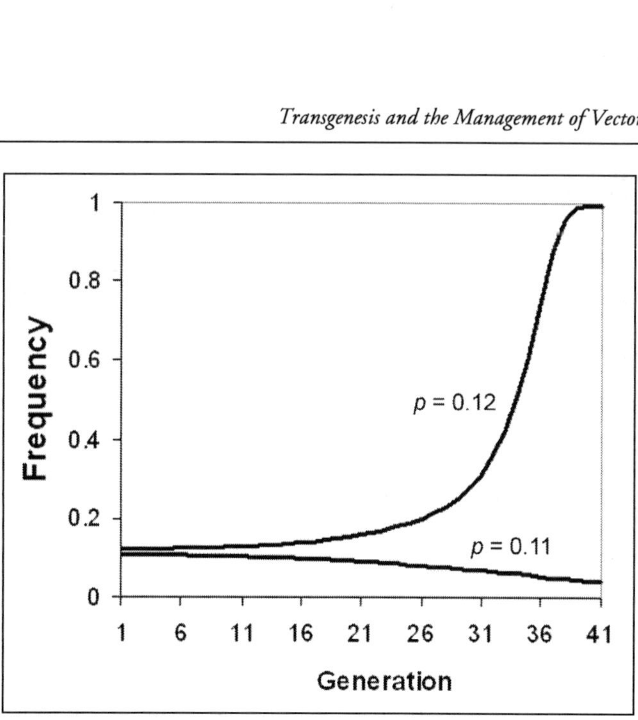

Figure 1. Dynamics of *Wolbachia* spread predicted by iterative solution of Eq. 1, where $\mu = 0.05$, $H = 0.1$ and $F = 0.95$. *Wolbachia* is predicted to spread if infection exceeds the unstable equilibrium (*pu*; Eq. 3) of 0.115. Two introduction scenarios are shown; 0.12 (which exceeds *pu* and spreads, and 0.11 (which does not exceed *pu* and is lost).

Root p_s is a stable equilibrium, while p_u is unstable. In this context, p_s represents the frequency infection will reach in the population after a successful invasion, while p_u represents the frequency that must be exceeded for infection to spread (i.e., the introduction threshold). If the initial release frequency is above p_u, infection frequency will increase until it reaches p_s. If the release frequency is less than p_u, infection will be lost from the population (Fig. 1). Accurate estimation of p_u is especially important from an applied perspective because it relates directly to the overall feasibility of potential releases (i.e., how large must the initial release be?). The time for invasion can be calculated by iterative solution of Eq. 1 (Fig. 1).

Can these models be used to predict the outcome of applied *Wolbachia* invasions into vector populations for disease control? For theoretical predictions to have any practical validity, models must be parameterized with field data and validated for the particular vector species of interest. There have been extensive theoretical and empirical studies on the dynamics of *Wolbachia* spread and model validation in natural *Drosophila* populations.[8,11-12] It was found that when values for μ, F and H were estimated under field conditions, the model closely predicted the behavior of infection spread in wild populations.[8] Data from these studies have been used to make inferences about how infection would be likely to behave in mosquito populations. However, until recently, there have been no studies attempting parameter estimation and model validation in natural vector populations. Differences in the biology between *Drosophila* and mosquitoes make firm conclusions problematic.

The only vector insect system where *Wolbachia* infection dynamics have been adequately investigated under field conditions is in the California *Culex pipiens* (L.) species complex.[13-14] This species complex consists of two subspecies (*Cx. pipiens pipiens* and *Cx. pipiens quinquefasciatus*) which hybridize freely where they come into contact in California.[15] During two years of sampling, *Wolbachia* infection frequency was found to be close to fixation throughout the state along a north-south transect. There was no molecular variation observed in *Wolbachia* surface protein (*wsp*) sequences within or between populations, and mosquitoes collected from widely separated populations were fully compatible when crossed in either direction. The findings suggested that a single strain of *Wolbachia* was present at very high frequency in all populations, and that infection prevalence levels were stable over time.[13-14]

The parameters that govern the dynamics of *Wolbachia* spread in insect populations (μ, H and F) were estimated under both laboratory and field conditions for the *Wolbachia* strain infecting the CA *Cx. pipiens* species complex.[13] Estimates for μ under field conditions ranged from $0.025 \geq \mu \geq 0.0077$, with a mean value of 0.014 (98.6% transmission). Hatch rates in incompatible crosses were 100% sterile, with no significant reduction in CI expression as males aged ($H = 0$). There were no detectable fitness effects due to infection ($F = 1$). Parameter estimates showed close agreement between lab and field experiments. Using these parameter values, Eq. 1 can be simplified to

$$p_{t+1} = \frac{p_t(1-\mu)}{1-p_t+p_t^2(1-\mu)} . \tag{4}$$

The stable equilibrium point p_s for Eq. 4 always equals 1.0 for any value of μ indicating that if infection successfully invades the population, it is expected to reach fixation. This theoretical prediction agreed closely with empirically observed infection frequencies in nature (99.4% combined statewide infection frequency). The unstable equilibrium for Eq. 4 (p_u) can be found by

$$p_u = \frac{\mu}{1-\mu} \tag{5}$$

If the initial introduction is below the level predicted by Eq. 5, infection will be eliminated from the population. Using the mean value of μ estimated from the field (0.014), infection is expected to spread if frequency exceeds a threshold level of 0.0142. Infection is predicted to reach 100% in approximately 30 generations with an initial introduction of $p = 0.05$ (Fig. 2).

Wolbachia with the characteristics of the CA *Cx. pipiens* strain would have potential for application in vector-borne disease control programs. Important characteristics include

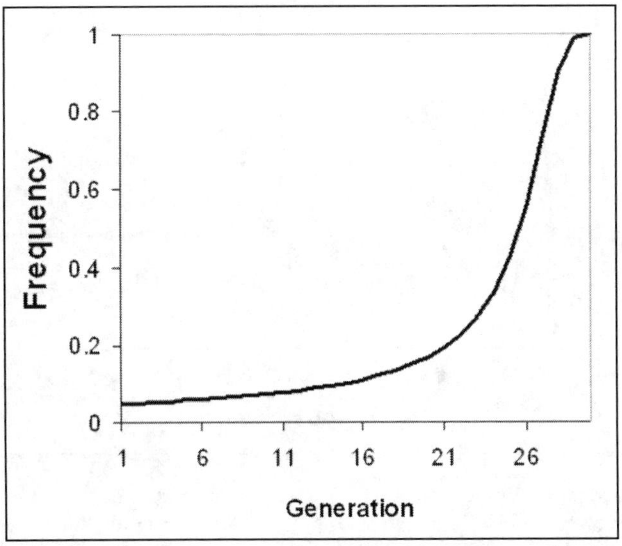

Figure 2. Dynamics of *Wolbachia* spread (Eq. 4), using field estimated parameters for *Cx. pipiens* ($\mu = 0.014$, $H = 0$, $F = 1$). Infection increases to fixation in approximately 30 generations from an initial introduction level of 0.05.

near-perfect maternal transmission, no detectable fitness effects and strong CI. These characteristics lead to the prediction that an economically and logistically feasible introduction (< 1.5%) will result in infection reaching fixation relatively rapidly.[13]

More Realistic Population Dynamics of *Wolbachia* Infections

For simplicity and mathematical tractability, the model outlined in equations 1-3 makes numerous simplifying assumptions that may not be ecologically realistic. Previous experimental and theoretical studies have shown that this simple model is quite robust in giving fairly accurate estimates of the *Wolbachia* stable equilibrium point p_s.[8,13] However, the simple model may not give accurate predictions of *Wolbachia* introduction thresholds if ecologically realistic assumptions are included in simulations. It has been shown, for instance, that conceptually simple factors such as population age structure and overlapping generations can greatly affect predictions of *Wolbachia* introduction thresholds.[16] Population age structure can result in up to a 10-fold or greater increase in *Wolbachia* introduction thresholds as compared to predictions of the simple model, depending on *Wolbachia* transmission, cytoplasmic incompatibility and fecundity effects.[16] This effect is due primarily to the death of released mosquitoes prior to completion of their first oviposition cycle, necessitating much larger initial releases (Fig. 3). Releasing gravid females is much more efficient because this period of preovipositional vector mortality is diminished, but it is logistically much more difficult to release gravid vs. teneral mosquitoes. Simulations also show that an accurate knowledge of the initial population age structure at the time of introduction is critical to predicting the success of the invasion, as deviations in the initial age structure can have a large positive or negative effect on the required magnitude of the initial transgenic release (Fig. 4).[16]

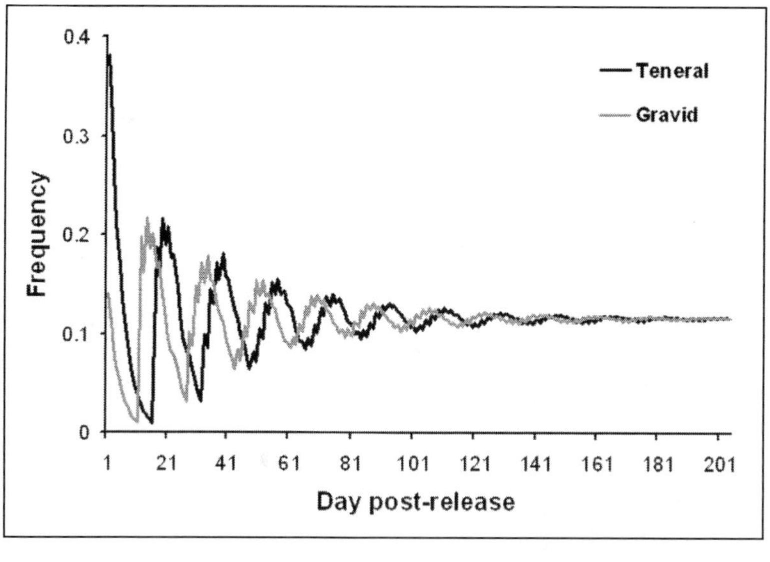

Figure 3. Dynamics of *Wolbachia* infection in adult mosquitoes in an age-structured population with overlapping generations. Infection was introduced into a population at the stable age distribution. *Wolbachia* parameters are: $\mu = 0.05$, $H = 0.1$, $F = 0.95$. Simulations were conducted using a "minimal release" (the minimum adult release required for infection to surpass the introduction threshold. Gray represents gravid release, black represents teneral release.

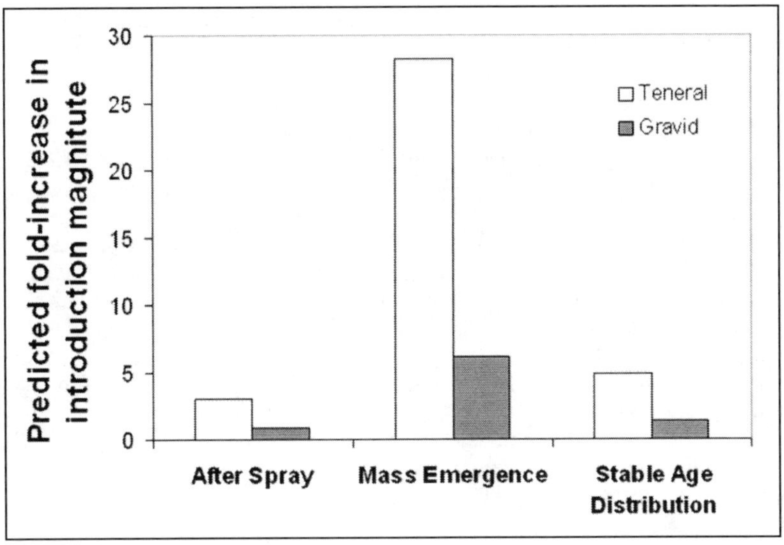

Figure 4. Deviation from the stable age distribution and effect on *Wolbachia* introduction thresholds. *Wolbachia* parameters are: $\mu = 0.05$, $H = 0.1$, $F = 0.95$. "Mass emergence" represents a population after an overwintering period where all mosquitoes are represented as young adults. "After spray" represents population where adult population has been eliminated by an adulticide application without affecting immature populations. Predicted fold-increase is relative to predictions of Eq. 3. White bars represent teneral release, gray bars represent gravid release.

Can Modeling Highlight a Better Way to Control Disease Using *Wolbachia* Infections?

The use of mathematical models to understand the dynamics of vector-borne diseases has a long history. Vectorial Capacity (C) (Eq. 6), defined as the number of new inoculations resulting from one infectious host entering a completely susceptible population per day, is often used to estimate the potential of a vector population to maintain pathogen transmission.[17] Understanding the parameters that make up C can also give insight to the most effective methods to control a vector-borne disease. Vectorial capacity is calculated as:

$$C = \frac{ma^2Vp^n}{-\ln p} \tag{6}$$

where m = mosquito density (female mosquitoes/host), a = daily biting rate, V = vector competence (proportion of mosquitoes ingesting pathogen that become competent to transmit), p = daily probability of vector survival, n = pathogen extrinsic incubation period (days).[17]

Most population replacement strategies focus on attempting to reduce arthropod vector competence (V in Eq. 6) , defined as the proportion of mosquitoes feeding on an infected host that become capable of transmitting a pathogen.[18-21] The idea that since vector-borne pathogens are dependent on arthropods for transmission, vertebrate infection and disease can be modulated by replacing susceptible vector populations with those that are genetically refractory to transmission, is intuitively attractive but not well supported theoretically or empirically. Transgenic vectors must be close to 100% refractory to pathogen transmission under field conditions to result in significant reductions in pathogen transmission.[21] There are many cases where refractory arthropods were efficient vectors because other aspects of their biology compensated for poor vector competence.[22-24]

According to the Vectorial Capacity equation, the most sensitive component of a vector's role in pathogen transmission is p, its daily probability of survival.[17] Disease control strategies that increase vector mortality are expected to be more efficient in reducing pathogen transmission than altering vector competence because small changes in daily survival can result in large changes in the number of new vertebrate host infections.

For a vector to transmit a pathogen, it must ingest the pathogen during bloodfeeding and survive until the pathogen can be transmitted to a vertebrate host. This time period, known as the extrinsic incubation period (EIP), can vary from days to weeks depending on ambient temperature, the vector species and the pathogen in question. Altering vector insects to reduce their lifespan would decrease vector survival through the EIP and their expectation of infective life; i.e., the number of days a vector is expected to live after becoming infectious. However, a trait that shortens vector lifespan will induce a major fitness cost and would not be expected to spread spontaneously.

A virulent *Wolbachia* strain (called *popcorn* or *w*MelPop) has been shown to kill adult *Drosophila* by over-replicating in the central nervous system. The average life span of adult infected flies is approximately one-half that of uninfected flies.[25] If a similar *Wolbachia* strain were transferred into disease vectors such as mosquitoes, CI might counteract the fitness disadvantages conferred by infection and spread the virulent infection through the population.[26-28]

Several modeling frameworks have been developed to examine the potential for virulent *Wolbachia* infections to shift the age structure of populations and thus affect disease transmission. The first model was outlined by Fine.[29] Fine incorporated a parameter β, representing the relative survival rate of infected individuals from hatching to reproduction. In this modeling framework, β has the same relationship to *Wolbachia* invasion dynamics as the fecundity parameter F; i.e., the relative fitness of infected vs. uninfected individuals equals βF, which would be substituted for F in Eqs. 1-3. The Fine model was specifically examined in the context of Dengue virus transmission of by Brownstein and colleagues.[28] Assuming perfect transmission ($\mu = 0$), Brownstein et al showed that virulent *Wolbachia* infections could invade populations and potentially affect pathogen transmission dynamics as long as introduction thresholds were relatively large (≥ 40%).

Rasgon and colleagues used a more detailed modeling framework to examine this question in age-structured populations.[27] Their analysis incorporated imperfect *Wolbachia* transmission, and used *Wolbachia* and mosquito life-table data empirically estimated from laboratory and field experiments in *Cx. pipiens*. They also developed an age-stratified extension of the classic Garrett-Jones Vectorial Capacity equation to measure theoretical changes in pathogen transmission in concert with virulent *Wolbachia* spread, for pathogens with extrinsic incubation periods ranging from 3 to 25 days. Using this approach, they were able to specifically examine the effect of the pattern of induced mortality on *Wolbachia* dynamics and disease potential.

Rasgon et al noted that the lowest introduction levels and greatest reduction in Vectorial Capacity were obtained when the onset of elevated mortality was delayed until infected individuals had an opportunity to reproduce.[27] In other words, there was a window of time where the infected individuals must live long enough to reproduce and pass on the infection, yet not survive through the extrinsic incubation period and transmit the pathogen. Interestingly, this is exactly the phenomenon observed in natural *popcorn* infected *Drosophila*—elevated mortality is delayed until adult flies are approximately 6-7 days old, which allows them to mate and oviposit before dying.[25] Under certain conditions, they showed that pathogen transmission could be essentially eliminated from populations with relatively low introduction levels (>0.15) depending on the pathogen extrinsic incubation period and pattern of induced mortality (Fig. 5).

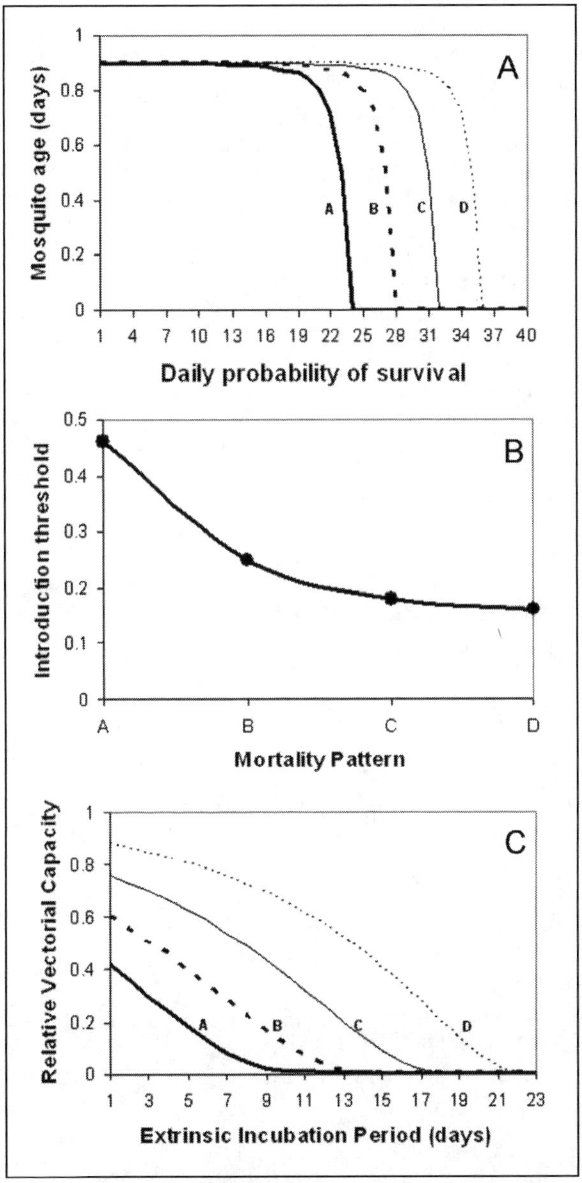

Figure 5. Introduction of virulent *Wolbachia* into populations and theoretical effect on pathogen transmission dynamics. *Wolbachia* parameters are: $\mu = 0.05$, $H = 0.1$, $F = 0.95$. A) Age-dependent daily survival patterns for pathogenic *Wolbachia*-infected mosquitoes. All age-dependent patterns have the same initial survival (0.9) that declines to 0 after varying latent periods; A: 100% mortality at day 14 of adult life, B: 100% mortality at day 18 of adult life, C: 100% mortality at day 22 of adult life, D: 100% mortality at day 26 of adult life. B) Pathogenic *Wolbachia* introduction threshold levels for virulent *Wolbachia* strains that affect mosquito survival according to the patterns in 5A. C) Reduction in Vectorial Capacity across a range of pathogen extrinsic incubation periods (EIP) after invasion of pathogenic *Wolbachia* that affect mosquito survival according to the patterns in 5A, expressed as percent of initial Vectorial Capacity before invasion.

Using *Wolbachia* to Drive Nuclear Traits?

The strategies outlined above rely on 100% linkage between *Wolbachia* and the trait of interest (i.e., the trait must be maternally inherited with perfect fidelity). Even slightly imperfect maternal transgene inheritance will result in disassociation between the transgene and the *Wolbachia* driver, resulting in elimination of the transgene from the population (Fig. 6).[7] Perfect linkage could be accomplished either by inserting the transgene into the *Wolbachia* genome or by placing the gene on separate maternally inherited construct. However, *Wolbachia* transformation protocols have not yet been developed and there are no current maternally-inherited constructs that satisfy the perfect transmission requirement.[6,30]

To get around this obstacle, some have suggested that if the *Wolbachia* genes responsible for CI were identified and cloned, they could be inserted into the host nuclear chromosomes and spread in a manner similar to an under-dominant trait.[7,31] If tightly linked to the anti-parasite gene, this would negate the need for a perfectly-inherited maternal construct.

The theoretical dynamics of single-locus nuclear incompatibility genes was investigated by Turelli and colleagues,[7,31] where allele A represents presence of the CI gene and allele a represents absence of the gene. It was further assumed that a single gene was responsible for sperm modification in the male and fertilization rescue in the female, and that the gene was completely dominant. H = the relative hatch rate of an incompatible cross and F represents the relative fitness of individuals carrying an A allele (AA and Aa have the same relative fitness). $s_h = (1-H)$ and $s_f = (1-F)$. P equals the frequency of AA individuals, Q equals the frequency of Aa individuals, and R = 1-(P + Q) equals the frequency of aa individuals (wild-type). Changes in the frequencies of adult genotypes can be calculated as

$$P' = \frac{F\left[P(P+Q)+0.25Q^2\right]}{1-s_f(1-R)-s_hR(1-R)} \quad (7)$$

and

$$Q' = \frac{0.5PQF+FPR+0.5FQ+HPR+0.5HQR}{1-s_f(1-R)-s_hR(1-R)} \quad (8)$$

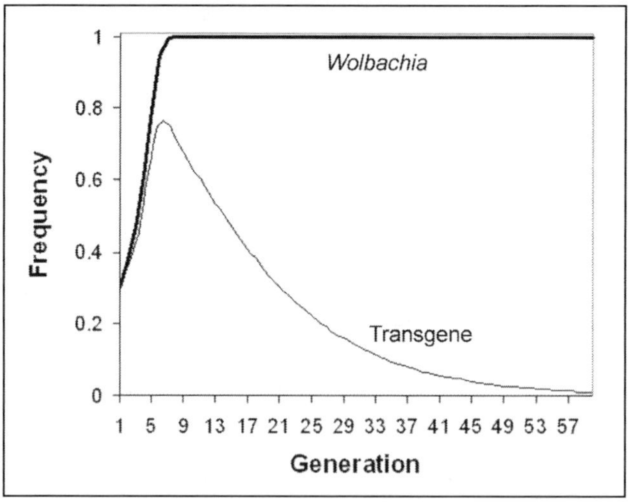

Figure 6. Breakdown of linkage between *Wolbachia* and an imperfectly transmitted maternally-inherited transgene (95%). *Wolbachia* spreads through the population, but the transgene is eliminated.

The model predicts that nuclear CI genes can invade populations, but not easily. The predicted efficiency of a nuclear-locus CI gene drive system is much lower than maternally-inherited CI gene drive, due to large reductions in transgene frequency caused by CI expression in the initial generations following a release. The number of transgenic insects that must be released (Introduction Threshold) and the time for the transgene to spread into the population are much greater for nuclear CI genes relative to maternally-inherited systems. The unstable equilibrium point is difficult to calculate analytically, but Turelli and Hoffmann showed that it can be approximated by the expression.[7]

$$p_u = 1 - \sqrt{0.5\left(1 - \frac{s_f}{s_h}\right)} \qquad (9)$$

Even under ideal conditions of no fitness cost ($F = 1.0$) and perfect CI ($H = 0$), the CI gene must be introduced at an introduction threshold greater than 0.36 (Fig. 7). High introduction levels may make this particular strategy unfeasible for many vector systems.

Turelli and Hoffmann further examined this scenario by relaxing the dominance assumption.[7] When nuclear CI genes act in a recessive manner, the unstable equilibrium point can be approximated by

$$p_u = \sqrt{0.5\left(1 + \frac{s_f}{s_h}\right)} \qquad (10)$$

Even under ideal conditions ($F = 1.0$, $H = 0$), recessive nuclear CI genes are not predicted to spread unless introduced at exceptionally high rates (>0.7), suggesting that if even if traits of this nature were successfully introduced at high levels and established in a local population, they would tend not to spread to other populations.

Sinkins and Godfray have outlined an alternative strategy that may get around this problem. They show theoretically that a nuclear "rescue" gene capable of restoring fertilization in an incompatible cross is predicted to spread in populations already infected with *Wolbachia*, or in populations where the infection is currently spreading, as long as *Wolbachia* is either imper-

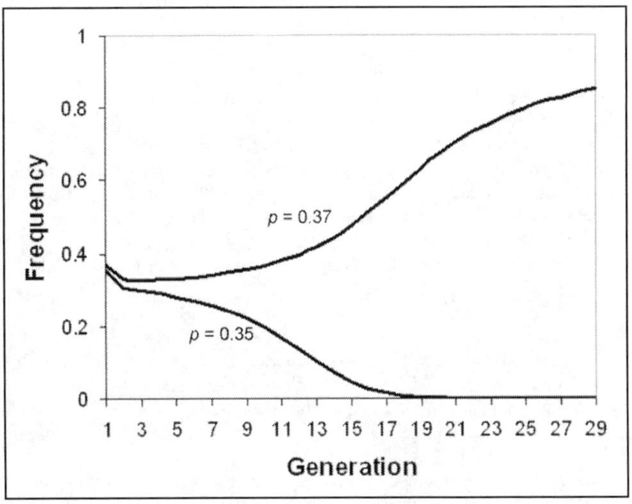

Figure 7. Dynamics of dominant nuclear CI genes, where $H = 0$ and $F = 1$. Genes invade the population if the frequency exceeds an unstable equilibrium of 0.36. Genes are lost below this frequency.

fectly transmitted or restores compatibility less efficiently than the nuclear rescue gene.[32] A gene of interest that is linked to the rescue gene will in theory be driven into the population at the same time. If *Wolbachia* is transmitted perfectly, completely restores fertility in a compatible cross and exhibits no fitness effects, the nuclear rescue gene will not increase deterministically in frequency and will behave as a neutral trait, or be lost if it imposes a fitness cost of its own on the mosquitoes. Imperfect transmission or CI is often observed in natural infections, so this caveat may not be limiting in practice.[8,13]

Interestingly, Sinkins and colleagues have identified nuclear factors in *Culex* mosquitoes that seem to restore compatibility between mosquitoes infected with putatively "incompatible" *Wolbachia* strains.[33] Similar nuclear factors capable of modulating CI expression have also been observed in other taxa.[34-37] Since we already have the ability to manipulate mosquito nuclear genomes using standard transgenesis technology,[5-6] the identification of nuclear rescue genes may offer a viable method to drive traits into populations in the absence of a *Wolbachia* transformation system.

Conclusions

The development of models to investigate the outcome of releasing transgenic insects for disease control are an absolutely critical first step before any actual release is ever considered. These modeling efforts must, whenever possible, be coupled with data from field populations. Ideally, the strategy (or strategies) that are most likely to result in a positive outcome will be identified and potential problems recognized by theoretical analyses prior to release. As models (and their underlying computer hardware and software) grow more sophisticated, we will ultimately be able to develop efficacious and cost-effective vector-borne disease control strategies based on transgenesis technology.

References

1. Hemingway J, Ranson H. Insecticide resistance in insect vectors of human disease. Annu Rev Entomol 2000; 45:371-391.
2. Shiff CJ. Can roll back malaria achieve its goal? A challenge. Parasitol Today 2000; 16:271-272.
3. Talisuna AO, Bloland P, Alessandro U. History, dynamics, and public health importance of malaria parasite resistance. Clin Microbiol 2004; 17:235-254.
4. Townson H, Nathan MB, Zaim M et al. Exploiting the potential of vector control for disease prevention. Bull World Health Organ 2005; 83:942-7.
5. Beaty BJ. Genetic manipulation of vectors: A potential novel approach for control of vector-borne diseases. Proc Natl Acad Sci USA 2000; 97:10295-10297.
6. James AA. Gene drive systems in mosquitoes: Rules of the road. Trends Parasitol 2005; 21:64-67.
7. Turelli M, Hoffmann AA. Microbe-induced cytoplasmic incompatibility as a mechanism for introducing transgenes into arthropod populations. Insect Mol Biol 1999; 8:243-255.
8. Turelli M, Hoffmann AA. Cytoplasmic incompatibility in Drosophila simulans: Dynamics and parameter estimates from natural populations. Genetics 1995; 140:1319-1338.
9. Gould F, Schliekelman P. Population genetics of autocidal control and strain replacement. Annu Rev Entomol 2004; 49:193-217.
10. Stouthamer R, Breeuwer JA, Hurst GD. Wolbachia pipientis: Microbial manipulator of arthropod reproduction. Annu Rev Microbiol 1999; 5371-102.
11. Hoffmann AA, Clancy DJ, Merton E. Cytoplasmic incompatibility in Australian populations of Drosophila melanogaster. Genetics 1994; 136:993-999.
12. Hoffmann AA, Hercus M, Dagher H. Population dynamics of the Wolbachia infection causing cytoplasmic incompatibility in Drosophila melanogaster. Genetics 1998; 148:221-231.
13. Rasgon JL, Scott TW. Wolbachia and cytoplasmic incompatibility in the California Culex pipiens mosquito species complex: Parameter estimates and infection dynamics in natural populations. Genetics 2003; 165:2029-2038.
14. Rasgon JL, Scott TW. An initial survey for Wolbachia (Rickettsiales: Rickettsiaceae) infections in selected California mosquitoes (Diptera: Culicidae). J Med Entomol 2004; 41:255-257.
15. Cornel A, McAbee R, Rasgon J et al. Differences in extent of genetic introgression between sympatric Cx. pipiens and Cx. quinquefasciatus in California and South Africa. J Med Entomol 2003; 40:125-132.

16. Rasgon JL, Scott TW. Impact of population age structure on Wolbachia transgene driver efficacy: Ecologically complex factors and release of genetically-modified mosquitoes. Insect Biochem Mol Biol 2004; 34:707-713.
17. Garrett-Jones C. The human blood index of malaria vectors in relations to epidemiological assessment. Bull World Health Organ 1964; 30:241-261.
18. Powers AM, Kamrud KI, Olson KE et al. Molecularly engineered resistance to California serogroup virus replication in mosquito cells and mosquitoes. Proc Natl Acad Sci USA 1996; 93:4187-4191.
19. Higgs S, Rayner JO, Olson KE et al. Engineered resistance in insect vectors of human disease. Am J Trop Med Hyg 1998; 58:663-670.
20. Ito J, Ghosh A, Moreira LA et al. Transgenic anopheline mosquitoes impaired in transmission of a malaria parasite. Nature 2002; 417:452-455.
21. Boete C, Koella JC. A theoretical approach to predicting the success of genetic manipulation of malaria mosquitoes in malaria control. Malar J 2002; 1:1-7.
22. Miller BR, Monath TP, Tabachnick WJ et al. Epidemic yellow fever caused by an incompetent mosquito vector. Trop Med Parasitol 1989; 40:396-399.
23. Walker ED, Torres EP, Villanueva RT. Components of the vectorial capacity of Aedes poicilius for Wuchereria bancrofti in Sorsogon province, Philippines. Ann Trop Med Parasitol 1998; 92:603-614.
24. Mellor PS, Boorman J, Baylis M. Culicoides biting midges: Their role as arbovirus vectors. Annu Rev Entomol 2000; 45:307-340.
25. Min KT, Benzer S. Wolbachia, normally a symbiont of Drosophila, can be virulent, causing degeneration and early death. Proc Natl Acad Sci USA 1997; 94:10792-10796.
26. Sinkins SP, O'Neill SL. Wolbachia as a vehicle to modify insect populations. In: Handler AM, James AA, eds. Insect Transgenesis: Methods and Applications. New York: CRC Press, 2000:271-287.
27. Rasgon JL, Styer LM, Scott TW. Wolbachia-induced mortality as a mechanism to modulate pathogen transmission by vector arthropods. J Med Entomol 2003; 40:125-132.
28. Brownstein JS, Hett E, O'Neill SL. The potential of virulent Wolbachia to modulate disease transmission by insects. J Invertebr Pathol 2003; 84:24-29.
29. Fine PEM. On the dynamics of symbiote-dependent cytoplasmic incompatibility in Culicine mosquitoes. J Invertebr Pathol 1978; 30:10-18.
30. Sinkins SP. Wolbachia and cytoplasmic incompatibility in mosquitoes. Insect Biochem Mol Biol 2004; 34:723-729.
31. Sinkins SP, Curtis CF, O'Neill SL. The potential application of inherited symbiont systems to pest control. In: O'Neill SL, Hoffmann AA, Werren JH, eds. Influential Passengers. Oxford: Oxforn University Press, 1997:155-175.
32. Sinkins SP, Godfray HC. Use of Wolbachia to drive nuclear transgenes through insect populations. Proc Biol Sci 2004; 271:1421-1426.
33. Sinkins SP, Walker T, Lynd AR et al. Wolbachia variability and host effects on crossing type in Culex mosquitoes. Nature 2005; 436:257-260.
34. Poinsot D, Bourtzis K, Markakis G et al. Wolbachia transfer from Drosophila melanogaster into D. simulans: Host effect and cytoplasmic incompatibility relationships. Genetics 1998; 150:227-237.
35. Bordenstein SR, Werren JH. Effects of A and B Wolbachia and host genotype on interspecies cytoplasmic incompatibility in Nasonia. Genetics 1998; 148:1833-1844.
36. Sasaki T, Ishikawa I. Transinfection of Wolbachia in the mediterranean flour moth, Ephestia kuehniella, by embryonic microinjection. Heredity 2000; 85:130-135.
37. Sakamoto H, Ishikawa Y, Sasaki T et al. Transinfection reveals the crucial importance of Wolbachia genotypes in determining the type of reproductive alteration in the host. Genet Res 2005; 85:205-210.

Chapter 11

Modifying Insect Population Age Structure to Control Vector-Borne Disease

Peter E. Cook, Conor J. McMeniman and Scott L. O'Neill*

Abstract

Age is a critical determinant of the ability of most arthropod vectors to transmit a range of human pathogens. This is due to the fact that most pathogens require a period of extrinsic incubation in the arthropod host before pathogen transmission can occur. This developmental period for the pathogen often comprises a significant proportion of the expected lifespan of the vector. As such, only a small proportion of the population that is oldest contributes to pathogen transmission. Given this, strategies that target vector age would be expected to obtain the most significant reductions in the capacity of a vector population to transmit disease. The recent identification of biological agents that shorten vector lifespan, such as *Wolbachia*, entomopathogenic fungi and densoviruses, offer new tools for the control of vector-borne diseases. Evaluation of the efficacy of these strategies under field conditions will be possible due to recent advances in insect age-grading techniques. Implementation of all of these strategies will require extensive field evaluation and consideration of the selective pressures that reductions in vector longevity may induce on both vector and pathogen.

Introduction

Most pathogens vectored by arthropods, such as *Plasmodium* and dengue virus, must undergo an extrinsic incubation period (EIP) in the vector, before they can be transmitted to a new host. During this time arborviruses and parasites like *Plasmodium*, penetrate the vector's midgut, replicate in various host tissues and infect the salivary glands prior to transmission during subsequent bloodfeeding. The duration of the EIP is variable and is heavily influenced by ambient temperature. For many pathogens the duration of the EIP consumes a significant proportion of the vector's lifespan. As a consequence, only a small fraction of the vector population that is oldest is of epidemiological importance.

The significance of vector age as a target for disease control was first realized by MacDonald through his quantifications of basic case reproductive rate for malaria.[1] Garrett-Jones[2,3] devised the concept of vectorial capacity by isolating the entomological components of MacDonald's[1] formulation. Essentially, it is a statement that expresses the intensity of transmission due to physiological, behavioral and ecological aspects of the vector population. While vectorial capacity increases linearly with factors such as vector competence and vector density, it increases exponentially with increases in mosquito longevity. Consequently, vector lifespan is the most sensitive determinant of a vector population's capacity for pathogen transmission. It is not surprising then, that very successful control strategies, such as indoor surface spraying of houses, effectively target older mosquitoes in the vector population.

*Corresponding Author: Scott L. O'Neill—School of Integrative Biology, The University of Queensland, Brisbane, Queensland, Australia. Email: scott.oneill@uq.edu.au

Transgenesis and the Management of Vector-Borne Disease, edited by Serap Aksoy.

Due to a number of complex social and biological factors, vector-borne diseases such as dengue fever are currently reemerging throughout the tropics.[4] With few exceptions, current control methods are not proving effective in stopping this alarming trend. In recent years a number of new approaches have been proposed that use biological agents to modify the age structure of mosquito populations. These include; (i) the use of virulent strains of the common bacterial endosymbiont *Wolbachia*,[5,6] (ii) entomopathogenic fungi[7] and (iii) densonucleosis viruses (densoviruses).[8] These biological agents induce mortality effects late in adult life, and as a result skew vector population age structure towards younger individuals, thereby potentially limiting pathogen transmission without eradicating the mosquito population. As these strategies target mosquito longevity, even minimal reductions in lifespan could yield significant reductions in vectorial capacity.

Entomological Components of Pathogen Transmission

After a mosquito consumes a blood meal containing a pathogen, there is a period during which the pathogen replicates and disseminates in the mosquito's body before it can be transmitted. This period from pathogen ingestion to potential infectivity is termed the extrinsic incubation period and lasts between 10-14 days for dengue virus[9] and 9-14 days for malaria.[10] Consequently, a mosquito disease vector must live longer than the combined duration of the initial nonfeeding period (approx. 2 days) plus the EIP of the pathogen before contributing to disease transmission. This combined period is long relative to adult mosquito lifespan, therefore, the majority of disease transmission is due to the feeding activity of old individuals. Consequently, the vectorial capacity of an insect population is closely linked to its demography.[11]

Vectorial capacity is defined as the daily rate at which future human inoculations arise from a single infective case.[12] Mathematically presented, vectorial capacity (V) is:

$$V = \frac{ma^2 p^n b}{-\ln p}$$

where:

m = vector density in relation to host density

a = probability a vector feeds on a host in 1 day (= host preference x feeding frequency)

p = probability of vector surviving 1 day

n = duration of extrinsic incubation period (EIP) in days

b = vector competence

$1/(-\ln p)$ = duration of a vector's life, in days after surviving EIP.[13]

The probability of contact between vector and host depends on the vector population size (term m) and vector feeding preference and frequency (term a). Vector feeding behavior, specifically host preference and feeding frequency, combine to form term a in the vecterial capacity equation. Feeding behavior can have significant implications for epidemiological risk as it influences the probability of contact between vector and host. For example, female *Ae. aegypti* feed almost exclusively on humans[14] and take multiple blood meals during each gonotrophic cycle.[15,16] Consequently, *Ae. aegypti* is the most epidemiologically important dengue vector even though *Aedes albopictus* is a more competent vector for dengue viruses.

Vector competence (term b) refers to a vector's ability to acquire and transmit a pathogen. Extensive research efforts are currently being directed towards genetic manipulation of vector competence to block disease transmission.[17-23] Vector competence is a linear component of vectorial capacity and as such, theoretical predictions suggest that large reductions in vector competence would be required to generate significant reductions in disease transmission.[24]

Vectorial capacity is most responsive to reductions in daily survivorship (term p). This is a consequence of the duration of the EIP of a pathogen being long relative to vector life expectancy. The duration of the EIP of dengue has been estimated to be between 10 to 14 days from early research using human volunteers.[9] Long-lived vectors contribute most to pathogen transmission and small decreases in vector life expectancy can cause large reductions in transmission rates.[2,11] This observation has attracted researchers to the possibility of utilizing different biological agents to shift the age structure of vector populations as a way to reduce disease transmission.

Wolbachia pipientis

Wolbachia pipientis is an obligate intracellular bacteria that was first observed in the reproductive tissues of the mosquito *Culex pipiens*[25] and later described in the same insect by Hertig.[26] Since its original description, *Wolbachia* has been found in a diverse range of invertebrate taxa including insects, mites, spiders, terrestrial crustaceans and nearly all filarial nematode species. Several PCR-based surveys have indicated that *Wolbachia* chronically infects more than 20% of all insect species.[27,28]

Wolbachia are vertically inherited by transovarial transmission within host populations and only very rarely appear to move horizontally and infect new species.[29] They induce a number of intriguing reproductive abnormalities that account for their success. These include parthenogenesis,[30] feminization,[31] male-killing,[32] and cytoplasmic incompatibility (CI; the developmental arrest of insect embryos that result when females are mated to males that have a different infection status).[33] All these reproductive effects enhance *Wolbachia* transmission at the expense of the arthropod population that is not infected. This in turn results in the active spread of this endosymbiont into uninfected host populations even if it confers a fitness cost.[34]

The ability of *Wolbachia* to invade host populations using the most common of these phenotypes, CI, has been quantified and modeled in several arthropod species including *Drosophila simulans*[35,36] and *Ae. albopictus*.[37,38] Some strains of *Wolbachia* induce high CI, which leads to rapid population invasion. For instance, the *Wolbachia* strain *w*Ri has been shown to spread geographically at a rate of 100km/year into uninfected *D. simulans* populations through the action of CI.[35] While CI has been proposed as a major mechanism allowing certain strains of *Wolbachia* to invade and persist in host populations, mechanisms that underlie the spread of other *Wolbachia* strains are less well understood. Several *Wolbachia* strains, such as *w*Au from *D. simulans* and *w*Mel from *D. melanogaster*, have invaded natural populations even though they induce little to no CI under field conditions.[39] It is hypothesized that these strains may provide an as yet undetermined fitness benefit to their hosts.[34]

Life-Shortening *Wolbachia*

Recently, virulent *Wolbachia* strains that replicate to unusually high densities and shorten adult host lifespan have been reported.[40,41] For example, *Wolbachia* infection in the parasitoid wasp *Leptopilina heterotoma* induces a significant reduction in adult survival, locomotor activity, and fecundity.[40] Despite causing detrimental effects to their hosts, these life-shortening strains apparently persist because hosts are able to reproduce before death. Consequently the life-shortening strain spreads due to the reproductive fitness benefit that CI imparts to individuals carrying the infection.

To date, the best characterized and most virulent *Wolbachia* strain is *popcorn* (*w*MelPop) from *D. melanogaster*. Min and Benzer[41] discovered *w*MelPop during a screen for gene mutations that cause brain degeneration in *D. melanogaster*. They noticed that a particular X-chromosome deficiency strain; $Df(1)^{ct4b1}\ y^{1}/Binsn$ originally isolated by Hannah in 1947,[42] had drastically reduced lifespan compared with normal flies. The basis for this reduced lifespan was the presence of the *w*MelPop infection. Interestingly, *w*MelPop is present in low numbers during embryonic, larval and pupal stages, but in adults density increases in several tissues including the brain, retina and flight muscle. With progressing age, cells in these tissues become filled with bacteria causing overt host cell pathology. This effect culminates in early death of adults, with the lifespan of infected flies being reduced by about 50% relative to uninfected controls. At 29°C this equates to 100% mortality of infected flies by 14 days.

The degree of life-shortening induced by *w*MelPop is influenced by several factors, including temperature and host genetic background. Several reports have noted that the virulence of *w*MelPop increases at higher temperatures,[41,43,44] suggesting an active link between temperature and the life-shortening phenotype. Renyolds and coworkers[44] documented slight variations in *D. melanogaster* lifespan when *w*MelPop was placed over different genetic backgrounds at 25°C. In addition, they found no significant difference between the lifespan of infected and uninfected flies at 19°C.[44]

When transferred from *D. melanogaster* to *D. simulans*, *w*MelPop still caused life-shortening at comparable rates to those observed in *D.* melanogaster, indicating the over-replication of *Wolbachia* was a property of the microbe's genome.[43] Furthermore, when transferred to *D. simulans*, *w*MelPop induces strong CI and had little fecundity cost after adaptation to its new host.[43,45] The ability of *w*MelPop to induce strong CI and not induce large fecundity costs in a novel host indicates that this strain has the potential to be introduced into different insect species where it may induce significant life-shortening, yet potentially still invade uninfected populations.

One possible application of *w*MelPop would be to alter the age structure of *Ae. aegypti* populations to reduce dengue transmission.[6] By introducing life-shortening strains of inherited *w*MelPop into mosquito populations, old mosquitoes could be selectively eliminated, preferentially removing the segment of the vector population responsible for most disease transmission. Predictions from recent theoretical models indicate that this strategy could result in significant reductions (80-100%) in disease transmission.[5,46]

Experimental Transfer of *Wolbachia* into Disease Vectors

Wolbachia occurs naturally in several medically important species of sandflies,[47] tsetse flies,[48] and mosquitoes[49,50] including several vectors of dengue such as *Ae. albopictus*,[51] *Aedes polynesiensis*,[52] and *Aedes scutellaris*.[53] *Ae. albopictus* is naturally infected with two *Wolbachia* strains, *w*AlbA and *w*AlbB, and both of these strains are capable of inducing strong CI.[37,38,54-58] However, some of the major disease vectors are not naturally infected, including all anopheline mosquitoes sampled to date, and *Ae. aegypti*, the primary vector of dengue.[59-61] Transinfection of these species is a major research priority.

The first experiment specifically designed to move *Wolbachia* between two host species involved the transfer of *w*Ri from *D. simulans* to an uninfected strain of *D. melanogaster* by injecting infected cytoplasm between presyncitial blastoderm stage embryos of these species.[62] Since then, *Wolbachia* has been successfully transferred between several insect species[43,63-67] and crustaceans,[68,69] including transfer attempts between phylogenetically distant insect orders.[70,71] Transfer experiments that generate novel infection types have been used to answer questions relating to *Wolbachia*-host interactions,[43,45] as well as to develop population replacement and population suppression strategies for controlling harmful insect species.[72,73]

As *Wolbachia* are extremely fastidious microorganisms and cannot be maintained in cell-free media, methods typically used to transfer *Wolbachia* between hosts include the direct-transfer of *Wolbachia*-infected cytoplasm and transfer of embryo homogenate via embryonic microinjection.[62,63,70,74] With both techniques, *Wolbachia* is microinjected into the posterior end of early embryos, with the goal of infecting embryonic pole cells that will develop into germ tissues. More recently, successful transfer of *Wolbachia* has been achieved using adult microinjection where *Wolbachia* purified from hemolymph or embryo homogenate has been microinjected into the thorax or abdomen of adults. *Wolbachia* then proceeds to disseminate throughout the body and infect the ovaries.[75] This technique has been successful in establishing *Wolbachia* infections in arthropods where embryonic microinjection techniques are not well developed including isopods,[69,76] parasitoid wasps,[65] planthoppers,[77] and mosquitoes.[78]

A complicating factor associated with transferring *Wolbachia* between species is the stable maintenance of the new association. In some cases, transferred strains are extremely stable and maternally inherited at very high rates.[67] In other cases the new infection appears poorly adapted to its new host and shows variable degrees of maternal transmission efficiency.[71,73,79] Several advances in the area of *Wolbachia* transfer and culture over the past few years may allow researchers to overcome these barriers. The ability to maintain *Wolbachia* in vitro in cell culture systems,[80-83] and the availability of cell lines from many recipient target species provides an ideal system to preadapt *Wolbachia* strains to the intracellular environment of novel host species, prior to transfer via embryonic or adult microinjection. It is possible that this preadaptation strategy may lead to higher initial levels of maternal transmission efficiency, than by directly transferring *Wolbachia* into a novel intracellular environment. An additional advantage of the use of cell culture systems as a source of *Wolbachia* for microinjection, is the ability to obtain large quantities of *Wolbachia* to optimize transinfection protocols and infective doses. Further, the suspension of purified *Wolbachia* in injection buffers that maintain viability and increase transinfection efficiency such as sucrose-phosphate glutamate buffer,[74] may allow researchers to establish infections in species previously thought refractory to *Wolbachia* infection.

Although *Ae. aegypti* does not naturally harbor *Wolbachia* in nature, the successful transfer of *w*AlbB from *Ae. albopictus* to *Ae. aegypti* using cytoplasm transfers between presyncitial blastoderm stage embryos indicates this species is not refractory to *Wolbachia* infection.[67] Further, *Ae. aegypti* has also been transinfected with a *w*AlbA and *w*AlbB double infection from *Ae. albopictus* using adult microinjections.[78] Following on from this initial success, *w*MelPop has recently been transferred to *Ae. aegypti* via embryonic microinjection. (Conor McMeniman, pers. comm.). The method used to establish this infection differs from all *Wolbachia* transfers to date in that it used *w*MelPop passaged for several years in the *Ae. aegypti* cell line RML-12, as a source of bacteria for microinjection. Several previous attempts in our laboratory to directly transfer *w*MelPop between *D. melanogaster* and *Ae. aegypti* using cytoplasm transfers and adult microinjections were unsuccessful, with infection being lost after one or two generations after establishment of the infection due to poor maternal transmission efficiency. In an attempt to preadapt *w*MelPop to the *Ae. aegypti* intracellular environment, to maximize the possibility that the new association would be stable, we established an in vitro infection in RML-12 before transfer to *Ae. aegypti*. Initial data indicates that maternal transmission efficiency of *w*MelPop in several newly generated lines is high, with the effect of *w*MelPop on parameters such as life-shortening and cytoplasmic incompatibility still to be determined. Importantly, pilot transfers of *w*MelPop from the RML-12 cell line back into its native host *D. melanogaster* indicate *w*MelPop has not attenuated in virulence during in vitro maintenance. (Conor McMeniman, pers. comm.).

Temperature and the Impact of Life-Shortening *Wolbachia*

As *w*MelPop virulence is temperature-dependent, the range of environmental temperatures encountered by *Ae. aegypti* is important for understanding the efficacy of this approach. In the field, temperature fluctuates on both a daily and seasonal basis. During summer when temperatures increase, the effect of *w*MelPop on *Ae. aegypti* longevity is expected to be most pronounced. Significantly, summer also corresponds to the period of peak dengue transmission in many regions.[84,85] During periods of sustained cooler temperatures, when dengue transmission risk is lowest, virulence of *w*MelPop in *Ae. aegypti* populations would also be predicted to decrease. Importantly, temperature will also influence the infection dynamics of *w*MelPop in *Ae. aegypti* populations. It is well documented that a trade-off exists between fitness costs generated by *Wolbachia*, and the unstable equilibrium point that must be exceeded for *Wolbachia* to actively spread into a population under the action of CI.[34] It would be expected that this point would be lower in cooler months and during this season the infection would spread more readily.

Layered upon the effect of seasonal changes in temperature on *w*MelPop in *Ae. aegypti*, are the daily thermal preferences of *Ae. aegypti* itself. Although little is known about the thermal ecology of mosquitoes, preference of *Ae. aegypti* for microhabitats such as vegetated and dark sheltered areas are well documented.[86] Preference for different microhabitats have the potential to alter the range of temperatures mosquitoes experience during the day. As such, to fully evaluate the impact of such environmental heterogeneity on the efficacy of this approach contained semi-field trails will need to be conducted. Quantitative measures from these field cages coupled with modeling will allow researchers to make biologically relevant predictions about the potential efficacy of this strategy.

Molecular Basis of Life-Shortening in *w*MelPop

Identification of putative loci that modulate life-shortening in *w*MelPop would be useful from an applied perspective. These loci may involve point mutations, insertions or deletions that may have occurred in genes or promoter regions of *w*MelPop involved in regulation of bacterial cell division or density, causing it to over-replicate in a temperature sensitive manner. It may be desirable to genetically modulate virulence of *Wolbachia* strains that naturally infect (or can be transinfected) into disease vectors. Currently, transformation technologies are still being developed for *Wolbachia.* Encouragingly, homologous recombination-based techniques have been used successfully to introduce point mutations and foreign DNA into the *Wolbachia* chromosome. (Iñaki Iturbe-Ormaetxe, pers. comm.). This alternative strategy may be useful if attempts to transinfect disease vectors with *w*MelPop are unsuccessful, or cytoplasmic incompatibility and fecundity effects of *w*MelPop in these host species are less than optimal.

To determine the molecular basis of the life-shortening phenotype the genome of *w*MelPop has been mapped on a number of levels. Initial studies focused on comparisons of the genome organization of *w*MelPop to the closely related, yet nonvirulent strain *w*Mel. Using southern hybridization with probes designed from the recently sequenced *w*Mel genome,[87] Sun et al[88] arranged restriction fragments of the *w*MelPop chromosome to create a comparative map of the *w*MelPop genome. Results indicated the genome organization of *w*MelPop was identical to *w*Mel with the exception of a single 150 kb inversion of its chromosome. This region has been further characterized using a finer scale genetic map of the *w*MelPop genome. This map, created by PCR-tiling across the entire genome has revealed that break points flanking the inversion occurred in noncoding regions. (Markus Riegler, pers. comm.). Therefore, it is unlikely that the inversion has disrupted operons that in turn influence the life-shortening phenotype. Mapping has also indicated minor size differences between the genomes of *w*MelPop and *w*Mel. When PCR fragments, 2-10 kb in length, were resolved on agarose gels, two differential IS5 transposon insertion sites and two separate variable number tandem repeats (VNTRs) in the *w*MelPop genome were identified. These minor genomic differences between *w*Mel and *w*MelPop were considered potentially informative until it was discovered that *w*MelPop shares these differences, including the orientation of the 150 kb chromosomal inversion, with a previously undescribed avirulent *w*Mel strain, called *w*MelCS.[89]

As these two strains share identical genome arrangements it is likely that *w*MelPop is a substrain of *w*MelCS, rather than *w*Mel. The *D. melanogaster* laboratory stock from which *w*MelPop was identified, $Df(1)^{ct4b1}$, $y^1/Binsn$, was isolated in 1947.[42] From the available literature concerning the origin of this strain, it is not possible to know whether *w*MelPop arose from *w*MelCS during the mutagenesis experiment used to generate this fly strain, or beforehand. Most other *D. melanogaster* stocks established around this time were infected with *w*MelCS.[89] To create the X-chromosome deficiency line $Df(1)^{ct4b1}$, $y^1/Binsn$, males from ring chromosome stock X^{c2} were exposed to an X-ray radiation source and the line selected by crosses to females from an un-specified X-chromosome marker stock.[42,90] Unless *w*MelPop was paternally inherited from males used is this study, a circumstance very rare in nature,[36] it is most likely that *w*MelPop arose in females from the marker stock used in this study. As no stock name for this marker line was specified the circumstances surrounding the origin of *w*MelPop remain a mystery.

Currently, the genome of *w*MelPop is being sequenced in our laboratory. Once completed, the genome sequences of *w*Mel and *w*MelPop will be compared and possible point mutations in gene and promoter regions identified. These differences will then be verified by comparative

PCR in *w*MelCS, and possible candidate genes for life-shortening identified. These candidates will then be functionally tested using site-directed mutagenesis via homologous recombination to introduce mutations in *w*MelCS or *w*Mel genome to test these putative loci for their role in life-shortening.

Entomopathogenic Fungi

An alternative strategy that reduces vector lifespan using entomopathogenic fungi has recently been proposed. Reports have shown that several disease vectors, including tsetse flies and mosquitoes, are susceptible to a range of fungal entomopathogens.[91-96] Mosquitoes infected with fungal spores have increased mortality rates, with most individuals surviving to 10-14 days post infection. More recent studies have illustrated the potential of using commercially available oil-based formulations of these fungi to alter the daily probability of survival of anopheline mosquitoes for malaria control.[7,97]

In a laboratory-based setting, Blanford and coworkers[7] examined the survival and sporozoite burden of *Anopheles stephensi* exposed to an isolate of the fungus *Beauveria bassiana.* Results indicated that short periods of exposure of mosquitoes to cage mesh sprayed with oil-based formulations of *Beauveria* were sufficient to cause > 90% mortality by day 14 after contact (the approximate EIP for malaria). Importantly, exposure of mosquitoes infected with the rodent malaria *Plasmodium chabaudi* to surfaces sprayed with *Beauveria* spores reduced the transmission risk by a factor of 80.[7] At day 14 post-exposure, 31% of malaria-infected control mosquitoes were alive and able to transmit, compared with only 0.4% mosquitoes in the *Beauvaria* and malaria treatment. The potential of this approach for malaria control was further demonstrated by a field-based study in a Tanzanian village by Scholte and coworkers.[97] In their study, the ability of a different fungus, *Metarhizium anisopliae,* to infect and reduce the lifespan of wild *An. gambiae* mosquitoes after resting on fungus-treated cotton sheets suspended from the ceiling of huts was evaluated. Over the three weeks of the trial, 580 female *An. gambiae* were collected from treated huts and 132 of these mosquitoes were infected with the fungus (23% infection rate). Significantly, infected mosquito lifespan was reduced to 11 days post-contact. Using these results, a model to estimate the intensity of malaria transmission indicated that a 75% reduction in transmission could be achieved using this strategy. In addition, the authors noted that increasing the coverage of treated mosquito resting sites to infect 50% of mosquitoes would reduce transmission by 96%. Interestingly, further experiments to characterize the effect of *M. anisopliae* on *An. gambiae* have indicated that infection with this fungus reduces the propensity of females to blood-feed, as well as decreasing fecundity and total lifetime productivity of mosquitoes.[98]

The documented effects of entomopathogenic fungi reducing the probability of daily survival[7,97] and feeding frequency[98] of malaria vectors are extremely encouraging. Initial studies suggest that implementation of this method in an applied setting may cause large decreases in vectorial capacity in mosquito populations. However, like residual insecticides, the gradual loss of viability of entomopathogenic fungal spores on treated surfaces over time[97] will necessitate their repeated application for sustained disease control.

Densonucleosis Viruses

Mosquito densoviruses (Family *Parvoviridae*) are also being investigated as potential control agents for mosquito-borne diseases due to the reductions in adult mosquito lifespan they induce.[8] Densovirus infected mosquitoes show increased mortality during both immature and adult life stages.[99-102] The level of pathogenicity of mosquito densoviruses appears to be highly dependent on densovirus strain and infective viral dose.[101,103] Female mosquitoes infected with densovirus strains may also have reduced fecundity and egg viability.[103,104]

Mosquito densoviruses were initially isolated from *Ae. aegypti* colony material sourced from Southeast Asia.[8] In addition, densovirus strains have been isolated from mosquito colonies[105] lines,[106,107] and from natural populations.[100,108] Currently little is known about the maintenance of densoviruses in natural mosquito populations. It is believed that densovirus transmission occurs within larval habitats as infected larvae excrete virus particles in the water.[99,101] Vertical transmission of densoviruses has been reported, yet the mode of transmission is unknown.[100,103,104] Initial results from population cage studies suggest that vertical transmission will allow the virus to spread to new larval habitats. This may mitigate the need for repeated applications of densovirus formulations to larval habitats.

It is unclear to what extent densovirus-induced mortality will have on disease transmission. Modeling of the mortality effects of densovirus infection in *Ae. aegypti* predicts a reduction of approximately 76% of infectious mosquito days.[103] While preliminary, this result suggests that densovirus-based vector control strategies may lead to significant reductions in dengue transmission. However, extensive research is needed to develop mosquito densovirus formulations for widespread application to control mosquito populations.

Evaluating of the Efficacy of Strategies Targeting Vector Longevity

One challenge associated with the development of control strategies that reduce vector longevity relates to how the efficacy of these approaches will be evaluated under field conditions. Researchers will need to accurately monitor changes in age structure of mosquito populations with overlapping generations. As such, tools that provide accurate estimates of mosquito survival are needed. Various methods of varying utility exist that enable the estimation of mosquito longevity.

Daily survivorship can be assessed through horizontal life tables constructed by following a cohort of individuals. Mark-release-recapture techniques (MRR), reviewed by Service[109] and Hagler and Jackson,[110] are central to this approach. Periodic collections give an indication of the number of marked individuals surviving within the population. Reliable estimates of daily survival rate require numerous marked individuals per recapture, which requires large numbers of mosquitoes at the initial release. Consequently the approach is labor intensive and researchers undertaking MRR studies in disease endemic areas need to be mindful of contributing to epidemiological risk which may limit the application of this approach.[111,112]

Daily survival rates can alternatively be inferred directly from the age structure of a vector population. Russian scientists pioneered the first mosquito age-grading techniques during the 1950's (reviewed by Detinova[113]). Their work identified consistent changes in female ovarian anatomy with age. As a female mosquito ages and passes through successive gonotrophic cycles, permanent changes occur in the ovaries and associated structures. The simplest reproductive age-grading technique differentiates nulliparous (has not oviposited; <4 days of age) and parous (one or more ovipositions; >4 days of age) females on the basis of ovarian tracheations.[113] Tracheoles are small tubules of the insect respiratory system that form tightly wound structures (skeins) on the ovaries of nulliparous mosquitoes. In parous females the tracheoles have stretched to accommodate the growth of the ovary during the first gonotrophic cycle and form a tracheal net. The limitation of this method is that it only differentiates two age classes and this provides limited information of epidemiological importance.

An alternative reproductive age-grading method counts individual dilatations in the ovariolar pedicel that correspond directly to multiple gonotrophic cycles.[113] This allows more age classes to be determined, as dilatations result from distension and incomplete contraction of ovariolar walls with the passage of a mature egg through the ovariole.[113] This method requires delicate dissection and examination of multiple undamaged ovarioles to ensure an accurate estimate of parity.[114] Improved techniques of preparing ovarioles for examination have been developed,[115,116] however, the method remains technically challenging and time consuming, and its application to large-scale studies is limited. Further, ovariolar dilatations may not be appropriate for age-grading all mosquitoes of medical importance, such as *Ae. aegypti*.[114,117]

More recent approaches for age-grading insects have focused on age-related changes in the abundance of pteridines[118-120] and cuticular hydrocarbons.[121-124] Pteridines are fluoresent pigments that are synthesized in the fat body of insects and transported to various parts of the body where they accumulate. Pteridines have been used to determine chronological age in several Dipterans.[119,125] Application of this method to mosquito age-grading was initially promising as laboratory studies observed inverse linear trends between pteridine abundance and age in *An. gambiae* and *An. stephensi*.[120] This association was, however, not apparent in *Anopheles albiminus* collected from the field.[119] Additional studies of pteridine fluorescence in *Aedes* and *Culex* spp. showed pteridine levels were below the detection limit of standard spectrofluorometers.[118]

The abundance of cuticular hydrocarbons (CHC) has been used to age-grade *Ae. aegypti*. Relative changes in the abundance of specific CHCs from *Ae. aegypti* legs were measured using gas chromatography and showed a linear association with age up to 15 calendar days.[122] A MRR experiment undertaken to validate this age-grading method under field conditions in Thailand demonstrated seasonal variation in CHC abundance.[124] Application of CHC profiles for age-grading several other mosquito species has shown promise.[121,126,127]

The most recent advance in mosquito age-grading has been the development of a transcriptional age-grading technique for *Ae. aegypti*.[128] This method quantifies age-related transcriptional changes of eight genes using quantitative reverse transcriptase PCR (qRT-PCR). Age predictions are derived for adult female *Ae. aegypti* using calibration data, transcriptional profiles quantified from mosquitoes reared to known ages under field conditions.[128] Field studies used to validate this method estimated the age of individual mosquitoes using both transcriptional and CHC age-grading methods. Mosquito legs were used for CHC analysis, while transcript abundances were quantified from the head and thorax of the same mosquito. This study demonstrated that transcriptional age-grading gave more accurate age estimates than those derived from CHC analysis.[128]

Cook et al[128] presented additional results showing age predictions derived from the three most informative genes. The subsequent calibration model resulted in a slight overestimation of age across most age classes, however this bias may be acceptable given the considerable simplification of the assay. It is expected that identifying additional genes of interest that are transcriptionally active in later age classes will remove this bias and increase the accuracy of age predictions in individuals older that 13 days of age. Preliminary findings suggest that appropriate candidates are available. Field-based evaluations of the transcriptional age-grading technique in different geographical locations and across different seasons are currently being undertaken.

Research is currently underway to transfer the transcriptional age-grading technique into other mosquito species. Transcriptional studies using *An. gambiae*[129] and *Drosophila*[130,131] microarrays have shown similar age-related trends in orthologs in two of the most informative genes used for *Ae. aegypti* age determination.[128] This may indicate the potential of the transcriptional age-grading method to be broadly applicable to other medically important Diptera.

Evolutionary Consequences of Strategies That Reduce Vector Longevity

Virulent *Wolbachia*, entomopathogenic fungi and densoviruses offer new and potentially effective vector control strategies that target mosquito lifespan. Given the rapid and widespread evolution of insecticide resistance, it would seem prudent to consider potential selection pressures that any of these life-shortening strategies may impose on both the pathogen and vector populations.

One concern is the possible development of resistance to these life-shortening agents by the mosquito vectors. At the present time little data is available to indicate how quickly mosquitoes might develop resistance. In the case of life-shortening *Wolbachia*, no signs of resistance have emerged in laboratory cultures of *D. melanogaster* since its initial description in 1997. It has also been suggested that mosquitoes may evolve mechanisms that limit the ability of entomopathogenic fungi to penetrate through the cuticle or replicate within the host.[132] However, to date there has been no reported cases of resistance to fungi used in insect pest control.

This is thought to be in part because fungi use several different effector molecules to attack the insect such as chitinases, proteases and toxins.[133]

Alternatively, reductions to vector lifespan may select for pathogens with faster development rates and shorter extrinsic incubation periods. Due to the limited lifespan of most disease vectors, pathogens must rapidly infect, proliferate and/or disseminate in the vector to ensure their transmission. Given this, it would be expected that most vector-borne pathogens would be under significant evolutionary pressure to maximize their ability to infect and proliferate within the vector or alternatively, adopt strategies that would increase vector survival.[134] This acceleration in development rate, within the vector, is not readily observed in natural vector-borne pathogen systems. As such, it likely that the duration of a pathogen's EIP is evolutionary constrained. Little is known about factors that may cause this limitation, however it would seem plausible that selective pressures imposed by the vector and vertebrate host may act in opposing directions. Research undertaken using murine malaria has demonstrated that higher parasite virulence can be selected relatively quickly when parasites are sequentially passaged through just a mammalian host.[135] However, selection is much less effective when parasites are alternated between mammalian and insect hosts.

Conclusion

Vector age is one of the most sensitive parameters influencing the epidemiology of vector-borne disease. Strategies that aim to reduce adult mosquito lifespan are expected to be very effective in reducing pathogen transmission. This is because vectorial capacity is most sensitive to changes in vector longevity. Entomopathogenic fungi, virulent *Wolbachia* strains and mosquito densoviruses all show the potential to modify insect vector population age structure to such a degree that significant reductions in disease transmission may result. Indeed, using these agents synergistically as part of an integrated approach reducing mosquito population age structure may yield the most effective results.

Acknowledgements

We would like to thank Elizabeth McGraw, Jeremy Brownlie, Iñaki Iturbe-Ormaetxe and Leon Hugo for their comments on earlier versions of this chapter. This work was supported by Australian Research Council Grant LP0455732 and a grant from the Foundation for the National Institutes of Health through the Grand Challenges in Global Health Initiative.

References

1. Macdonald G. The epidemiology and control of malaria. London: Oxford University Press, 1957.
2. Garrett-Jones C. Prognosis for interruption of malaria transmission through assessment of the mosquito's vectorial capacity. Nature 1964; 204:1173-5.
3. Garrett-Jones C. The human blood index of malaria vectors in relation to epidemiological assessment. Bull World Health Organ 1964; 30:241-61.
4. Gratz NG. Emerging and resurging vector-borne diseases. Annu Rev Entomol 1999; 44:51-75.
5. Brownstein JS, Hett E, O'Neill SL. The potential of virulent Wolbachia to modulate disease transmission by insects. J Invertebr Pathol 2003; 84(1):24-9.
6. Sinkins SP, O'Neill SL. Wolbachia as a vehicle to modify insect populations. In: Handler AM, James AA, eds. Insect Transgenesis: Methods and Applications. London: CRC Press, 2000:271-87.
7. Blanford S, Chan BH, Jenkins N et al. Fungal pathogen reduces potential for malaria transmission. Science 2005; 308(5728):1638-41.
8. Carlson J, Suchman E, Buchatsky L. Densoviruses for control and genetic manipulation of mosquitoes. Adv Virus Res 2006; 68:361-92.
9. Siler JF, Hall MW, Hitchens AP. Dengue: Its history, epidemiology, mechanism of transmission, etiology, clinical manifestations, immunity and prevention. Philipp J Sci 1926; 29(1-2):1-304.
10. Gilles HM, Warrell DA. Essential malariology. 4th ed. London: Arnold, 2002.
11. Dye C. The analysis of parasite transmission by bloodsucking insects. Annu Rev Entomol 1992; 37:1-19.
12. Garrett-Jones C, Grab B. Assessment of insecticidal impact on malaria mosquito's vectorial capacity from data on proportion of parous females. Bull World Health Organ 1964; 31(1):71-86.

13. Black IVth WC, Moore CG. Population biology as a tool for studying vector-borne diseases. In: Marquardt WC, ed. Biology of Disease Vectors. 2nd ed. Boston: Elsevier Academic Press, 2005:187-206.
14. Edman JD, Strickman D, Kittayapong P et al. Female Aedes aegypti (Diptera: Culicidae) in Thailand rarely feed on sugar. J Med Entomol 1992; 29(6):1035-8.
15. Scott TW, Chow E, Strickman D et al. Blood-feeding patterns of Aedes aegypti (Diptera: Culicidae) collected in a rural Thai village. J Med Entomol 1993; 30(5):922-7.
16. Scott TW, Naksathit A, Day JF et al. A fitness advantage for Aedes aegypti and the viruses it transmits when females feed only on human blood. Am J Trop Med Hyg 1997; 57(2):235-9.
17. Olson KE, Adelman ZN, Travanty EA et al. Developing arbovirus resistance in mosquitoes. Insect Biochem Mol Biol 2002; 32(10):1333-43.
18. Alphey L, Beard CB, Billingsley P et al. Malaria control with genetically manipulated insect vectors. Science 2002; 298(5591):119-21.
19. Kokoza V, Ahmed A, Cho WL et al. Engineering blood meal-activated systemic immunity in the yellow fever mosquito, Aedes aegypti. Proc Natl Acad Sci USA 2000; 97(16):9144-9.
20. Aultman KS, Beaty BJ, Walker ED. Genetically manipulated vectors of human disease: A practical overview. Trends Parasitol 2001; 17(11):507-9.
21. Land KM. Transgenic mosquitoes in controlling malaria transmission. Trends Parasitol 2002; 18(9):383.
22. Jacobs-Lorena M. Interrupting malaria transmission by genetic manipulation of anopheline mosquitoes. J Vector Borne Dis 2003; 40(3-4):73-7.
23. Travanty EA, Adelman ZN, Franz AW et al. Using RNA interference to develop dengue virus resistance in genetically modified Aedes aegypti. Insect Biochem Mol Biol 2004; 34(7):607-13.
24. Boete C, Koella JC. A theoretical approach to predicting the success of genetic manipulation of malaria mosquitoes in malaria control. Malar J 2002; 1(1):3.
25. Hertig M, Wolbach SB. Studies on rickettsia-like micro-organisms in insects. J Med Res 1924; 44:329-74.
26. Hertig M. The rickettsia, Wolbachia pipientis (gen. et sp. n.) and associated inclusions in the mosquito, Culex pipiens. Parasitology 1936; 28(4):453-86.
27. Werren JH, Windsor D, Guo LR. Distribution of Wolbachia among neotropical arthropods. Proc R Soc Lond B Biol Sci 1995; 262:197-204.
28. Jeyaprakash A, Hoy MA. Long PCR improves Wolbachia DNA amplification: Wsp sequences found in 76% of sixty-three arthropod species. Insect Mol Biol 2000; 9(4):393-405.
29. Heath BD, Butcher RD, Whitfield WG et al. Horizontal transfer of Wolbachia between phylogenetically distant insect species by a naturally occurring mechanism. Curr Biol 1999; 9(6):313-6.
30. Stouthamer R, Breeuwer JAJ, Luck RF et al. Molecular identification of microorganisms associated with parthenogenesis. Nature 1993; 361:66-8.
31. Rousset F, Bouchon D, Pintureau B et al. Wolbachia endosymbionts responsible for various alterations of sexuality of arthropods. Proc R Soc Lond B Biol Sci 1992; 250:91-8.
32. Hurst GD, Jiggins FM, Graf von der Schulenberg JH et al. Male killing Wolbachia in two species of insects. Proc R Soc Lond B Biol Sci 1999; 266:735-40.
33. O'Neill SL, Giordano R, Colbert AM et al. 16S rRNA phylogenetic analysis of the bacterial endosymbionts associated with cytoplasmic incompatibility in insects. Proc Natl Acad Sci USA 1992; 89(7):2699-702.
34. Hoffmann AA, Turelli M. Cytoplasmic incompatibility in insects. In: O'Neill SL, Hoffmann AA, Werren JH, eds. Influential Passengers: Inherited Microorganisms and Arthropod Reproduction. Oxford: Oxford University Press, 1997:42-80.
35. Turelli M, Hoffmann AA. Rapid spread of an inherited incompatibility factor in California Drosophila. Nature 1991; 353(6343):440-2.
36. Turelli M, Hoffmann AA. Cytoplasmic incompatibility in Drosophila simulans: Dynamics and parameter estimates from natural populations. Genetics 1995; 140(4):1319-38.
37. Dobson SL, Marsland EJ, Rattanadechakul W. Mutualistic Wolbachia infection in Aedes albopictus: Accelerating cytoplasmic drive. Genetics 2002; 160(3):1087-94.
38. Dobson SL, Rattanadechakul W, Marsland EJ. Fitness advantage and cytoplasmic incompatibility in Wolbachia single- and superinfected Aedes albopictus. Heredity 2004; 93(2):135-42.
39. Hoffmann AA, Clancy D, Duncan J. Naturally-occurring Wolbachia infection in Drosophila simulans that does not cause cytoplasmic incompatibility. Heredity 1996; 76:1-8.
40. Fleury F, Vavre F, Ris N et al. Physiological cost induced by the maternally-transmitted endosymbiont Wolbachia in the Drosophila parasitoid Leptopilina heterotoma. Parasitology 2000; 121(5):493-500.

41. Min KT, Benzer S. Wolbachia, normally a symbiont of Drosophila, can be virulent, causing degeneration and early death. Proc Natl Acad Sci USA 1997; 94(20):10792-6.
42. Hannah AM. Radiation-mutations involving the cut-locus in Drosophila. Proc 8th Intl Congr Genet 1948 Hereditas (Lund) 1949; 588-9.
43. McGraw EA, Merritt DJ, Droller JN et al. Wolbachia density and virulence attenuation after transfer into a novel host. Proc Natl Acad Sci USA 2002; 99(5):2918-23.
44. Reynolds KT, Thomson LJ, Hoffmann AA. The effects of host age, host nuclear background and temperature on phenotypic effects of the virulent Wolbachia strain popcorn in Drosophila melanogaster. Genetics 2003; 164(3):1027-34.
45. McGraw EA, Merritt DJ, Droller JN et al. Wolbachia-mediated sperm modification is dependent on the host genotype in Drosophila. Proc R Soc Lond B Biol Sci 2001; 268(1485):2565-70.
46. Rasgon JL, Styer LM, Scott TW. Wolbachia-induced mortality as a mechanism to modulate pathogen transmission by vector arthropods. J Med Entomol 2003; 40(2):125-32.
47. Ono M, Braig HR, Munstermann LE et al. Wolbachia infections of Phlebotomine sand flies (Diptera: Psychodidae). J Med Entomol 2001; 38(2):237-41.
48. Cheng Q, Ruel TD, Zhou W et al. Tissue distribution and prevalence of Wolbachia infections in tsetse flies, Glossina spp. Med Vet Entomol 2000; 14:44-50.
49. Ruang-Areerate T, Kittayapong P, Baimai V et al. Molecular phylogeny of Wolbachia endosymbionts in Southeast Asian mosquitoes (Diptera: Culicidae) based on wsp gene sequences. J Med Entomol 2003; 40(1):1-5.
50. Sinkins SP. Wolbachia and cytoplasmic incompatibility in mosquitoes. Insect Biochem Mol Biol 2004; 34(7):723-9.
51. Kittayapong P, Baimai V, O'Neill SL. Field prevalence of Wolbachia in the mosquito vector Aedes albopictus. Am J Trop Med Hyg 2002; 66(1):108-11.
52. Dean JL, Dobson SL. Characterization of Wolbachia infections and interspecific crosses of Aedes (Stegomyia) polynesiensis and Ae. (Stegomyia) riversi (Diptera: Culicidae). J Med Entomol 2004; 41(5):894-900.
53. Behbahani A, Dutton TJ, Davies N et al. Population differentiation and Wolbachia phylogeny in mosquitoes of the Aedes scutellaris group. Med Vet Entomol 2005; 19(1):66-71.
54. Dobson SL, Marsland EJ, Rattanadechakul W. Wolbachia-induced cytoplasmic incompatibility in single- and superinfected Aedes albopictus (Diptera: Culicidae). J Med Entomol 2001; 38(3):382-7.
55. Sinkins SP, Braig HR, O'Neill SL. Wolbachia pipientis: Bacterial density and unidirectional cytoplasmic incompatibility between infected populations of Aedes albopictus. Exp Parasitol 1995; 81(3):284-91.
56. Sinkins SP, Braig HR, O'Neill SL. Wolbachia superinfections and the expression of cytoplasmic incompatibility. Proc R Soc Lond B Biol Sci 1995; 261(1362):325-30.
57. Kittayapong P, Baisley KJ, Sharpe RG et al. Maternal transmission efficiency of Wolbachia superinfections in Aedes albopictus populations in Thailand. Am J Trop Med Hyg 2002; 66(1):103-7.
58. Kittayapong P, Mongkalangoon P, Baimai V et al. Host age effect and expression of cytoplasmic incompatibility in field populations of Wolbachia-superinfected Aedes albopictus. Heredity 2002; 88(4):270-4.
59. Kittayapong P, Baisley KJ, Baimai V et al. Distribution and diversity of Wolbachia infections in Southeast Asian mosquitoes (Diptera: Culicidae). J Med Entomol 2000; 37(3):340-5.
60. Rasgon JL, Scott TW. An initial survey for Wolbachia (Rickettsiales: Rickettsiaceae) infections in selected California mosquitoes (Diptera: Culicidae). J Med Entomol 2004; 41(2):255-7.
61. Ricci I, Cancrini G, Gabrielli S et al. Searching for Wolbachia (Rickettsiales: Rickettsiaceae) in mosquitoes (Diptera: Culicidae): Large polymerase chain reaction survey and new identifications. J Med Entomol 2002; 39(4):562-7.
62. Boyle L, O'Neill SL, Robertson HM et al. Interspecific and intraspecific horizontal transfer of Wolbachia in Drosophila. Science 1993; 260(5115):1796-9.
63. Chang NW, Wade MJ. The transfer of Wolbachia pipientis and reproductive incompatibility between infected and uninfected strains of the flour beetle, Tribolium confusum, by microinjection. Can J Microbiol 1994; 40:978-81.
64. Clancy DJ, Hoffmann AA. Behavior of Wolbachia endosymbionts from Drosophila simulans in Drosophila serrata, a novel host. Am Nat 1997; 149(5):975-88.
65. Grenier S, Pintureau B, Heddi A et al. Successful horizontal transfer of Wolbachia symbionts between Trichogramma wasps. Proc R Soc Lond B 1998; 265:1441-5.
66. Sasaki T, Kubo T, Ishikawa H. Interspecific transfer of Wolbachia between two lepidopteran insects expressing cytoplasmic incompatibility: A Wolbachia variant naturally infecting Cadra cautella causes male killing in Ephestia kuehniella. Genetics 2002; 162(3):1313-9.

67. Xi Z, Khoo CCH, Dobson SL. Wolbachia establishment and invasion in an Aedes aegypti laboratory population. Science 2005; 310:326-8.
68. Rigaud T, Juchault P. Success and failure of horizontal transfers of feminizing Wolbachia endosymbionts in woodlice. J Evol Biol 1995; 8:249-55.
69. Rigaud T, Pennings PS, Juchault P. Wolbachia bacteria effects after experimental interspecific transfers in terrestrial isopods. J Invertebr Pathol 2001; 77(4):251-7.
70. Braig HR, Guzman H, Tesh RB et al. Replacement of the natural Wolbachia symbiont of Drosophila simulans with a mosquito counterpart. Nature 1994; 367(6462):453-5.
71. Van Meer MM, Stouthamer R. Cross-order transfer of Wolbachia from Muscidifurax uniraptor (Hymenoptera: Pteromalidae) to Drosophila simulans (Diptera: Drosophilidae). Heredity 1999; 82(2):163-9.
72. Xi Z, Dean JL, Khoo C et al. Generation of a novel Wolbachia infection in Aedes albopictus (Asian tiger mosquito) via embryonic microinjection. Insect Biochem Mol Biol 2005; 35(8):903-10.
73. Zabalou S, Riegler M, Theodorakopoulou M et al. Wolbachia-induced cytoplasmic incompatibility as a means for insect pest population control. Proc Natl Acad Sci USA 2004; 101(42):15042-5.
74. Xi Z, Dobson SL. Characterization of Wolbachia transfection efficiency by using microinjection of embryonic cytoplasm and embryo homogenate. Appl Environ Microbiol 2005; 71(6):3199-204.
75. Frydman HM, Li JM, Robson DN et al. Somatic stem cell niche tropism in Wolbachia. Nature 2006; 441(7092):509-12.
76. Bouchon D, Rigaud T, Juchault P. Evidence for widespread Wolbachia infection in isopod crustaceans: Molecular identification and host feminization. Proc R Soc Lond B Biol Sci 1998; 265(1401):1081-90.
77. Kang L, Ma X, Cai L et al. Superinfection of Laodelphax striatellus with Wolbachia from Drosophila simulans. Heredity 2003; 90(1):71-6.
78. Ruang-Areerate T, Kittayapong P. Wolbachia transinfection in Aedes aegypti: A potential gene driver of dengue vectors. Proc Natl Acad Sci USA 2006; 103(33):12534-9.
79. Riegler M, Charlat S, Stauffer C et al. Wolbachia transfer from Rhagoletis cerasi to Drosophila simulans: Investigating the outcomes of host-symbiont coevolution. Appl Environ Microbiol 2004; 70(1):273-9.
80. Noda H, Miyoshi T, Koizumi Y. In vitro cultivation of Wolbachia in insect and mammalian cell lines. In Vitro Cell Dev Biol Anim 2002; 38(7):423-7.
81. Dobson SL, Marsland EJ, Veneti Z et al. Characterization of Wolbachia host cell range via the in vitro establishment of infections. Appl Environ Microbiol 2002; 68(2):656-60.
82. O'Neill SL, Pettigrew MM, Sinkins SP et al. In vitro cultivation of Wolbachia pipientis in an Aedes albopictus cell line. Insect Mol Biol 1997; 6(1):33-9.
83. Kubota M, Morii T, Miura K. In vitro cultivation of parthenogenesis-inducing Wolbachia in an Aedes albopictus cell line. Entomol Exp Appl 2005; 117:83-7.
84. Bartley LM, Donnelly CA, Garnett GP. The seasonal pattern of dengue in endemic areas: Mathematical models of mechanisms. Trans R Soc Trop Med Hyg 2002; 96(4):387-97.
85. Wearing HJ, Rohani P. Ecological and immunological determinants of dengue epidemics. Proc Natl Acad Sci USA 2006; 103(31):11802-7.
86. Vezzani D, Rubio A, Velazquez SM et al. Detailed assessment of microhabitat suitability for Aedes aegypti (Diptera: Culicidae) in Buenos Aires, Argentina. Acta Trop 2005; 95(2):123-31.
87. Wu M, Sun LV, Vamathevan J et al. Phylogenomics of the reproductive parasite Wolbachia pipientis wMel: A streamlined genome overrun by mobile genetic elements. PLoS Biol 2004; 2(3):E69.
88. Sun LV, Riegler M, O'Neill SL. Development of a physical and genetic map of the virulent Wolbachia strain wMelPop. J Bacteriol 2003; 185(24):7077-84.
89. Riegler M, Sidhu M, Miller WJ et al. Evidence for a global Wolbachia replacement in Drosophila melanogaster. Curr Biol 2005; 15(15):1428-33.
90. Valencia JI, Muller HJ. The mutational potentialities of some individual loci in Drosophila. Hereditas, Lund: Proc 8th Intl Congr Genet 1948 Hereditas (Lund) 1949:681-3.
91. Kaaya GP, Munyinyi DM. Biocontrol potential of the entomogenous fungi Beauveria bassiana and Metarhizium anisopliae for testse flies (Glossina spp.) at developmental sites. J Invertebr Pathol 1995; 66:237-41.
92. Kaaya GP. Glossina morsitans morsitans: Mortalities caused in adults by experimental infection with entomopathogenic fungi. Acta Trop 1989; 46:107-14.
93. Clark TB, Kellen W, Fukuda T et al. Field and laboratory studies on the pathogenicity of the fungus Beauveria bassiana to three genera of mosquitoes. J Invertebr Path 1968; 11:1-7.
94. Soares Jr GG. Pathogenesis of infection by the hyphomycetous fungus Tolyplcladium cylindrosporum in Aedes sierrensis and Culex tarsalis (Dipt.: Culicidae). Entomophaga 1982; 27:283-300.

95. Scholte EJ, Njiru BN, Smallegange RC et al. Infection of malaria (Anopheles gambiae s.s.) and filariasis (Culex quinquefasciatus) vectors with the entomopathogenic fungus Metarhizium anisopliae. Malar J 2003; 2:29.
96. Scholte EJ, Takken W, Knols BGJ. Pathogenicity of six East African entomopathogenic fungi to adult Anopheles gambiae s.s. (Diptera: Culicidae) mosquitoes. Proc Exp Appl Entomol NEV Amsterdam 2003; 14:25-9.
97. Scholte EJ, Ng'habi K, Kihonda J et al. An entomopathogenic fungus for control of adult African malaria mosquitoes. Science 2005; 308(5728):1641-2.
98. Scholte EJ, Knols BG, Takken W. Infection of the malaria mosquito Anopheles gambiae with the entomopathogenic fungus Metarhizium anisopliae reduces blood feeding and fecundity. J Invertebr Pathol 2006; 91(1):43-9.
99. Barreau C, Jousset FX, Bergoin M. Pathogenicity of the Aedes albopictus parvovirus (AaPV), a denso-like virus, for Aedes aegypti mosquitoes. J Invertebr Pathol 1996; 68(3):299-309.
100. Kittayapong P, Baisley KJ, O'Neill SL. A mosquito densovirus infecting Aedes aegypti and Aedes albopictus from Thailand. Am J Trop Med Hyg 1999; 61(4):612-7.
101. Ledermann JP, Suchman EL, Black WC et al. Infection and pathogenicity of the mosquito densoviruses AeDNV, HeDNV, and APeDNV in Aedes aegypti mosquitoes (Diptera: Culicidae). J Econ Entomol 2004; 97(6):1828-35.
102. Suchman E, Carlson J. Production of mosquito densonucleosis viruses by Aedes albopictus C6/36 cells adapted to suspension culture in serum-free protein-free media. In Vitro Cell Dev Biol Anim 2004; 40(3-4):74-5.
103. Suchman E, Kononko A, Plake E et al. Effects of AeDNV infection on Aedes aegypti lifespan and reproduction. Biol Control 2006; 39(3):456-473.
104. Barreau C, Jousset FX, Bergoin M. Venereal and vertical transmission of the Aedes albopictus parvovirus in Aedes aegypti mosquitoes. Am J Trop Med Hyg 1997; 57(2):126-31.
105. Jousset FX, Baquerizo E, Bergoin M. A new densovirus isolated from the mosquito Culex pipiens (Diptera: Culicidae). Virus Res 2000; 67(1):11-6.
106. Jousset FX, Barreau C, Boublik Y et al. A parvo-like virus persistently infecting a C6/36 clone of Aedes albopictus mosquito cell-line and pathogenic for Aedes aegypti larvae. Virus Res 1993; 29(2):99-114.
107. O'Neill SL, Kittayapong P, Braig HR et al. Insect densoviruses may be widespread in mosquito cell lines. J Gen Virol 1995; 76(Pt 8):2067-74.
108. Rwegoshora RT, Baisley KJ, Kittayapong P. Seasonal and spatial variation in natural densovirus infection in Anopheles minimus S.L. in Thailand. Southeast Asian J Trop Med Public Health 2000; 31(1):3-9.
109. Service MW. Mosquito ecology: Field sampling methods. 2nd ed. London: Elsevier Applied Science Publishers, 1993.
110. Hagler JR, Jackson CG. Methods for marking insects: Current techniques and future prospects. Annu Rev Entomol 2001; 46:511-43.
111. Muir LE, Kay BH. Aedes aegypti survival and dispersal estimated by mark-release-recapture in northern Australia. Am J Trop Med Hyg 1998; 58(3):277-82.
112. Gillies MT. Methods for assessing the density and survival of blood-sucking Diptera. Annu Rev Entomol 1974; 19:345-62.
113. Detinova TS. Age-grouping methods in Diptera of medical importance with special reference to some vectors of malaria. WHO Monograph No 47. Geneva: World Health Organization, 1962:216.
114. Tyndale-Biscoe M. Age-grading methods in adult insects: A review. Bull Entomol Res 1984; 74:341-77.
115. Hoc TQ, Charlwood JD. Age determination of Aedes cantans using the ovarian oil injection technique. Med Vet Entomol 1990; 4(2):227-33.
116. Hoc TQ, Schaub GA. Improvement of techniques for age grading hematophagous insects: Ovarian oil-injection and ovariolar separation techniques. J Med Entomol 1996; 33(3):286-9.
117. Mondet B. Application of the Polovodova's method to the determination of the physiological age of Aedes (Diptera: Culicidae) transmitting yellow fever. Ann Soc Entomol Fr 1993; 29(1):61-76.
118. Lardeux F, Ung A, Chebret M. Spectrofluorometers are not adequate for aging Aedes and Culex (Diptera: Culicidae) using pteridine fluorescence. J Med Entomol 2000; 37(5):769-73.
119. Penilla RP, Rodriguez MH, Lopez AD et al. Pteridine concentrations differ between insectary-reared and field-collected Anopheles albimanus mosquitoes of the same physiological age. Med Vet Entomol 2002; 16(3):225-34.
120. Wu D, Lehane MJ. Pteridine fluorescence for age determination of Anopheles mosquitoes. Med Vet Entomol 1999; 13(1):48-52.

121. Brei B, Edman JD, Gerade B et al. Relative abundance of two cuticular hydrocarbons indicates whether a mosquito is old enough to transmit malaria parasites. J Med Entomol 2004; 41(4):807-9.
122. Desena ML, Clark JM, Edman JD et al. Potential for aging female Aedes aegypti (Diptera: Culicidae) by gas chromatographic analysis of cuticular hydrocarbons, including a field evaluation. J Med Entomol 1999; 36(6):811-23.
123. Desena ML, Edman JD, Clark JM et al. Aedes aegypti (Diptera: Culicidae) age determination by cuticular hydrocarbon analysis of female legs. J Med Entomol 1999; 36(6):824-30.
124. Gerade BB, Lee SH, Scott TW et al. Field validation of Aedes aegypti (Diptera: Culicidae) age estimation by analysis of cuticular hydrocarbons. J Med Entomol 2004; 41(2):231-8.
125. Hayes EJ, Wall R. Age-grading adult insects: A review of techniques. Physiol Entomol 1999; 24(1):1-10.
126. Caputo B, Dani FR, Horne GL et al. Identification and composition of cuticular hydrocarbons of the major Afrotropical malaria vector Anopheles gambiae s.s. (Diptera: Culicidae): Analysis of sexual dimorphism and age-related changes. J Mass Spectrom 2005; 40(12):1595-604.
127. Hugo LE, Kay BH, Eaglesham GK et al. Investigation of cuticular hydrocarbons for determining the age and survivorship of Australasian mosquitoes. Am J Trop Med Hyg 2006; 74(3):462-74.
128. Cook PE, Hugo LE, Iturbe-Ormaetxe I et al. The use of transcriptional profiles to predict adult mosquito age under field conditions. Proc Natl Acad Sci USA 2006; 103(48):108060-5.
129. Marinotti O, Calvo E, Nguyen QK et al. Genome-wide analysis of gene expression in adult Anopheles gambiae. Insect Mol Biol 2006; 15(1):1-12.
130. Arbeitman MN, Furlong EEM, Imam F et al. Gene expression during the life cycle of Drosophila melanogaster. Science 2002; 297(5590):2270-5.
131. Pletcher SD, Macdonald SJ, Marguerie R et al. Genome-wide transcript profiles in aging and calorically restricted Drosophila melanogaster. Curr Biol 2002; 12(9):712-23.
132. Kanzok SM, Jacobs-Lorena M. Entomopathogenic fungi as biological insecticides to control malaria. Trends Parasitol 2006; 22(2):49-51.
133. Scholte EJ, Knols BGJ, Samson RA et al. Entomopathogenic fungi for mosquito control: A review. J Insect Sci 2004; 4.
134. Paul REL, Ariey F, Robert V. The evolutionary ecology of Plasmodium. Ecol Lett 2003; 6(9):866-80.
135. Mackinnon MJ, Bell A, Read AF. The effects of mosquito transmission and population bottlenecking on virulence, multiplication rate and rosetting in rodent malaria. Int J Parasitol 2005; 35(2):145-53.

Chapter 12

Technological Advances to Enhance Agricultural Pest Management

Thomas A. Miller,* Carol R. Lauzon and David J. Lampe

Abstract

Biotechnology offers new solutions to existing and future pest problems in agriculture including, for the first time, possible tools to use against insect transmitted pathogens causing plant diseases. Here, we describe the strategy first described as Autocidal Biological Control applied for the development of conditional lethal pink bollworm strains. When these strains are mass-reared, the lethal gene expression is suppressed by a tetracycline repressor element, which is activated by the presence of chlorotetracycline, a normal component of the mass-rearing diet. Once removed from the tetracycline diet, the lethal genes are passed on to offspring when ordinary lab-reared pink bollworms mate with special lethal strains. Lethality is dominant (one copy sufficient for lethality), expressed in the egg stage and affects all eggs (100% lethal expression). The initial investment by the California Cotton Pest Control Board is an outstanding example of research partnerships between agriculture industry, the USDA and land grant universities.

Introduction

The control of pest insect populations by genetic means came into its own during the last century as noted by Davidson.[1] Transgenic insect technology stands to impact agricultural pest management in several areas. The first obvious applications as suggested by Ashburner et al[2] are use of transgenic insects in Sterile Insect Technique programs by improving the genetic control mechanism.

Another application of transgenic technology is the use of female-killing factors in mass-rearing colonies. It was suggested by Heinrich and Scott[3] that releasing only sterile males is much more efficient and cost effective than rearing, sterilizing and releasing both sexes. One way to switch mass-rearing to a males-only production is by applying transgenic methods, which are faster than waiting to find appropriate chance mutations.

Still another opportunity presented by advances in biotechnology is disruption of transmission of plant pathogens by insects. First conceived of in human disease (Chagas disease) protection by Frank Richards and colleagues,[4] paratransgenesis methods are suitable for delivery of anti-disease reagents to a variety of pathogens.

A number of plant disease complexes are potential targets for control. Some of the older and more established plant pathogens and diseases are rice stripe virus, Tristeza virus, Citrus canker, Citrus Greening and Citrus Variegated Chlorosis of citrus, Pierce's disease and Grape Yellows of grapevines, a variety of scorch diseases of ornamentals and crop plants, Curly Top Virus

*Corresponding Author: Thomas A. Miller—Entomology Department, University of California, Riverside, California, USA. Email: thomas.miller@ucr.edu

Transgenesis and the Management of Vector-Borne Disease, edited by Serap Aksoy.

transmitted by beet leafhopper and Cotton and tomato leaf crumple and curl diseases caused by viruses transmitted by insects (whiteflies).

There are no cures for any of the diseases mentioned above and treatments include removal of the ailing plants as inoculation sources. The vector insects may be treated with insecticides as an indirect method to prevent the spread of the pathogen, but this can put a burden on nontarget organisms and might disrupt Integrated Pest Management (IPM) schemes. While biotechnology offers hope for crop protection from these and other diseases, the Pew Foundation[5] reported in early 2004 that the regulatory apparatus in the United States lacks experience and procedures for approving the new methods.

The symbiotic control of Chagas disease mentioned above employs recombinant tactics, but Peter Cotty[6] reported a simpler form of symbiotic control in which he selected a strain of *Aspergillus flavus* that did not secrete aflatoxin and used it as a product (AF-36) to treat soil ahead of planting time to competitively displace the *A flavus* responsible for aflatoxin contamination of cotton seed. One treatment of AF-36 reportedly protects one field for several years and natural dispersal protects surrounding fields downwind from the treatment area.

Another symbiotic control method that was granted permits for field trials is protection against dental caries offered by a selected strain of *Streptococcus mutans*.[7] Jeffrey Hillman cofounded a company (Oragenics, Inc)[8] partly to develop *S. mutans* as a treatment against tooth decay. Instead of symbiotic control, this application is called by the developers, replacement therapy.

Other symbiotic control applications include possible treatments for inflamed bowel disease (IBD)[9] and protection against HIV.[10] Thus applications span the field of agriculture and medicine. Whenever major technological breakthroughs occur, opportunities abound.

Sterile Insect Technique

Successful use of genetic control methods was first applied by mass-rearing target insect pests, irradiating to produce sterility and releasing overwhelming numbers daily to drastically reduce chances of mating between members of fertile wild-type populations. The method, known as Sterile Insect Technique (SIT) was developed by Edward Knipling.[11] SIT is available only for the most economically compelling of pest insect complexes with compatible biology because of the high cost of operations. The biology of the target pest must allow mass-rearing, transportation, handling and the target population must be in a defined area where migration does not dilute the effectiveness of the sterile release insects.

The SIT operations to control pink bollworm, *Pectinophora gossypiella* (Saunders), in California were established by the Cotton Pest Control Board in 1968.[12] Exposure of pink bollworm pupae to gamma radiation from ^{60}cobalt sources in the SIT program has fitness costs as described by Van Steenwyk et al[13] and Miller et al.[14]

The California Cotton Pest Control Board supported a project in the 1980s to produce sterile insects using the modern ability to make conditional lethal transgenic pink bollworms. As described in Miller,[15] the elements necessary to achieve this goal included finding a transformation protocol, a marker gene to use for selection and a conditional lethal gene designed in such a manner as to allow mass-rearing but capable of passing on dominant lethal genes to any offspring from mating between released insects and wild types. A single copy of the gene (in the heterozygote progeny) must be fully lethal in the egg stage for the strategy to work.

The most difficult part of this process, finding a lethal gene, was actually done first when Carl Fryxell[16] realized that the mutant *Notch* gene he was studying in *Drosophila melanogaster* had conditional lethal properties. The other necessary elements were reported independently about the same time including use of fluorescent marker genes such as green fluorescent protein (GFP) by Doug Prasher[17] and the discovery of the piggyBac element by Mac Fraser.[18]

After all of these separate components were identified, Steve Thibault[15,19] put them together with a specially designed *Bm*A3 actin promoter from *Bombyx mori* to make a plasmid. Upon injection into pink bollworm eggs along with a piggyBac helper plasmid supplied by Al

Handler of the USDA, Steve and John Peloquin achieved transformation on the first try in February of 1998. Luke Alphey[20] found the ideal combination of transcription factors and lethal genes from another *Drosophila* lethal gene candidate that eventually proved to be the winning combination and was used to develop a working strain of pink bollworm with single genes proving 100% lethal in eggs (Greg Simmons, personal communication, 2005).

Symbiosis and Pierce's Disease

The principles of symbiotic control were described by the Frank Richards and colleagues[4] early on. As newer applications are developed, the principles are affirmed and modified to fit each case. The key feature of symbiotic control is that the delivery agent must come from the natural cycle as established in the field. Thus in the Chagas disease application, the microbial symbiont, *Rhodococcus rhodnii*, resides in the hindgut of the vector insect, *Rhodnius prolixus*, and therefore has access to the protozoan pathogen, *Trypanosoma cruzi*, occupying the same space. Moreover, *R. rhodnii* is known to supply nutrients to the vector insect and therefore has developed a symbiotic relationship with the host insect. The host insect vector has a natural tendency to selectively retain the symbiont. A coprophagic habit by triatomine insects ensures that feces are sampled on a regular basis repeatedly supplying the symbiont to the hindgut.

The dental caries example given above[7] also picks a microbial symbiont, *Streptococcus mutans*, which occupies and evolved to reside in the oral cavity of humans. Therefore there is a predisposed mutualism between the human and the symbiont. The tooth decay replacement therapy tactic removed from the symbiont by recombinant methods the genes responsible for supplying enzymes to catalyze the conversion of glucose to lactic acid, a major source of tooth decay.

The selection of symbiont in the aflatoxin contamination[6] application also mentioned above found a candidate symbiont in the natural ecosystem. AF-36 is named for the 36th isolate of *Aspergillus flavus* that Peter Cotty screened from agricultural fields in Arizona where the strategy was developed for the Arizona cotton industry.

An application for symbiotic control of Pierce's disease has been described.[21,22] Pierce's disease is caused by a specific strain of the pathogen, a bacterium, *Xylella fastidiosa*, that is transmitted physically by the vector insect, a type of leafhopper called a sharpshooter. The glassy-winged sharpshooter (GWSS), *Homalodisca coagulata*, a fairly recent arrival to California, is a far more serious threat to spread the pathogen than any of the native sharpshooters. Various strains of *Xylella* are acquired by GWSS or related sharpshooter insects by the act of xylem feeding from plants.

Several symbiotic bacteria are also acquired by GWSS from the xylem fluid of host plants during feeding. Both symbionts and *Xylella* strains form biofilms on the lining of the buccal cavity in the foregut of the insect. From here they detach and are swept into the xylem fluid of the next host plant during ordinary feeding and probing by the vector insects. A strain of *Xylella* becomes pathogenic to a given host plant when it physically clogs the xylem vessels causing "scorch" symptoms. One strain of *Xylella fastidiosa* causes Pierce's disease of grapevines; another strain causes oleander leaf scorch (OLS) in the ornamental oleanders. An unprecedented epidemic of OLS in southern California has destroyed thousands of these plants and threatens other ornamental plants due to the presence of still other strains of the pathogen all being cycled by the extremely active GWSS.

Carol Lauzon[21,22] identified symbiotic bacteria from extractions of the heads of GWSS collected from the endemic infestation area. The common endophyte, *Alcaligenes xylosoxidans* var. *denitrificans* (*Axd*) was found along with a few other bacteria and *Axd* has properties suitable as a delivery vehicle for anti-*Xylella* strategies.

Again, *Axd* must occupy exactly the same niche as the target pathogen. This means that *Axd* has access to *Xylella fastidiosa* and once introduced into the disease cycle could disrupt the transmission between vector insect and host plants.

Bacterial Transgenesis and the Suppression of Horizontal Gene Transfer

Axd is an attractive candidate bacterial species to deliver anti-*Xylella* factors in either plants or sharpshooter vectors, but very little work has been done on this species with regard to genetics or physiology. Its genome has also not been sequenced which limits the approaches one can take to modifying it genetically.

Fortunately, there are broad host range tools that can be employed to modify *Axd* in a sophisticated way despite our limited knowledge of its genetics. Since modified strains of *Axd* are meant for environmental release, concerns about drug markers and horizontal gene transfer must be incorporated into the design of transgenic *Axd*. We have employed the *Himar1 mariner* transposon carried on a suicide plasmid to introduce transgenes into the chromosome of *Axd*.[23] *Himar1* is a eukaryotic transposon of the *mariner* transposable element family that works very well in phylogenetically diverse organisms, including bacteria and archae.[23-25] We reasoned that since this element is not normally found in prokaryotes, the chances of it being mobilized from the *Axd* chromosome in the future are essentially zero. The transgenesis system that we have developed is simple to use, is easily mated into *Axd* from *E. coli* via the broad host range RP4 origin of transfer, and results in chromosomal insertions that are stable and easily isolated due to the transfer of kanamycin resistance (kanR) contained in the transposon to *Axd*. Because of concerns over widespread drug resistance in bacteria, the kanR gene can be removed later using FLP recombinase since the kanR gene is flanked by direct repeats of the recognition site of this enzyme.[26,27] The resulting strains of *Axd* carry no drug markers.

Although chromosomal insertions of DNA are inherently stable, bacteria do undergo lateral DNA transfer and thus concerns remain over the horizontal transfer of novel transgenes from *Axd* to other bacterial species. Indeed, this is one of the chief concerns expressed by regulators when evaluating transgenic bacterial species aimed at environmental release. We recently tested one genetic system in *Axd* that can suppress the transfer of DNA from *Axd* to other bacteria dramatically. This is the *colE3*/ *immE3* system from the plasmid ColE3-CA38 that encodes the antibacterial protein colicin and its immunity factor.[28] Colicin /*immE3* is a kind of toxin/antidote system commonly found on plasmids that helps to ensure plasmid maintenance by the bacterial cells that carry them.[29] In such systems, a long-lived toxin is produced in addition to a short-lived antidote to that toxin. As long as both are produced, the cell is viable. If the cell loses the antidote gene for any reason (carried naturally on the ColE3-CA38 plasmid), the cell will die since the toxin remains behind to kill the cell. Similarly, cells that only receive the toxin gene will die since they do not also receive an antidote. Colicin E3 targets 16S ribosomal DNA, cleaving it near its 3' end, thus interfering with ribosome synthesis. The product of the *immE3* gene blocks this function, allowing the cell to live. Linking transgenes to colicin is an ideal way to prevent the horizontal transfer of the transgenes since recipient cells will be killed by colicin if horizontal transfer were ever to occur.

We tested this system in *Axd* by creating strains that carried the *immE3* gene on the chromosome. To these strains we introduced either plasmid pVLT31 (a broad host range matable plasmid) or pEDF5 (pVLT31 carrying the *colE3* gene that produces colicin).[30] We then attempted to transfer pVLT31 or pEDF5 from *Axd* to *E. coli*. While pVLT31 was easily transferred from *Axd* to *E. coli*, pEDF5 was never recovered. Horizontal transfer of pEDF5 from *Axd* to *E. coli* was thus suppressed by the presence of the *colE3* gene at least by a factor of 3×10^7.

Anti-*Xylella* Factors

Progress on the development of anti-*Xylella* factors has been frustrated by the difficulty in culturing this bacterial species and our limited understanding of how it causes disease. Despite the availability of complete or partial genome sequences for four different *Xylella fastidiosa* strains (Temecula 1, 9a5c, Ann-1, and Dixon), our knowledge of how this bacterium functions is very incomplete.[31] Nevertheless, at least three promising avenues have appeared for anti-*Xylella* factors and more are likely to follow.

The first of these factors are antimicrobial peptides. These peptides are relatively short (<10 kDa) and have been isolated from a wide variety of living organisms where they form the basis of the innate immune system.[32] Most antibacterial peptides are thought to act by disrupting the cell membranes of pathogens leading to cell lysis. Importantly, some of these can have comparatively narrow specificities offering the possibility of isolating peptides that have anti-*Xylella* activity but not affecting the other bacteria that inhabit grape xylem. Several anti-*Xylella* antibacterial peptides have been recently reported.[33]

A second class of anti-*Xylella* factor is likely to be single chain antibodies (scFv's). Single chain antibodies are synthetic genes that unite the antigen binding domains of vertebrate antibody heavy and light chains into a single gene by means of a synthetic linker. These genes can be expressed and secreted from bacteria, and can be created as libraries of billions of different members that can be screened against virtually any antigen. Purified proteins and even entire cells can be used to screen such libraries which can allow the targeting of cell surface factors that are important in the growth and pathogenicity of *Xylella*. Moreover, scFv's can be linked to toxins or antibacterial peptides to deliver them directly and specifically to a particular target.[34]

Finally, factors that can interfere with cell-cell communication have been proposed as anti-*Xylella* factors. *Xylella* is known to form biofilms inside grapevine xylem and in its insect vector. Importantly, this biofilm formation has been implicated in its pathogenicity. *Xylella* biofilms are formed in response to a diffusible alpha, beta unsaturated fatty acid signal molecule. Interference with this signal molecule has been suggested as a means to control *X. fastidiosa*.[35]

Ecological Microbiology

A common and likely most appreciated descriptor of microorganisms in ecosystems is that of governor. Microbes govern many activities, such as material cycles, mediating the movement of organic and inorganic compounds on our planet. In doing so, they modulate pH balance and climate, and regulate fluxes. These activities not only occur on a global scale, as we often describe similar activities in an animal gut. Thus, introduction of a modified autochthonous or allocthonous microbe into the environment tends to elicit concerns by some that the natural order of a system or systems, or at least communities within a system, may be disrupted and result in a dysbiosis.

The use of *Axd* in the management or control of PD requires that *Axd* remain in ecosystems for limited but effective periods of time and cause minimal and reversible, or no disruption to a host or ecosystem. To begin to assess efficacy and risk associated with the use of *Axd* in the field, we conducted studies aimed to monitor the fate of *Axd* in soil, water, and plant ecosystems under semi-natural conditions. We also examined the potential of *Axd* to engage in horizontal gene transfer.

To assess the efficacy and risk of use for *Axd* in the field we employed Real Time- Polymerase Chain Reaction (RT-PCR) to semi-quantitative *Axd* growth in lake water under semi-natural conditions. We found that *Axd* grew better in autoclaved lake water than in lake water that contained indigenous microbial populations. Thus, competitive attributes associated with established microbial communities overrode the ability of *Axd* to establish within these communities.

Axd growth was also monitored in soil and on leaf surfaces under semi-natural conditions using microbiological and molecular techniques. *Axd* was not retrieved from soils containing indigenous microbial populations unless the soil was autoclaved. *Axd* was retrieved from leaf surfaces from citrus, strawberry, sage, and basil. We are currently examining the effect of introducing *Axd* to citrus leaf microbial communities using denaturing gradient gel electrophoresis and terminal restriction fragment length polymorphism.

We also initiated studies whereby *Axd* was screened for the presence of endogenous plasmids. Endogenous plasmids have been shown by Taghavi et al[36] to engage in horizontal gene transfer (HGT) to members of endophytic communities in poplar trees. We have found that *Axd* can be introduced and recovered viably from citrus xylem, therefore, we began our assessment of the propensity of *Axd* to engage in HGT. We first screened *Axd* for the presence of endogenous plasmids. A strain of *E. coli* containing a single copy plasmid was used as a control in our survey. Plasmid preparations were conducted to screen for "very low" copy, "medium-low," and "high copy" plasmids using agarose gels. We also conducted to pulse field gel electrophoresis (FIGE) to visualize plasmids that range in size from 50-200 kb and that would not be detected on standard agarose gels used in our survey. Some smeared material was detected on the FIGE gels near 200 kb size, however, this materials was likely genomic DNA. Thus, our data suggest that *Axd* does not contain any endogenous plasmids up to 150 kb that would be horizontally transferred to other bacteria in nature.

We subsequently examined the likelihood that *Axd* could acquire plasmids in nature by monitoring transfer and uptake of two plasmid vectors, DsRed (pIRES-DsRed Express, Invitrogen) and pTZ18r (Amersham Biotech). Transformation attempts included both chemical and electroporation protocols. *E. coli* was used as a control. In both cases, *Axd* resisted transformation while *E. coli* was successfully transformed.

To create a more natural environment whereby *Axd* may be transformed, we studied HGT potential between *Axd* and strains of *E. coli* and *Shigella* sp. that carry fluorescent and antibiotic-marked endogenous plasmids. In coculture studies, *Axd* did not acquire the plasmids from either *E. coli* or *Shigella* sp. where control organisms were transformed. This study is currently being examined with an *Alcaligenes* sp. that contains a plasmid. It is hypothesized that a related species is more likely to transfer genes to *Axd* than a distant relative.

A complete understanding of the risks associated with the introduction of a modified bacterium into an ecosystem includes knowing if the bacterium engages in quorum sensing, or the behavioral language of bacteria. Community members communicate with their own and other species by releasing and responding to the accumulation of signaling molecules known as auto-inducers in their local environment. These auto-inducers, or signaling molecules are assembled into bacterial messages that represent the response of bacteria sensing other bacteria in their local environment. This phenomenon of cell to cell communication in bacteria is the mechanism behind the monitor and coordinated activities bacteria exhibit in response to cell density. Quorum sensing regulates many bacterial behaviors such as symbiosis, sporulation, conjugation, virulence, antibiotic production, and biofilm formation.[37] Thus, we screened *Axd* for the production of acyl-homoserine lactones, autoinducers of Gram negative bacteria. We found that *Axd* does produce these compounds. While we do not know if *Axd* produces these compounds in nature to control its or other bacterial density, we did conduct a series of coculture experiments where *Axd* was grown in the presence of other bacteria, namely, *Staphylococcus aureus*, *Pseudomonas aerunginosa*, and *E. coli*. In all cases, *Axd* did not grow well in coculture with these bacteria and growth of the other bacterial species was also decreased.

While we continue to assess the impact, or risk of introducing *Axd* into natural ecosystems, we also must assess the same responses of *Axd* in terms of strain efficiency. Energetic costs associated with survival and nutrient acquisition may effect antibody gene expression, for example. Therefore, we first designed an assay to semi-quantitatively measure antibody gene expression using Reverse Transcription PCR and Real Time PCR. Studies are currently being conducted that measure antibody production by *Axd* when in different plant xylem. This information will assist us with optimizing our Symbiotic Control agent and provides a useful tool for risk assessment.

Regulatory Issues

The regulatory activities associated with development of transgenic pink bollworm for area-wide control strategies was described by Miller and Staten,[38] and Miller,[39] and are summarized in Table 1, taken from a PowerPoint presentation developed by Robert I. Rose, formerly a regulatory official with EPA and then USDA-APHIS before he retired December 2004. Dr. Rose is now a regulatory consultant specializing in biopesticides.

The regulatory activity associated with the Chagas disease symbiotic control project is largely anecdotal because there are no public outlets for documenting regulatory activity except for publication of environmental assessments in the Federal Register; but EAs summarize the biology of the regulatory object and do not necessarily describe the regulatory activities themselves. Thus researchers seeking permits and registrations are forced to reinvent the wheel each time a new case is posed unless they hire a consultant like Bob Rose.

When we first asked for permission from the Environmental Protection Agency to inject commercial grapevines with a genetically marked *Alcaligenes xylosoxidans* var. *denitrificans* (*Axd*) endophytic symbiont, they did not have a section that dealt specifically with symbionts, whether genetically modified or not. EPA did have a section that specifically dealt with the finished symbiont designed to secrete a reagent meant to control a pathogen, such as *Xylella fastidiosa.* EPA called our symbiont, *Axd*, a "microbial pesticide," even though the word pesticide is taken completely out of context in this application and perhaps microbial antibiotic is closer to the truth.

The EPA promptly responded to the first application for field trials, but required us to burn the grapevines as the conclusion of the trials.[39] We found this extremely odd since the Biosafety Committee at UC Riverside had already given us permission to use genetically modified *Axd* in out laboratory at BL-1 level. This level allows the bacteria to be used in High School Biology laboratories. Thus it is difficult to escape the impression that the requirement to burn the grapevines was overkill.

Results of the first year of field trials (2003) showed that genetically marked *Axd* (with a *Ds*Red gene inserted, therefore nicknamed "R*Axd*") did not survive in the xylem fluid of grapevines in commercial vineyards. We found,[40] in fact, that R*Axd* preferred to colonize the xylem of citrus far more than grapevines. Poor colonization of grapevines by *Axd* was independently confirmed by Steve Lindow (personal communication).

Since GWSS prefer citrus over grapevines, and because we normally collect GWSS on citrus, it made sense that the GWSS we use to isolate endophytes would reflect the normal complement of the xylem of citrus host trees. This would have to be taken into consideration for later application strategies, but for regulatory purposes, a poor colonization of grapevines from injections of R*Axd* would seem to introduce a further protection and safety layer. This was ignored by EPA who continued to insist on burning grapevines for a potential third year (not funded as it turned out) of field trials.

Table 1. Regulatory actions on transgenic pink bollworm

Action	Start	End	Time Lapsed	Result
Movement to Phx	1 Sep 1998	8 Mar 1999	6 months	Permit issued
Contained outdoors	29 Jan 2001	1 Oct 2001	8 months	Permit (and EA issued)
Movement to Phx	6 Dec 2002	22 Jan 2003	<2 months	Permit issued
Contained outdoors	14 Apr 2003	14 Jul 2003	3 months	Permit issued
Movement to Phx	1 Dec 2003	4 Feb 2004	2 months	Permit issued
Field release	5 Jan 2004 '	Cancelled	6 months	Withdrawn
Contained outdoors	27 Apr 2004	14 Jun 2004	2 months	Permit issued
Field release	8 Apr 2005	June 2006	14 months	Permit issued
Contained outdoors	25 Apr 2005	2 Aug 2005	3 months	Permit issued
Contained outdoors	28 Apr 2005	2 Aug 2005	3 months	Permit issued

*Environmental Assessment was posted (12 Feb 06) 8 months after it was written.

During the R*Axd* field trials from 2003-2005, the "Glofish"[41] was introduced. Glofish is a freshwater zebrafish, *Danio rerio*, originally found in the Ganges River in East India and Burma. Once genetically altered with a *Ds*Red gene, they appear red in normal room light and glow red under ultraviolet light. A Texas company asked for permission to sell the Glofish out of pet stores as an aquarium novelty. All of the federal regulatory agencies realized they did not have regulations that dealt directly with this case and waived review. California was the only state whose Fish and Game Commission denied permission to sell the Glofish. Their reason given was "I think selling genetically modified fish as pets is wrong."

California is not the only place to ban sale of the Glofish. There is no science in this conclusion by the California Fish and Game Commission, just a value judgment. While two of the three commissioners are entitled to their private opinion, they have made it public policy by this choice.

One might think that an aversion by the general public to the word "recombinant" or genetic "engineering" might be behind regulatory reticence. Transgenic crops are now an accepted part of agriculture, yet certain groups remain vocal in opposition. Mendocino, Marin and Trinity Counties in Northern California have voted to ban all transgenic crops (2005). Other Counties (Sonoma and Ventura) notably voted down a ban (also 2005). So there is clearly a difference of opinion in California amongst the voters.

It turns out that public influence of regulatory activities has more to do with scientific peer pressure than the public at large. Two pertinent studies[42,43] from the National Academy of Sciences appeared in 2004. The first was commissioned by the California Grape and Wine industry for the purpose of prioritizing funding.[42] On page 109 of that report the study group concluded that symbiotic control using recombinant had a limited chance of success due to the technical difficulty, operational difficulty and regulatory difficulty.

The head of the review committee, Jan Leach, captured the size of the threat to vineyards posed by the combination of GWSS and *Xylella* in the preface.[42] She admitted that her own experience suggested that "... breeding for resistance is the most economically feasible and environmentally sound approach to disease management." However, the report offered no clues to an eventual solution and at the same time the industry was warning researchers that transgenic grapevine solutions were not going to be tolerated. Indeed, if all of the research that was highly recommended in the report was fully funded and fully successful, in ten years California would still have Pierce's disease and the industry would still not know what to do about it. There seems to be a lack of appreciation for the enormity of the problem and lack of appreciation that traditional methods are inadequate.

Because existing technology is incapable of being used to stem the threat of Pierce's disease (or not allowed to bear, if the powerful tool of transgenic grapevines is off the table), one is forced to consider new technologies. Thus it seems obvious that the difficult work must be done to perfect new technology.

The second NAS study[43] largely concluded that nothing was known of the long-term consequences of release of genetically modified organisms other than plants. That seems fairly obvious since they have not been developed before. Caution in introducing new technology is always a wise course, however, laboratory studies never duplicate natural effects and no amount of laboratory data will anticipate all that might happen. At some point transgenic animals will have to be introduced simply because there are no other options.

Conclusions

Biotechnology offers new solutions to existing and future pest problems in agriculture including, for the first time, possible tools to use against insect transmitted pathogens causing plant diseases. Although the regulatory apparatus necessary to deal with the new strategies is in place, progress is slow partly because of the novelty of the application. Transgenic organisms developed through biotechnology are not the only examples facing difficulty; even the use of symbiotic bacteria that are selected by traditional nontransgenic methods is being delayed.

Conditional lethal pink bollworm strains are currently being held in quarantine at the USDA-APHIS laboratories in Phoenix, AZ. When these strains are mass-reared, the lethal gene expression is suppressed by a tetracycline repressor element that is activated by the presence of chlorotetracycline, a normal component of the mass-rearing diet. Once removed from the tetracycline diet, the lethal genes are passed on to offspring when ordinary lab-reared pink bollworms mate with special lethal strains. Lethality is dominant (one copy sufficient for lethality), expressed in the egg stage and affects all eggs (100% lethal expression). The strategy first described as Autocidal Biological Control[16] over ten years ago. This technology is very close to completion for use in controlling pink bollworm. The initial investment by the California Cotton Pest Control Board is an outstanding example of research partnerships between agriculture industry, the USDA and land grant universities.

Symbiotic control was borne from the fertile imagination of Frank Richards at Yale Medical school, called paratransgenesis by David O'Brochta (University of Maryland) and has been translated from medicine into agriculture as the latest example of biotechnology innovations. The possible uses of this new technology will grow as it becomes applied to pest and disease complexes.

Acknowledgements

The authors gratefully acknowledge Lloyd Wendell and Beth Stone-Smith of the USDA-APHIS whose unwavering support allowed the Pierce's disease project to begin and prosper under Cooperative Agreement No. 8500-0510-GR CA. Federal and state funds also supported the project from: Biotechnology Risk Assessment Program (BRAG) from CSREES [Award No. 2004-39454-15205]; and CDFA [grant No. 03-0335] from California Department of Food and Agriculture and the PD Board; Hatch Projects 5887-H and 7342-CG; and NSF grant 0091044 to DJL. The pink bollworm work was supported by the California Cotton Pest Control Board, by token support from Cotton, Inc. and by Cooperative Agreements from USDA-APHIS, and funds from the Arizona Cotton Research and Protection Council, and the Cotton State Support Committees of California and Arizona.

References

1. Davidson G. The genetic control of insect pests. New York: Academic Press Inc., 1974.
2. Ashburner M, Hoy MA, Peloquin JJ. Prospects for the genetic transformation of arthropods. Insect Mol Biol 1998; 7:201-213.
3. Heinrich JC, Scott MJ. A repressible female-specific lethal genetic system for making transgenic insect strains suitable for a sterile-release program. Proc Natl Acad Sci USA 2000; 97:8229-8232.
4. Beard, CF, Durvasula RV, Richards FF. Bacterial symbiosis in arthropods and the control of disease transmission. Emerg Infect Dis 1998; 4:581-591.
5. Anonymous. "Bugs in the System? Issues in the science and regulation of genetically modified insects." The Pew Initiative on Food and Biotechnology. Washington DC: 2004:109, (www.pewagbiotech.org).
6. Ehrlich KC, Cotty PJ. An isolate of Aspergillus flavus used to reduce aflatoxin contamination in cottonseed has a defective polyketide synthase gene. Appl Microbiol Biotechnol 2005; 65(4):473-8.
7. Hillman JD. Genetically modified Streptococcus mutans for the prevention of dental caries. Antonie van Leeuwenhoek 2002; 82:361-366.
8. Oragenics, Inc. www.oragenics.com.
9. Westendorfa AM, Gunzerb F, Deppenmeierc S et al. Intestinal immunity of Escherichia coli NISSLE 1917: A safe carrier for therapeutic molecules. FEMS Immunol Med Microbiol 2005; 43:373-384.
10. Rao S, Hu S, McHugh L et al. Toward a live microbial microbicide for HIV: Commensal bacteria secreting an HIV fusion inhibitor peptide Proc Natl Acad Sci USA 2005; 102:11993-11998.
11. Knipling EF. Possibilities of insect control or eradication through the use of sexually sterile males. Econ Entomol 1955; 48:459-462.
12. Staten RT, Rosander RW, Keaveny DF. Genetic control of cotton insects; The PBW as a working programme. Proc Internat Symp On Management of Insect Pests. Vienna: 1992:269-283, (Pub International Atomic Energy Agency, Vienna, 1993).
13. Van Steenwyk RA, Henneberry TJ, Ballmer GR et al. Mating competitiveness of laboratory-cultured and sterilized PBW for use in a sterile moth release program. J Econ Entomol 1979; (72):502-505.
14. Miller E, Staten RT, Jones E et al. Effects of 20 krad of gamma irradiation on reproduction of pink bollworm (Lepidoptera: Gelechiidae) and their F_1 progeny: Potential impact on the identification of trap catches. J Econ Entomol 1984; (77):304-307.

15. Miller TA. Control of pink bollworm. Pesticide Outlook 2001; 12:68-70.
16. Fryxell KJ, Miller T. Autocidal Biological Control; A general strategy for insect control based on genetic transformation with a highly conserved gene. J Econ Entomol 1995; 88:1221-1232.
17. Prasher DC. Using GFP to see the light. Trends Genet 1995; 11:320-323.
18. Fraser MJ, Ciszczon T, Elick T et al. Precise excision of TTAA-specific lepidopteran transposons piggyBac (IFP2) and tagalong (TFP3) from the baculovirus genome in cell lines from two species of Lepidoptera. Insect Mol Biol 1996; 5:141-151.
19. Peloquin JJ, Thibault ST, Staten R et al. Germ-line transformation of pink bollworm (Lepidoptera: Gelechiidae) mediated by the piggyBac transposable element. Insect Mol Biol 2000; 9:323-333.
20. Thomas DD, Donnelly CA, Wood RJ et al. Insect population control using a dominant, repressible, lethal genetic system. Science 2000; 287:2474-2476.
21. Miller TA, Lampe DJ, Lauzon CR. Transgenic and paratransgenic insects in crop protection. In: Ishaaya I, Nauen R, Horowitz R, eds. Insecticide Design Using Advanced Technologies. Heidelberg, Germany: Springer-Verlag, 2007:87-103.
22. Miller TA, Lauzon CR, Lampe D et al. Paratransgenesis applied to control insect-transmitted plant pathogens: The Pierce's disease case. In: Bourtzis K, Miller TA, eds. Insect Symbiosis 2. FL: Taylor and Francis, London/CRC Press, Boca Raton, 2006:247-263.
23. Rubin EJ, Akerley BJ, Novik VN et al. In vivo transposition of mariner-based elements in enteric bacteria and mycobacteria. Proc Natl Acad Sci USA 1999; 96:1645-1650.
24. Zhang JK, Pritchett MA, Lampe DJ et al. In vivo transposon mutagenesis of the methanogenic archaeon Methanosarcina acetivorans C2A using a modified version of the insect mariner-family transposable element Himar1. Proc Natl Acad Sci USA 2000; 97:9665-70.
25. Bextine B, Lampe D, Lauzon C et al. Establishment of a genetically marked insect-derived symbiont in multiple host plants. Curr Microbiol 2005; 50:1-7.
26. Hoang TT, Karkhoff-Schweizer RR, Kutchma AJ et al. A broad-host-range Flp-FRT recombination system for site-specific excision of chromosomally-located DNA sequences: Application for isolation of unmarked Pseudomonas aeruginosa mutants. Gene 1998; 212:77-86.
27. Chiang SL, Mekalanos JJ. Construction of a Vibrio cholerae vaccine candidate using transposon delivery and FLP recombinase-mediated excision. Infect Immun 2000; 68:6391-6397.
28. Masaki H, Ohta T. Colicin E3 and its immunity genes. J Mol Biol 1985; 182:217-227.
29. Engelberg-Kulka H, Glaser G. Addiction modules and programmed cell death and antideath in bacterial cultures. Ann Rev Microbiol 1999; 53:43-70.
30. Diaz E, Munthali M, de Lorenzo V et al. Universal barrier to lateral spread of specific genes among microorganisms. Mol Microbiol 1994; 13:855-61.
31. Van Sluys MA, de Oliveira MC, Monteiro-Vitorello CB et al. Comparative analyses of the complete genome sequences of Pierce's disease and citrus variegated chlorosis strains of Xylella fastidiosa. J Bacteriol 2003; 185:1018-26.
32. Reddy KV, Yedery RD, Aranha C. Antimicrobial peptides: Premises and promises. Int J Antimicrob Agents 2004; 24:536-47.
33. Kuzina LV, Miller TA, Cooksey DA. In vitro activities of antibiotics and antimicrobial peptides against the plant pathogenic bacterium Xylella fastidiosa. Lett Appl Microbiol 2006; 42:514-20.
34. Yoshida S, Ioka D, Matsuoka H et al. Bacteria expressing single-chain immunotoxin inhibit malaria parasite development in mosquitoes. Mol Biochem Parasitol 2001; 113:89-96.
35. Newman KL, Almeida RP, Purcell AH et al. Cell-cell signaling controls Xylella fastidiosa interactions with both insects and plants. Proc Natl Acad Sci USA 2004; 101:1737-42.
36. Taghavi S, Barac T, Greenberg B et al. Horizontal gene transfer to endogenous endophytic bacteria from poplar improves phytoremediation of toluene. Appl Environ Microbiol 2005; 71:8500-8505.
37. Bassler B. Small Talk: Cell-to-cell communication in bacteria. Cell 2002; 109:421-424.
38. Miller T, Staten RT. The pink bollworm. Tale of a Transgenic Tool. Agrichemical Environmental News, 2001, (www.aenews.wsu.edu/June01AENews/June01AENews.htm).
39. Miller TA. Rachel Carson and the adaptation of biotechnology to crop protection. Amer Entomol 2004; 50:194-198.
40. Bextine B, Lampe D, Lauzon C et al. Establishment of a genetically marked insect-derived symbiont in multiple host plants. Current Microbiol 2005; 50:1-7.
41. Hallerman E. Glofish, the first GM animal commercialized: Profits amid controversy. 2004, (http://www.isb.vt.edu/articles/jun0405.htm).
42. Anonymous California Agricultural Research Priorities, Pierce's Disease. National Research Council. Washington DC: The National Academy Press, 2004:109.
43. Anonymous Biological Confinement of genetically engineered organisms. National Research Council. Washington DC: The National Academy Press, 2004.

CHAPTER 13

Applications of Mosquito Ecology for Successful Insect Transgenesis-Based Disease Prevention Programs

Thomas W. Scott,* Laura C. Harrington, Bart G. J. Knols and Willem Takken

Introduction

Successful application of genetically modified mosquitoes (GMMs) for disease prevention requires close collaboration among scientists with a diverse spectrum of expertise. Perspectives ranging from theoretical to empirical-within the context of appropriate ethical, social, and cultural guidelines-will provide the essential insights that shape informed evaluation and implementation of GMMs by vector-borne disease specialists, public health officials, and policy makers. Ecologists and population biologists have key roles to play in this process.[1-6]

Various discussions of the importance of mosquito ecology and population biology to GMM strategies motivated the development of a list of broad issues and their most important implications. These essential issues merit increasingly detailed attention. They are best viewed as challenges, rather than barriers, that must be met for the GMM strategy to be safely and effectively applied. Five principal topics are (1) spread and stability of introduced genes; (2) evolutionary consequences of mosquito transformation; (3) entomological risk, pathogen transmission, and disease severity; (4) qualitative analyses of mosquito biology, disease, and GMM control; and (5) regulatory issues.[1,5] Without an improved understanding of these subjects, application of GMM technology will lack an appropriate conceptual and factual foundation. For example, although past genetic control failures were sometimes attributed to factors other than the technology applied, four topics were identified as particularly problematic: male biology, mating behavior, colonization and mass-production effects on mosquito fitness, and population biology.[1] Consequently, by its interdisciplinary nature the genetic control approach requires close collaboration between ecologists and molecular geneticists, recruitment of expertise from outside the vector-borne disease arena, greater involvement from scientists from disease endemic countries, training for young scientists, adequate funding, consideration of ethical, legal, and social issues, and a sustained effort. Ultimately, our goal should be to integrate genetic control into overall vector-borne disease prevention strategies, now and in the future.

In this chapter we discuss four topics that are highly relevant to any GMM strategy and require additional research attention or about which important new developments can be reported. All four topics fit squarely within the general framework presented above: (1) mosquito mating behavior and male biology, (2) assessing fitness, (3) population biology, and (4)

*Corresponding Author: Thomas W. Scott—Department of Entomology, University of California, Davis, CA, USA. Email: twscott@ucdavis.edu

Transgenesis and the Management of Vector-Borne Disease, edited by Serap Aksoy.

regulatory issues. There is an urgent need for greater understanding of these topics for the GMM strategy as well as vector-borne disease prevention in general. Our discussion focuses on the taxa for which the greatest body current knowledge exists; i.e., *Anopheles* vectors of human malaria parasites and *Aedes aegypti*, the world-wide vector of dengue and yellow fever viruses.

Mating Behavior and Male Biology

Knowledge of mating behavior is central to any successful genetic control strategy. With a few notable exceptions, mosquito biologists have not contributed as much as they could have to an improved understanding of insect mating biology.[1,2,7,8] Important unanswered questions concern how mates are acquired, how many mates an individual has, which individuals they mate with, and what differences exist among species, between the two sexes and among different populations in the same species. A central issue in this line of investigation concerns sex-specific differences that influence reproductive success and thus, fitness.[10-11] For example, in mosquitoes, and animals in general, females produce a few large gametes while males produce many small gametes. Differences in gamete size (anisogamy) drive differences in sexual strategies. Females can be inseminated by a single male, but males can inseminate multiple females. Females may increase their reproductive success by choosing the fittest males to inseminate their eggs and converting resources to eggs and offspring. Males increase their reproductive success by finding and mating with as many females as possible.

There are well-founded reasons to challenge current concepts in mosquito mating behavior.[11] For example, are mosquitoes always monandrous? Do male mosquitoes compete for access to females? Do females choose the males with whom they mate? The frequency at which females mate with different males has important implications for predictions concerning mosquito mating systems[9] and the effectiveness of GMMs. For instance, even if rare, polyandry can profoundly undermine genetic control strategies, particularly if genetic modification of a mosquito confers a fitness deficit compared to wild type conspecifics. There will be a need to determine circumstances under which different mating scenarios are most likely or least likely to occur and the ramifications of those processes on intrasexual competition and male fitness. A significantly improved understanding of male fitness, which has been largely ignored, should be identified as a research priority. It is expected that in most cases GMMs released into natural habitats will be males in order to prevent increased human biting.

Results from investigations of these kinds of topics will provide important insights into mosquito population biology, evolution, control, and disease prevention. An improved understanding of mating will explain patterns of reproductive isolation and the rate and direction of sexual selection. Is there variation in mate preference and a cost of choosing a mate? Are differences in mating behavior explained by factors associated with ecological conditions or morphological traits? Can costs be manipulated to examine their effect on mate choice? What are the underlying mechanisms of mate choice? Understanding these processes will be critical for genetic control of vectors and the pathogens they transmit.[1-2] If genetically-modified mosquitoes are not competitive with wild mosquitoes, assortative mating can undermine genetic control strategies. Similarly, fitness and competitive disadvantages for genetically modified mosquitoes could select for increased polyandry or influence the spread of important/desirable traits.

Mating experiments will need to be carried out in settings as close to natural circumstances as possible.[12-13] Because GMMs are competitive for mates in the laboratory does not necessarily mean they will mate competitively with wild-type conspecifics in the field.[7]

Below we review the state-of-knowledge of mating by taxa and identify research opportunities, with special emphasis on male mosquitoes.

Anopheles

The notion that incomplete understanding of male biology, and in particular mating behavior, is a crucial flaw in the design of genetic control programs dates back to the 1980s.[7]

Nevertheless, in spite of this, remarkably little research has been undertaken on the ecology of male mosquitoes over the past two decades. And if at all, most of this research was laboratory-based.[8] On a more pragmatic note, it has been argued that outcome measures (such as field-measured induced sterility) are sufficient to gauge the quality and competitiveness of released males and indeed, as long as the desired effects are observed, such measures are adequate. Some genetic control trials have gone beyond this point, to merely measure the benefits of the program; e.g., a reduction in disease incidence. However, once such effects are not observed, a myriad of possible causes may underpin failure. The New World Screwworm eradication program in Jamaica provides a good example of this. The dogmatic belief that sterile male flies would be superior in performance and highly competitive was taken for granted based on experiences elsewhere. Such assumptions proved wrong and were, among various operational issues, key reasons for lack of progress in this project.[14] Intricacies involved in the transfer of sterile or engineered sperm from mass-reared males to wild virgin females in the field thus remain high on the research agenda.[15]

Complexity

Most contemporary research on anophelines involves *Anopheles gambiae sensu stricto* as it target organism. After all, this species is considered the prime vector of malaria in sub-Saharan Africa. Although this is true, there is more complexity observed in nearly all of Africa endemic for malaria, where multi-vectorial systems are in place.[16-17] There are relatively few places in Africa where a single *Anopheles* species is present and vectors malaria. Examples include the islands of São Tomé and Principe where *An. gambiae sensu stricto* is the only taxon present. The island of La Réunion has a population of *An. arabiensis* only.[18] On mainland Africa *An. arabiensis* appears to be the exclusive vector in parts of the horn of Africa. Thus, beyond these confined single-vector settings, the importance of this biological diversity involved in malaria transmission is not to be underestimated and various consequences of the introduction of transgenes in such systems have been modeled. For instance, the introduction of refractory *An. gambiae s.s.* in a multi-vectorial system may influence the evolutionary relationships that *P. falciparum* maintains with other vector species and other malaria parasites.[19] The sympatric occurrence of more than one vector increases the complexity of malaria transmission dynamics, and is aggravated when temporal variation in contribution to disease transmission occurs. In much of Africa *An. arabiensis* is more resilient to harsh (drier and hotter) climatic conditions than *An. gambiae*, with the former species causing the bulk of transmission far into the dry season and the latter in the rainy season. Such observations have been made in West Africa with annual transmission being maintained by *An. funestus* with other vectors having seasonal contributions[20] as well as in East Africa.[21]

Beyond the *An. gambiae* species complex level, now consisting of seven sibling species with distinct genetic and eco-ethological differences, an additional level of complexity is found with regard to sympatric occurrence of sub-taxa with apparent reproductive isolation limiting gene flow. With the identification of five chromosomal forms within the *An. gambiae s.s.* taxon based on polytene chromosome studies[22-23] and subsequent further differentiation to molecular forms based on differentiation of the rDNA region on the X chromosome,[24-25] restrictions to gene flow indicate ongoing speciation within the complex, and possible effects on malaria epidemiology and control. Further subdivision of the M molecular form has recently been proposed.[26] Clearly, the mass rearing and introduction of a form that may not be compatible or only partially capable of driving transgenes across the various forms, this poses questions regarding the choice of mosquito for genetic engineering and subsequent release. Evidence that genetic 'leakage' may be severely limited is available from West Africa, where *kdr* resistance in some areas has been found to be restricted to one of the molecular forms. Recently, however, evidence for introgression of *kdr* resistance from the Savanna to Bamako form has been documented.[27]

The existence of strong assortative mating within the *An. gambiae* molecular forms begs the question as to how these taxa remain, at least partially, reproductively isolated and what this means for the design and implementation of genetic control strategies. A variety of mechanisms have been proposed[28] and will be discussed in the section below.

Male Swarming Behavior

It is generally accepted that anophelines aggregate in mating swarms and that this behaviour constitutes an integral part of their reproductive effort.[11,28-30] Mating systems based on aerial male aggregations that function as encounter sites for receptive virgin females have evolved repeatedly in various groups of insects.[31-32] In most swarming species, the bulk of the swarm consists of males; females typically approach a swarm, acquire a mate, and leave *in copula.* Pairs are formed within breeding aggregations, and considering the low number of females present at any given time, but relatively large number of males that compete, Shuster and Wade[33] classify this mating system as polygynous mating swarms. Examples include swarming species with limited sexual dimorphism (e.g., male claspers) to facilitate the grasp of males on females. The process of swarming is normally initiated at dusk, with reducing light intensity triggering the activity of resting males resulting in house exodus.[34] It is assumed that distinct features in the landscape, combined with visual input (light contrasts) serve to detect so-called swarm markers that are used to keep the swarm stationary.[11] Swarms may form above specific features in the landscape, and the importance of such features has been shown for *An. melas*, for which swarms would move along with an artificial swarm marker (a piece of cloth) when this was dragged along the ground.[35] Some markers are visited regularly,[34] others may only serve the purpose once. The nature of markers can vary dramatically, from small shrubs, to the edges of roofs, chimneys, or even the hood of a car.[34,36-39] The temporal component to swarming thus appears to be influenced by light intensity; the spatial component is less clearly understood and at least highly variable. The recruitment of males to swarms remains elusive too. Aggregation in a landscape where several objects can serve as potential markers, coupled with a limited time span in which swarming must take place, demands an effective aggregation mechanism. It is obvious that, given the fact that releases of GM mosquitoes will probably be restricted to males, knowledge of factors governing swarm site selection, swarm formation and swarm sustenance is vital.

Little is known about the age at which males engage in swarming activity, but their need to mature their sexual organs before being able to mate, prevents them from successfully engaging in reproductive behavior in the first few days of life.[11,15,40] First, inversion of the terminalia takes place within a day post-emergence. Maturation of the accessory glands requires an extra 1-2 days. In a laboratory study, Verhoek and Takken[40] observed maximum mating activity and insemination rates in 5-7 day old male *An. gambiae s.s.* and *An. arabiensis* when offered virgin females for a 24 hr period. This period is likely to be shorter under field conditions; males are likely to engage in mating activity two days after emergence. As it remains difficult to precisely age-grade male anophelines, the age distribution of males in swarms has not been studied in detail. However, following the development of age-grading techniques for *An. stephensi* by Mahmood and Reisen,[41-42] Huho et al[43] recently adapted the technique for the African *An. gambiae.* It was shown that the number of spermatocysts in male testes decreased with age, and the relative size of their sperm reservoir increased. The presence of a clear area around accessory glands was also linked to age and mating status. A quantitative model was able to categorize males from the blind trial into age groups of young (≤ 4 days) and old (> 4 days) with an overall efficiency of 89%. Using the parameters of this model, a simple table was compiled that can be used to predict male age. In contrast, mating history could not be reliably assessed because virgins could not be distinguished from mated males. Simple assessment of a few morphological traits which are easily collected in the field now allows accurate age-grading of male *An. gambiae.* Studies that evaluate demographic patterns and mortality in wild and released males in populations targeted by GM or sterile male-based control programes are now possible.

Of the many studies that have been undertaken on the energetics of mosquitoes, virtually all have focused on the female sex.[44-45] Yuval et al[46] studied the energy budget of male *An. freeborni* and concluded that a single night of swarming may consume up to 50% of a male's caloric reserves. Clearly, the ability of a male to subsequently replenish reserves through plant feeding[47-50] is a major contributor to reproductive success. Recent studies have shown that male feeding on nectar/sugar sources is common,[50] particularly for male *An. gambiae*, and plant feeding has been observed in the wild at night on plants such as *Lantana camara* (Hortance Manda *pers. comm.*). Yuval et al[51] concluded that stored sugars and glycogen provide the fuel for swarming flight in *An. freeborni*. A recent comparative field study in Tanzania on the reproductive potential and life-history of *An. gambiae s.s.* from both standardized laboratory conditions and from natural field settings showed that body size and lipid reserves of wild males were substantially greater than those reared under standard laboratory conditions. Higher levels of sugars and glycogen in the laboratory mosquitoes would enable swarming flight, but their lack of lipids would impair their resting metabolism. The authors cautioned that the energetic limitations of insects as identified in the laboratory may underestimate their resilience in the wild, and discussed the implications of this phenomenon with respect to vector-borne disease control programs based on genetic control of mosquitoes.[52] Thus it remains important to consider the nutritional status of mosquitoes earmarked for releases. A 'priming' meal to balance their reserves with those measured in field populations or even increase the nutritional balance in their favor, should be considered.

Frequency of Insemination

Very few investigators have studied anopheline mating frequency among field populations. Giglioli and Mason[53] examined mating plugs in *An. maculipennis*. Of 3,866 females with mating plugs, only 5 (0.13%) contained more than one plug. *An. gambiae,* males inseminated 5 or more females in rapid succession but could not produce more than two mating plugs.[53] In western Siberia, Novikov[54] concluded that *An. messeae* were generally monandrous. Similarly, in Thailand, Baimai and Green[55] examined *An. dirus* karyotypes and found the progeny of gravid females were all one of two karyotypes. Yuval and Fritz[51] reported that one of 25 *An. freeborni* females had mated with two males (4%). In contrast, Scarpassa et al[56] found evidence of 15% double mating from offspring of 40 field-collected *An. nuneztovari*. Tripet et al[57] detected 2.5% polyandry in 239 *An. gambiae s.s.* that were collected in Mali. They examined sperm removed from spermatheca for more than two alleles at any of 4 microsatellite loci.

Aedes aegypti

Male Swarming Behavior

Unlike many other mosquito species, *Ae. aegypti* do not form traditional, station-keeping swarms.[11] This strategy appears to yield little genetic gain for a low-density species that feeds and rests inside human dwellings. Instead, male *Ae. aegypti* aggregate around hosts who represent the primary female encounter site.[58] Gubler and Bhattachaya[59] described a similar behavior by *Ae. albopictus*, a closely related *Stegomyia* species. Males arrive at human hosts first and matings can be observed just before or after females feed on blood from the human host.

Markers for *Ae. aegypti* Genetic Fingerprinting

Details of *Ae. aegypti*'s mating biology can be investigated with heritable, co-dominant, single copy genetic markers. Such markers must be capable of distinguishing closely related individuals. Microsatellites from anophelines meet these criteria and have proven useful for studying polyandry.[57] In anophelines, microsatellites are composed of simple repeats that are abundant and polymorphic.[60] Unfortunately, *Ae. aegypti* microsatellites have not been as useful as genetic markers because too often they are not composed of simple repeats, not abundant, not polymorphic, and tend to be embedded in repetitive regions of the genome.[26,61-62] Studies of *Ae. aegypti*'s mating biology will require the capacity to distinguish genotypes of closely related individuals from the same population.

Frequency of Insemination

Female mosquitoes, including *Ae. aegypti*, are typically thought to be monandrous[63-64] while males can inseminate many females and are considered polygynous.[11] Results from several laboratory studies indicate that there may be times when females are polyandrous. Gwadz and Craig,[65] for example, reported up to 30% polyandry in female *Ae. aegypti* kept in small cages when males of two different genotypes were introduced simultaneously.

Two laboratory studies investigated female *Ae. aegypti* mating frequency beyond the first oviposition, which is an important limitation because a significant proportion of wild females undergo multiple gonotrophic cycles.[66] Young and Downe[67] examined mating frequency of *Ae. aegypti* over several gonotrophic cycles and reported that 75-90% of females mate again prior to the second gonotrophic cycle. Williams and Berger[68] similarly detected multiple mating by females that completed several ovipositions. Six percent of females mated again after 4 gonotrophic cycles, 22% after 5, 38% after 6 and 48% after 7 cycles.

Components of Mosquito Semen

Components of mosquito semen have not been carefully characterized.[11] Considerably more detailed information is available for *Drosophila*. *Drosophila* semen is comprised of several components, including sperm, accessory gland proteins, and ejaculatory duct and ejaculatory bulb products.[69-72] It is reasonable to expect that, as in *Drosophila*, mosquito semen is complex with components that have a variety of effects on female behavior and reproduction.

Role of Seminal Products

Semen deposited in the female bursa seminalis provides a medium for sperm to move into the opening of the spermathecal ducts (spermathecal vestibule). Despite limited knowledge of the components of mosquito semen and their impact on female behavior, components of semen appear to be essential for regulating mosquito post-mating behavior as they are for *Drosophila*.[89-90] Mating or direct injection of seminal substances affected female behaviors including flight,[91-96] response to host cues,[58,97-99] oviposition,[100-103] fertility and ovarian development,[88,104-107] blood digestion[108-109] and sexual refractoriness (discussed below). Many of these effects are reviewed by references 11 and 110.

Early work on *Ae. aegypti* seminal products identified a partially purified protein, named matrone,[63] that had a variety of effects on female reproductive and feeding behavior. Additional work identified a 7.6 kDa peptide that appeared to reduce female host-seeking behavior.[164] These substances have never been fully characterized. Sirot et al.[165] recently discovered, however, the presence of male reproductive proteins in *Ae. aegypti*. Using bioinformatic comparisons with *Drosophila melanogaster* accessory gland proteins and mass spectrometry of proteins from *Ae. aegypti* male accessory glands and ejaculatory ducts and female reproductive tracts, the authors identified 63 new *Ae. aegypti* proteins. Twenty-one of these proteins were found in mated females, but not in virgin females, suggesting that they were transfered from males to females during mating. Most of the male reproductive proteins are in the same class as male proteins from other organisms. Many of these proteins may be rapidly evolving because they lack identifiable homologs in *Culex pipiens*, *An. gambiae*, and *D. melanogaster*. The authors identified 15 proteins in the 30 kDa range and another two in the 60 kDa range, which are similar in size to the main protein bands in partially-purified matrone.[111] Male reproductive proteins do not appear to be limited to *Ae. aegypti*. Using bioinformatics searches, Dottorini et al.[166] identified 46 male accessory gland genes in *An. gambiae*. Results from these two recent studies open the door for new molecular investigations into female mosquito post mating responses.

Sperm Capacity and Depletion

Although the most successful reproductive strategy for male mosquitoes is likely to involve mating with as many females as possible, the number of females that males can inseminate and the rate and permanence of depletion of sperm and other seminal fluid components has not been clearly defined. Estimates of the number of *Ae. aegypti* females that can be inseminated by a male in a single day ranges from 3-4[65] to 5.8.[73] Over a males' lifetime the total ranged from 8-9.[74] Ponlawat and Harrington[75] investigated sperm capacity of laboratory and field strains of *Ae. aegypti.* The authors found increasing numbers of sperm in virgin males with age up to day 10. These results suggest that male mosquitoes undergo spermatogenesis as adults which may allow males to continue mating with females as they age.[75] GMM strategies should focus on releasing males at their peak sperm capacity (after 5 days of age).

Mosquito Body Size and Reproductive Success

Very few studies have assessed the competitiveness and reproductive success of male mosquitoes based on their body size. Benjamin and Bradshaw[76] failed to find differences in offspring number from small vs. large male *Wyeomyia smithii.* Dickinson and Klowden[77] measured the whole protein content of accessory glands in large vs. small *Ae. aegypti* males before and after mating to estimate the amount of accessory gland protein (AGP) transferred. They found significantly less protein in glands of small males (wing length = 1.66 mm) than large males (wing length = 2.20 mm) and small males transferred significantly less protein to females. This is the only paper we are aware of that directly addressed the effect of different male body sizes on female sexual receptivity. They did not examine the range of male body sizes found in wild *Ae. aegypti* populations; i.e., wing length = 1.67 to 3.0 mm. Ponlawat and Harrington[75] found a significant increase in overall sperm capacity of virgin males with increasing male body size. The effect of body size on fertilization success has not been addressed.

In other insects, correlation of male body size with pre and post-copulatory success revealed complex relationships with, in some cases, compensatory responses. In certain insects large body size makes males more physically competitive and attractive to females.[78] However, where flight agility is important or scramble competition is the predominant strategy, small size is advantageous.[79-80] In some insect species such as the yellow dung fly, body size is compensated for during copulatory events. Larger males have greater fertilization success during copulation due to a higher rate of ejaculate transferred,[81] but smaller males compensate by copulating longer.[82-83] Bangham et al[84] demonstrated that larger male *D. melanogaster* had greater post copulatory success than smaller males and males with larger accessory glands mated more frequently. The effect of female body size on reproductive success is well documented for *Ae. aegypti*. There is a direct relationship between female size and total oocyte number.[66] In a study with *Wy. smithii,* female size was positively correlated to, and the primary indicator of, eggs produced (Bradshaw and Holzapfel 1992, 1996).[85-86] Okanda et al[87] reported that male *An. gambiae* held in laboratory cages preferentially mated with large females. Large females were more likely than small ones to successfully oviposit and laid more eggs. Additional studies are needed to directly examine the effects of male and female size on reproduction.

Few investigators have directly examined the impact of male mosquito nutritional status on their reproductive success. Klowden and Chambers[88] reported that colonized male *Ae. aegypti* held in the laboratory without sugar for 3 days were less likely to stimulate oogenesis than sugar-fed males. Whether this applies to free-ranging wild *Ae. aegypti* is unknown.

Sexual Competition

Intersexual selection and its ramifications have been investigated considerably more thoroughly in other flies than for mosquitoes. For *Drosophila,* mating appears to decrease female lifespan.[114] Several important factors are involved including the cost of reproduction and male harassment.[115-116] Some evidence supports the idea that male seminal fluid is toxic to female *Drosophila.*[117] These intersexual effects have not been studied in mosquitoes. Intrasexual male mosquito competition has only been examined in a few instances; e.g., in the context of swarming efficiency. Two potentially very important parameters in *Ae. aegypti*'s mating system, male competition and female choice, merit significantly more research attention.

Assessing Fitness

A critical component for successful development of GMMs will be to minimize detrimental fitness affects from transgenesis, or to at least to be sure that the strength of the transgene driver exceeds fitness costs associated with transformation.[5] Although it is possible that genetic modification could increase fitness,[118] most investigators assume that it will result in a fitness costs. Reduced fitness could undercut a population reduction strategy by making GMMs non-competitive for wild type mates. For population replacement, it is proposed that genetic drive mechanisms will be used to spread desirable genes into a target population.[119] If transformed mosquitoes and their offspring are less fit than wild mosquitoes and the transgene driver is not strong enough to overcome fitness effects, desirable genes will be lost rapidly from the target population. Whenever possible, both strategies should aim to minimize adverse fitness effects due to genetic modification. Because fitness is a complex and controversial concept[120] it is important that we define our use of the term and as a research community we should strive to reach consensus on how to predict GMM fitness relative to their wild type targets. Below we (1) define our view of fitness, (2) review complexities associated with measuring fitness of GMM, (3) review previous fitness evaluations of GMMs, and (4) propose a methodology for predicting the fitness of GMMs released into a natural environment. For more detailed discussion of these topics see reference 5.

A generally accepted definition of fitness is "success in producing offspring, irrespective of the causes of that success".[121] It is important to recognize, however, that this is a complex and dynamic concept. Fitness of organisms with the same genotype can vary across different environments and structures of populations. Differences in fitness are not due solely to differences in transgenes or genotype. Environment can have profound affects on fitness. The genetic background into which a transgene is inserted is similarly important because it can modify fitness outcomes due to affects from the parental genotype, inbreeding depression, and/or epistasis and epigenetics. For these reasons, it is best to avoid use of mosquitoes from laboratory colonies. Colonization is a founding event that can result in genotypes not representative of the natural population as well as inbreeding depression.[7,122-123] Depending on the GMM approach taken, outcrossing with wild-type mosquitoes or transgene introgression into a wild-type background is recommended before carrying out fitness assessments. Care should always be taken not to assume that results from laboratory studies will automatically extrapolate to a natural, field setting.[124] When moving from the lab to the field potential sources of fitness variation, even among mosquitoes with identical genotypes, are potentially extensive and can be difficult to precisely define.

Because of these complexities there is considerable debate concerning how best to measure fitness.[120,125] Frequently it is expressed as the average contribution of individuals, genotypes or alleles to the next or succeeding generations. Two commonly used measures of fitness are net replacement rate (R) and per capita instantaneous growth rate (r). R is a product of offspring production across all ages ($R = \Sigma\, l_x m_x$) and r accounts for the potentially important effects of generation time. In most practical cases, relative fitness is assessed;[125-126] that is, comparisons of survival, reproduction, and population expansion are made across different genotypes.

Previous efforts to assess GMM fitness revealed valuable insights. From a series of laboratory and field studies with mosquito vectors of dengue, lymphatic filariasis, and malaria carried out during the 1970s[3,7,127-132] researchers concluded that

- although some species and strains of GMMs can mate competitively with their wild-type counter parts in the laboratory and field, for others assortative mating prevents random mating and thus can be an important source of failure in population reduction and population replacement control programs;
- successful population reduction programs can release larval mosquito populations from density dependent survival constraints, producing more adults than if no control was applied, and thus undermine a seemingly successful mosquito control program; and
- unrelated genetic background of released mosquitoes can adversely affect their competitiveness for mating with local, wild type conspecific mosquitoes.

More recent studies carried out in the laboratory during 2003 and 2004 with genetically modified *An. stephensi* and *Ae. aegypti* examined the fitness effects of transgenesis compared to parental, laboratory mosquito strains.[118,133-135] Two general conclusions were drawn (1) genetic transformation did not necessarily confer a fitness cost and (2) when a fitness reduction was detected it appeared to be due to insertional mutagenesis, detrimental expression of transgenes or inbreeding depression. In these studies, little attention was paid to simulating natural environmental conditions.

Considering the goal of developing a practical approach and the applied disease prevention objective of genetic control, high through-put competition experiments are the recommended approach for assessing GMM relative fitness.[5] Competition experiments are an efficient way to rapidly assess relative fitness of different genotypes compared to field strains. Life table studies are too labor intensive, but if they must be carried out it is best that appropriate methodology is rigorously applied (see refs. 135-138). For operationally amenable evaluations, phased competition experiments are recommended starting in the laboratory and concluding in large outdoor semi-field enclosures. The first phase, conducted in the laboratory, is a direct comparison at equal frequencies of transgenic strains versus wild-type mosquitoes from the target population. As with all phases, transgenic strains with major fitness affects are eliminated. The second phase is identical to phase 1 except it is conducted at the proposed field site in order to account for environmental affects. In phase 3, GMMs are competed against local wild-type mosquitoes in large outdoor semi-field enclosures, like the ones described below. Environmental conditions in phase 3 should be as close as is possible to the natural habitat of wild type, target population mosquitoes. Replication of comparisons will be important to avoid error associated with genetic drift. Relative fitness can be estimated in subsequent generations from the observed frequencies of transgenic genotypes; i.e., transgenic homozygotes, heterozygotes and wild type homozygotes. To estimate components of fitness and predict genotype frequencies, a subset of offspring can be randomly selected over multiple generations.[139-142] It will be critical to keep track of genotypes instead of strains; monitoring transgene frequencies should be the primary objective. The intent of this kind of rapid fitness assessment is to facilitate rapid identification of constructs and GMM strains with the greatest likelihood of successful application in the field. A drawback is that unlike life table analyses competition experiments do not allow one to examine complex processes such as mortality and how it varies among different genotypes. On the other hand, competition experiments allow meaningful evaluation (i.e., replicate cages with large study populations) in a relatively short time, all mosquito life stages can be examined, and different genotypes are directly compared. In a defined research environment, it is possible to determine whether a GMM strain is neutral (no frequency change), less fit (frequency decrease) or more fit (frequency increase) compared to wild type mosquitoes. It is expected that this process should begin with many candidate GMM strains and end with one or a few lines that show serious potential for disease prevention.

Population Biology

Natural variation in population biology of mosquito vectors can potentially affect carefully planned GMM strategies. Laboratory cultures of mosquitoes usually express reduced fitness compared to wild type, which is caused by the uncontrolled loss of specific genes during the colonization process.[122]

Much work on population biology of malaria vectors has focused on the *Anopheles gambiae* complex in Africa. Members of this complex belong to the most efficient malaria vectors and are principal candidates under consideration for genetic control. Lehmann et al[143] described the heterogeneity of *An. gambiae sensu stricto* along a cline from East to West Africa. These authors demonstrated that within country heterogeneity among members of this species was greater than among populations from East and West Africa. It was also found that kinship among anopheline mosquitoes was greatest between individuals collected within the same village, and that over the relatively short distance of 7 km kinship relationships were no longer evident.[144] Chromosomal and molecular genetic studies on *An. gambiae s.s.* revealed the existence of at least 5 chromosomal forms

(SAVANNA, MOPTI, BAMAKO, BISSAU and FOREST) and two molecular forms; forms M and S.[145-146] Pairing of these genetic forms is asynchronous, several M forms belonging to different chromosomal forms, and most S forms belonging to one chromosomal form only (the FOREST form). Although the molecular forms frequently co-existent in parts of West Africa, they are reproductively isolated, based on the near absence of hybrids in wild populations.[23,146-147] Interestingly, *An. funestus*, another prominent and important malaria vector in Africa, was found to express greater genetic heterogeneity compared to *An. gambiae*.[148-149]

Although much is known about the genetic distribution of these important malaria vectors in Africa, there is until now little evidence for ecological associations of several members of these complexes. This means, for instance, that although the population structure of malaria vectors in Africa is sometimes known in great detail, the significance of these findings for mosquito behavior, reproduction and survival suffers from the incomplete data that are available on these topics. One notable exception to this is the detailed information about the *An. gambiae* complex in West Africa.[145] Here, detailed genetic analysis of field populations has been complemented by ecological information such as genetic exchange, seasonal phenology, biting behaviour and vectorial capacity.[23,26,150-151] Such studies revealed that the chromosomal form *An. gambiae* MOPTI is favored by man-made irrigation agriculture and the form BAMAKO is selected against.[152] Introduction of a GMM made from only one chromosomal form could relax environmental pressure on that or a closely related form. Tripet et al[151] found a low level of genetic exchange between chromosomal forms BAMAKO and MOPTI, which suggests that ecological factors may favor the introgression of the genetic construct into only one chromosomal form. The finding of incipient speciation in *An. gambiae s.s.*[22-23] is further proof that GMM strategies should consider the occurrence of such processes and establish a strategy for affecting all members belonging to this genetic group.

Geographic distribution of the *An. gambiae s.s.* molecular forms M and S[146] revealed an unequal spread of these forms across Africa. Molecular form S appears to be present across much of the African continent, while form M has a much more restricted distribution. Both forms share a large geographic range, where they occur sympatrically. Whereas there is some evidence of ecotypic differentiation between both forms, there is no specific ecological feature that favors one form above the other.[153] Current research on molecular forms of *An. gambiae s.s.* in Cameroon may provide more data on the ecological determinants that favoring one or the other form (G. Lanzaro pers. comm.).

Throughout its range *An. gambiae s.s.* is considered synanthropic, expressing a high degree of anthropophily coupled with endophily and endophagy.[154-155] To-date, none of the genetic forms has been associated with variations in this behavior, suggesting that these behavioral traits are inherited independently from the chromosomal and molecular characters discussed above. However, under specific conditions *An. gambiae s.s.* can adopt an opportunistic feeding behavior and be more zoophilic than at other times.[156] Whether this switching is of transient nature or being selected because of environmental change needs to be established. Across much of its range, *An. gambiae s.s.* shares many of its breeding sites with *An. arabiensis*,[155] and genetic modification of only one species may provide selective advantages to one of the two species with hitherto unpredictable consequences.

Unlike *An. gambiae s.s.*, *Ae. aegypti* has a near-cosmopolitan, subtropical, worldwide distribution, being present on five continents. Results from a series of studies over the past two decades revealed local and global patterns in its population structure that appear to have been fashioned by relatively recent patterns of migration.[157] Tabachnick, Powell, and colleagues carried out analyses of more than 80 populations worldwide using isozyme loci. Geographic differences related to variation in susceptibility to yellow fever virus infection were viewed as complex, variable, and dynamic. More recently, detailed analyses based on random amplified polymorphic DNA loci, NADH dehydrogenase subunit 4 mitochondrial DNA gene (ND4), and random amplified polymorphic DNA loci were carried out with geographically stratified samples of *Ae. aegypti* from Mexico. Results indicated that population structure was different in different regions of the country, but that in general populations are isolated by distance and that free gene flow occurs among collections within ~100-200 km. Genetic differentiation in southern Mexico appears to be due to greater habitat diversity than in the northeast. The authors concluded that contemporary dispersal via human commerce significantly effects patterns of natural gene flow.[158-160] To compare *Ae. aegypti* population structure in South East Asia to the detailed understanding that is unfolding in Mexico, Bosio et al[161] collected 4th instar larvae and pupae from a variety of containers in and around homes at 19 sites in Thailand along a ~1,400 km transect from Chang Mai in the north to Songkhla Province along the southern border with Malaysia. Analysis of SSCP variation in the ND4 gene was used to describe population structure. From a total of 1,346 individuals, 7 haplotypes were divided into 2 lineages. There was no isolation by distance across the regions sampled, but there was genetic structuring at different geographic scales. These results are consistent with the hypothesis that *Ae. aegypti* populations in Thailand have experienced substantial bottlenecks as they were established, most probably through human movement of mosquitoes. Variation among collections from within large cities (collections ≥25 km apart) was significant, indicating a high degree of genetic differentiation at a small spatial scale, which is indicative of genetic drift. Two unique haplotypes were detected from a single population in Bangkok and 2 rare haplotypes in collections from Hat Yai. Conversely, suburban/rural sites exhibited more restricted gene flow and an absence of rare haplotypes. Depending on the genetic strategy selected for implementation, results from studies like these will have important implications for the deployment of genetically modified *Ae. aegypti*. The recently publication of *Ae. aegypti* genome sequence[162] will be helpful as a baseline for future comparisons between among local populations, regions and continents.

Regulatory Issues

To-date, no field trial or other experiment outside of a rigidly contained indoor facility has been conducted using transgenic disease vectors. It should be realized, therefore, that for any activity that involves the use of genetically-modified vectors, a lengthy pathway must be followed in order to satisfy the internationally-accepted standards on biosafety (see below). In contrast to the now widely-applied technology of GM crops (e.g., corn, rice, cotton), there is no established precedent for government agency, international organization, covenant, or treaty regulation of GM vectors, other than public human and animal health import permit and quarantine regulations intended to prevent the incursion of new pests and diseases into countries where they are not indigenous. GM vectors are new or unique in their nature and science, and thus are unfamiliar to agencies with responsibilities to regulate their importation, distribution, contained testing, and release or use in the environment. As of this writing, the only regulatory precedent for transgenic insects is in the United States. It is based on a combination of import permit procedures, a transparent documentation process, and a law that mandated preparation and public comment on two Environmental Assessments and an Environmental Impact Statement conducted under the National Environmental Policy Act.[†] Agencies or authorities charged with regulation of GM mosquitoes likely will not have established processes or personnel to regulate them. For early trials involving GM mosquitoes, they may be inclined therefore to rely on experience with conventional organisms that are typically quarantined or transgenic plants. Because the impact of human vector-borne disease is often not measured in economic terms, in contrast to the impact of pests and diseases in agriculture, the costs incurred with establishing national, regional and international regulation and internationally-accepted treaties can be considerable and may prevent the execution of field trials and implementation of a proposed GM vector strategy for disease prevention.

Recently, the PEW Initiative on Food and Biotechnology[163] issued a helpful list of issues that should be considered at the planning stage of a GM insect control strategy. Although

---

[†]Example of Environmental Assessments for transgenic pink bollworm cotton pests may be found at http://www.aphis.usda.gov/brs/arthropod.assess.html

disease vectors are addressed in this publication, it is clear that because of the particularly sensitive issue of disease transmission, strategies for the use of GM insects for vector-borne disease control require regulations that are currently not available. This issue has been further addressed in the Daegu protocol on Enhancing Regulatory Communication for Microbial Pesticides and Transgenic Insects, which was completed in August 2007 (http://biopesticide.ucr.edu/daegu/daegu.html). In this protocol, disease vectors are explicitly addressed. Both documents provide information on regulatory issues that need to be addressed in the application of GM technologies for insects. It is expected that in the near future the World Health Organization will develop guiding principles for the application of GM vectors, including regulatory and biosafety issues (Y.T. Touré, pers. comm.). This might be a helpful development as the United Nations is widely considered a respected international platform through which international treaties are being established and from which national governments can develop their country-specific (national) regulations.

Currently, movement and use of transgenic organisms is arranged through the Cartagena protocol to the International Convention on Biodiversity (http://www.biodiv.org/biosafety/default.aspxon). This protocol arranges in principle transboundary movements of living organisms, including transgenic organisms. Both transport and containment guidelines are provided, however, the protocol does not address the specific use of GM vectors.

Selection of a field site for any trial with a GM vector depends on the ecology, geographic distribution and presence of the specific disease (and its vectors) under consideration. In addition to these environmental factors, ethical, cultural and social issues should also be considered, as the introduction of GM vectors, even under strictly contained conditions, may be considered as a potential health and/or ecological risk (Lavery, Harrington, Scott, pers. comm.). Not only national authorities, but also regional and local residents must be involved in discussions leading to a decision for the application of GM vector studies. Although at this stage the use of GM vectors is only considered in contained facilities, these discussions should address the accidental release of a GM vector, status of abortive actions (in case of undesired escapes), and future developments concerning the use of transgenic insect technologies for disease prevention.

Once a field site for a contained study with a GM vector has been identified, and local and national support for the study has been obtained, an environmental impact assessment (EIA) may be required, detailing the potential risks associated with the proposed study. Here, the existence of a national protocol for the EIA will be helpful, and this protocol can then be used paying particular attention to the use of disease vectors. In the absence of EIA regulations, the standard EPI regulations used in the USA may be considered as these are generally considered to provide an objective assessment procedure. The assessors should make a special provision for the impact of accidental release of the GM vector under study, as this is expected to be the principle concern of the local and national authorities.

In the (likely) absence of effective regulatory issues concerning the biosafety for the use of GM vectors, research programs may experience considerable delay unless these issues are addressed at an early stage. It is anticipated that the collective experience of pilot programs working with GM vectors will facilitate future experiments and application of the GM technology for disease prevention.

Conclusions

Three general conclusions can be drawn concerning applications of mosquito ecology for successful GMM disease prevention. First, application of GMM technology will lack an appropriate conceptual and factual foundation without an improved understanding of ecological topics like the ones we reviewed above. Without a proper knowledge base prospects for successful genetic-based disease prevention will be greatly diminished. To achieve a meaningful reduction in disease, it is essential that we increase our understanding of mosquito ecology and effectively apply that knowledge to interfering with a mosquito's role in pathogen transmission. Second, mating behavior, male biology, assessing fitness, and population biology should be priorities for research programs on mosquito vectors and funding agencies. Third, effective development and application of genetic control will require a long-term commitment from scientists, public health officials, and funding agencies. Greater and sustained collaboration among people with diverse backgrounds will be necessary to reach this goal.

For genetic control programs, and vector-borne disease prevention in general, there is growing recognition of the critical need for a significant increase in ecological and behavioral research on target mosquito vector species and that this must be done, as much as possible, in natural field settings. Laboratory-based research has the advantage of being well controlled, but suffers from the potential of producing ecologically or epidemiologically irrelevant or misleading results. Continued recognition of the importance of field-based vector ecology will be the foundation for revealing insights into the topics outlined above and ultimately to rigorous assessment of genetic control programs.

References

1. Scott TW, Takken W, Knols BGJ et al. The ecology of genetically modified mosquitoes. Science 2002; 298:117-9.
2. Takken W, Scott TW. Ecological aspects for application of genetically modified mosquitoes. Springer Dordrecht: Frontis, 2003.
3. Gould F, Schliekelman P. Population genetics of autocidal control and strain replacement. Ann Rev Entomol 2004; 49:193-217.
4. Knols BGJ, Louis C. Strategic plan to bridge laboratory and field research in disease vector control. Springer Dordrecht: Frontis, 2006.
5. Scott TW. Current thoughts about the integration of field and laboratory sciences in genetic control of disease vectors. In: Knols BGJ, Louis C, eds. Strategic Plan to Bridge Laboratory and Field Research in Disease Vector Control. Springer Dordrecht: Frontis, 2005:67-76.
6. Mukabana RW, Kannady K, Kiama GM et al. Ecologists can enable communities to implement malaria vector control in Africa. Malar J 2006; 5:9.
7. Reisen W. Lessons from the past: Historical studies by the University of Maryland and the University of California, Berkeley. In: Takken W, Scott TW, eds. Ecological Aspects for Application of Genetically Modified Mosquitoes. Springer Dordrecht: Frontis, 2003:25-32.
8. Ferguson HM, John B, Ng'habi K et al. Redressing the sex imbalance in knowledge of vector biology. Trends Ecol Evol 2005; 20:202-9.
9. Thornhill RA, Alcock J. Insect Mating Systems. Cambridge: Harvard University Press, 1983.
10. Partridge L. Lifetime reproductive success in Drosophila. In: TH CB, ed. Reproductive Success. Chicago: University of Chicago Press, 1988:11-23.
11. Clements AN. The biology of mosquitoes; sensory reception and behaviour. 1999.
12. Knols BGJ, Ng'habi K, Mathenge EM et al. MalariaSphere: A greenhouse-enclosed simulation of a natural Anopheles gambiae (Diptera: Culicidae) ecosystem in western Kenya. Malar J 2002; 1:19.
13. Knols BGJ, Nijru BN, Mukabana RW et al. Contained semi-field environments for ecological studies on transgenic African malaria vectors: Benefits and constraints. In: Scott TW, ed. Ecological Aspects for Application of Genetically Modified Mosquitoes. Springer Dordrecht: Frontis, 2003:91-106.
14. Dyck VA, Hendrichs J, Robinson AS. Sterile insect technique: Principles and practices in area-wide pest management. New York: Springer, 2005.
15. Takken W, Costantini C, Dolo G et al. Bringing laboratory and field research for genetic control of disease vectors: Springer/wageningen UR frontis series, 2006.
16. White GB. Anopheles gambiae complex and disease transmission in Africa. Trans R Soc Trop Med Hyg 1974; 68:278-301.
17. Coetzee M, Craig M, le Sueur D. Distribution of African malaria mosquitoes belonging to the Anopheles gambiae complex. Parasitol Today 2000; 16:74-7.
18. Girod R, Salvan M, Simard F et al. Evaluation of the vectorial capacity of Anopheles arabiensis (Diptera: Culicidae) on the island of Reunion: An approach to the health risk of malaria importation in an area of eradication. Bull Soc Pathol Exot 1999; 92:203-9.
19. Chevillon C, Paul RE, Meeus Td et al. Thinking transgenic vectors in a population context: Some expectations and many open questions. In: Boete C, ed. Genetically Modified Mosquitoes for Malaria Control. Georgetown: Landes Bioscience, 2006:117-31.
20. Fontenille D, Lochouarn L, Diagne N et al. High annual and seasonal variations in malaria transmissions by anopheline and vector species composition in Dielmo, a holoendemic area in Senegal. Am J Trop Med Hyg 1997; 56:247-53.

21. Charlwood JD, Kihonda J, Sama S et al. The rise and fall of Anopheles arabiensis (Diptera: Culicidae) in a Tanzanian village. Bull Entomol Res 1995; 85:37-44.
22. Favia G, Della Torre A, Bagayoko M et al. Molecular identification of sympatric chromosomal forms of Anopheles gambiae and further evidence of their reproductive isolation. Insect Mol Biol 1997; 6:377-83.
23. Della Torre A, Fanello C, Akogbeto M et al. Molecular evidence of incipient speciation within Anopheles gambiae s.s. in West Africa. Insect Mol Biol 2001; 10:9-18.
24. Favia G, Lanfrancotti A, Spanos L et al. Molecular characterization of ribosomal DNA polymorphisms discriminating among chromosomal forms of Anopheles gambiae s.s. Insect Mol Biol 2001; 10:19-23.
25. Della Torre A, Costantini C, Besansky NJ et al. Speciation within Anopheles gambiae- the glass is half full. Science 2002; 298:115-7.
26. Slotman MA, Mendez MM, Della Torre A et al. Genetic differentiation between the Bamako and Savanna chromosomal forms of Anopheles gambiae as indicated by amplified fragment length polymorphism analysis. Am J Trop Med Hyg 2006; 74:641-8.
27. Tripet F, Wright J, Cornel AJ et al. Longitudinal survey of knockdown resistance to pyrethroid (kdr) in Mali, West Africa, and the evidence of its emergence in the Bamako form of Anopheles gambiae s.s. Am J Trop Med Hyg 2007; 76:81-7.
28. Yuval B. Mating systems of blood-feeding flies. Ann Rev Entomol 2006; 51:413-40.
29. Downes JA. The swarming and mating flight of Diptera. Ann Rev Entomol 1969; 14:271-98.
30. Takken W, Knols BGJ. Odor-mediated behavior of Afrotropical malaria mosquitoes. Ann Rev Entomol 1999; 44:131-57.
31. Sullivan RT. Insect swarming and mating. Florida Entomologist 1981; 64:44-65.
32. Cooter RJ. Swarm flight behaviour in flies and locusts. In: Goldsworthy GJ, Wheeler CH, eds. Insect Flight. Boca Raton: CRC Press, 1989:165-203.
33. Shuster SM, Wade MJ. Mating systems and mating strategies. Monographs in Behavior and Ecology. Princeton: Princeton University Press, 2003:533.
34. Marchand RP. Field observations on swarming and mating in Anopheles gambiae mosquitoes in Tanzania. Neth J Zool 1984; 34:367-87.
35. Charlwood JD, Jones MDR. Mating behaviour in the mosquito Anopheles gambiae II swarming behaviour. Physiol Entomol 1980; 5:315-20.
36. Charlwood JD, Pinto J, Sousa CA et al. The swarming and mating behaviour of Anopheles gambiae (Diptera: Culicidae) from Sao Tome Island. J Vector Ecol 2002a; 27:178-83.
37. Yuval B, Bouskila A. Temporal dynamics of mating and predation in mosquito swarms. Oecologia 1993; 95:65-9.
38. Charlwood JD, Thompson R, Madsen H. Observations on the swarming and mating behaviour of Anopheles funestus from southern Mozambique. Malar J 2003; 2:2.
39. Yuval B, Wekesa JW, Washino RK. Effects of body size on swarming behavior and mating success of male Anopheles freeborni (Diptera: Culicidae). J Insect Behavior 1993; 6:333-42.
40. Verhoek B, Takken W. Age effects on insemination rate of Anopheles gambiae s.l. in the laboratory. Entomol Exp Appl 1994; 72:167-72.
41. Mahmood F, Reisen WK. Anopheles stephensi (Diptera: Culicidae): Changes in male mating competence and reproductive system morphology associated with aging and mating. J Med Entomol 1982; 19(5):573-88.
42. Mahmood F, Reisen WK. Anopheles culicifacies: Effects of age on the male reproductive system and mating ability of virgin adult mosquitoes. Med Vet Entomol 1994; 8:31-7.
43. Huho BJ, Ng'habi KR, Killeen GF et al. A reliable morphological method to assess the age of male Anopheles gambiae. Malar J 2006; 5:62.
44. Briegel H. Physiological bases of mosquito ecology. J Vector Ecol 2003; 28:1-11.
45. Fernandes L, Briegel H. Reproductive physiology of Anopheles gambiae and Anopheles atroparvus. J Vector Ecol 2005; 30:11-26.
46. Yuval B, Holliday-Hanson ML, Washino RK. Energy budget of swarming male mosquitoes. Ecol Entomol 1994; 19:74-8.
47. Foster WA. Mosquito sugar feeding and reproductive energetics. Ann Rev Entomol 1995; 40:443-74.
48. Impoinvil DE, Kongere JO, Foster WA et al. Feeding and survival of the malaria vector Anopheles gambiae on plants growing in Kenya. Med Vet Entomol 2004; 18:108-15.
49. Gary RE, Foster WA. Anopheles gambiae feeding and survival on honeydew and extra-floral nectar of peridomestic plants. Med Vet Entomol 2004; 18:102-7.
50. Gary RE, Foster WA. Diel timing and frequency of sugar feeding in the mosquito Anopheles gambiae, depending on sex, gonotrophic state and resource availability. Med Vet Entomol 2006; 20:308-16.

51. Yuval B, Fritz GN. Multiple mating in female mosquitoes—evidence from a field population of Anopheles freeborni (Diptera: Culicidae). Bull Entomol Res 1994; 84:137-49.
52. Huho BJ, Ng'habi KR, Killeen GF et al. Nature beats nurture: A case study of the physiological fitness of free-living and laboratory-reared male Anopheles gambiae s.l. J Exp Biol 2007; 210:2939-2947.
53. Giglioli MEC, Mason GF. The mating plug in anopheline mosquitoes. Proc R Ent Soc Lond Dordrecht:1966; A41(7-9):123-9.
54. Novikov YM. Monogamy of Anopheles messeae under natural conditions. Zool Zh 1981; 60:214-20.
55. Baimai V, Green CA. Monandry (monogamy) in natural populations of anopheline mosquitoes. J Am Mosq Contr Assoc 1987; 3(3):481-4.
56. Scarpassa VM, Tadei WP, Kerr WE. Biology of Amazonian anopheline mosquitoes. XVI. Evidence of multiple insemination (polyandry) in Anopheles nuneztovari Gabaldon, 1940 (Diptera: Culicidae). Rev Brasil Genet 1992; 15:51-64.
57. Tripet F, Toure Y, Dolo G et al. Frequency of multiple inseminations in field-collected Anopheles gambiae females revealed by DNA analysis of transferred sperm. Am J Trop Med Hyg 2003; 68:1-5.
58. Hartberg WK. Observations on the mating behaviour of Aedes aegypti in nature. Bull World Health Organ 1971; 45:847-50.
59. Gubler DJ, Bhattacharya NC. Swarming and mating of Aedes albopictus subgenus Stegomyia in nature. Mosq News 1972; 32:219-23.
60. Norris DE, Shurtleff AC, Toure YT et al. Microsatellite DNA polymorphism and heterozygosity among field and laboratory populations of Anopheles gambiae s.s. (Diptera: Culicidae). J Med Entomol 2001; 38:336-40.
61. Fagerberg AJ, Fulton RE, Black WC. Microsatellite loci are not abundant in all arthropod genomes: Analyses in the hard tick, Ixodes scapularis, and the yellow fever mosquito, Aedes aegypti. Insect Mol Biol 2001; 10:225-36.
62. Huber K, Loan LL, Hoang TH et al. Genetic differentiation of the dengue vector, Aedes aegypti (Ho Chi Minh City, Vietnam) using microsatellite markers. Mol Ecol 2002; 11:1629-35.
63. Craig G. Mosquitoes: Female monogamy induced by male accessory gland substance. Science 1967; 156:1499-500.
64. Spielman A, Leahy MG, Skaff V. Seminal loss in repeatedly mated female Aedes aegypti. Biol Bull 1967; 132:404-12.
65. Gwadz RW, Craig GG. Female polygamy due to inadequate semen transfer in Aedes aegypti. Mosq News 1970; 30:355-60.
66. Christophers SR. Aedes aegypti (L.), the yellow fever mosquito: Its life history, bionomics and structure. Cambridge: Cambridge University Press, 1960.
67. Young ADM, Downe AER. Renewal of sexual receptivity in mated female mosquitoes, Aedes aegypti. Physiol Entomol 1982; 7:467-71.
68. Williams RW, Berger A. The relation of female polygamy to gonotrophic activity in the ROCK stain of Aedes aegypti. Mosq News 1980; 40:597-604.
69. Wolfner MF. Tokens of love: Functions and regulation of Drosophila male accessory gland products. Insect Biochem Mol Biol 1997; 27:179-92.
70. Wolfner MF. The gifts that keep on giving: Physiological functions and evolutionary dynamics of male seminal proteins in Drosophila. Heredity 2002; 88:85-93.
71. Chapman T, Bangham J, Vinti G et al. The sex peptide of Drosophila melanogaster: Female post-mating responses analyzed by using RNA interference. Proc Natl Acad Sci USA 2003; 100:9923-8.
72. Kubli E. Sex-peptides: Seminal peptides of the Drosophila male. Cell Mol Life Sci 2003; 60:1689-704.
73. Foster WA, Lea AO. Renewable fecundity of male Aedes aegypti following replenishment of seminal vesicles and accessory glands. J Insect Physiol 1975; 21:1085-90.
74. Youngson JHAM, Welch HM, Wood RJ. Meitotic drive and the D(M^D) locus and fertility in the mosquito, Aedes aegypti (L). Genetica 1981; 54:335-40.
75. Ponlawat A, Harrington L. Age and body size influence male sperm capacity of the Dengue vector Aedes aegypti (Diptera: Culicidae). J Med Entomol 2007; 44(3):422-5.
76. Benjamin SN, Bradshaw WE. Body size and flight activity effects on male reproductive success in the pitcher plant mosquito (Diptera: Culicidae). Ann Entomol Soc Am 1994; 87(3):331-6.
77. Dickinson JM, Klowden MJ. Reduced transfer of male accessory gland proteins and monandary in female Aedes aegypti mosquitoes. J Vector Ecol 1997; 22:95-8.
78. Simmons LW. Male size, mating potential, and lifetime reproductive success in the field cricket Gryllus bimaculatus. Animal Behavior 1988; 36:372-9.
79. Goldsmith SK, Alcock J. The mating chances of small males of the cerambycid beetle Trachyderes mandibularis differ in different environments (Coleoptera, Cerambycidae). J Insect Behavior 1993; 6:351-60.

80. Vencl FV, Carlson AD. Proximate mechanisms of sexual selection in the firefly Photinus pyralis (Coleoptera: Lampyridae). J Insect Behavior 1998; 11:191-207.
81. Simmons LW, Stockley P, Jackson RL et al. Sperm competition or sperm selection: No evidence for female influence over paternity in yellow dung flies Scatophaga stecoraria. Behavioral Ecol Sociobio 1996; 38:199-206.
82. Simmons LW, Parker GA. Individual variation in sperm competition success in the field cricket Gryllus bimaculatus. Animal Behavior 1992; 34:1463-70.
83. Parker GA, Simmons L. Evolution of phenotypica optima and copula duration in dungflies. Nature 1994; 370:53-6.
84. Bangham J, Chapman T, Partridge L. Effects of body size, accessory gland, and testis size on pre- and postcopulatory success in Drosophila melanogaster. Animal Behavior 2002; 64:915-21.
85. Bradshaw WE, Holzapfel CM. Reproductive consequences of density-dependent size variation in the pitcher plant mosquito, Wyeomyia smithii (Diptera, Culicidae). Ann Entomol Soc Am 1992; 85:274-81.
86. Bradshaw WE, Holzapfel CM. Genetic constraints to life history evolution in the pitcher-plant mosquito Wyeomyia smithii. Evolution 1996; 50:1176-81.
87. Okanda FM, Dao A, Njiru BN et al. Behavioural determinants of gene flow in malaria vector populations: Anopheles gambiae males select large females as mates. Malar J 2002; 1:10.
88. Klowden MJ, Chambers GM. Male accessory gland substances activate egg development in nutritionally stressed Ae. aegypti mosquitoes. J Insect Physiol 1991; 37:721-6.
89. Kalb J, di Benedetto AJ, Wolfner MF. Probing the function of Drosophila melanogaster accessory glands by directed cell ablation. Proc Natl Acad Sci 1993; 92:10114-8.
90. Xue L, Noll M. Drosophila female sexual behavior induced by sterile males showing copulation complementation. Proc Nat Acad Sci 2000; 97:3272-5.
91. Taylor B, Jones MDR. The circadian rhythm of flight activity in the mosquito Aedes aegypti (L): The phase-setting effects of light on and light off. J Exp Biol 1969; 51:59-70.
92. Jones MDR, Gubbins SJ. Changes in the circadian flight activity of the mosquito Anopheles gambiae in relation to insemination, feeding, and oviposition. Physiol Entomol 1978; 7:281-89.
93. Jones MDR, Gubbins SJ. Modification of female circadian flight-activity by a male accessory gland pheromone in the mosquito Culex pipiens quinquefasciatus. Physiol Entomol 1979; 4:345-51.
94. Jones MDR. The programming of circadian flight activity in relation to mating and gonotrophic cycle in the mosquito, Aedes aegypti. Physiol Entomol 1981; 6:307-13.
95. Chiba Y, Shinkawa Y, Yamamoto Y. A comparative study on insemination dependency of circadian activity pattern in mosquitoes. Physiol Entomol 1992; 17:213-8.
96. Chiba Y, Yamamoto Y, Shimizu C. Insemination-dependent modification of circadian activity of the mosquito Culex pipiens pallens. Zool Sci 1990; 7:895-906.
97. Lavoipierre MMJ. Biting behavior of mated and unmated females of an African strain of Aedes aegypti. Nature 1958; 181:1781-2.
98. Judson CL. Feeding and oviposition behavior in the mosquito Aedes aegypti (L.) I. Preliminary studies of physiological control mechanisms. Biol Bull 1967; 133:369-77.
99. Klowden MJ, Lea AO. Humoral inhibition of host seeking in Aedes aegypti during oocyte maturation. J Insect Physiol 1979; 24:231-5.
100. Gillett JD. Variation in the hatching-responses of Aedes eggs (Diptera: Culicidae). Bull Entomol Res 1955; 46:241-54.
101. Leahy MG, Craig GG. Accessory gland substance as a stimulant for oviposition in Aedes aegypti and Ae. albopictus. Mosq News 1965; 25:448-52.
102. Hiss EA, Fuchs M. The effect of matrone on oviposition in the mosquito, Aedes aegypti. J Insect Physiol 1972; 18:2217-27.
103. Ramalingam S, Craig GG. Functions of the male accessory gland secretions of Aedes mosquitoes (Diptera: Culicidae): Transplantation studies. Can Entomol 1976; 108:995-60.
104. Freyvogel T, Hunter R, Smith E. Nonspecific esterases in mosquitoes. J Histochem 1968; 16:765-90.
105. Klowden MJ, Chambers GM. Ovarian development and adult mortality in Aedes aegypti treated with sucrose, juvenile hormone, and methoprene. J Insect Physiol 1989; 35:513-7.
106. Klowden MJ, Chambers GM. Reproductive and metabolic differences between Ae. aegypti and Ae. albopictus (Diptera: Culicidae). J Med Entomol 1992; 29:467-71.
107. Klowden MJ. Mating and nutritional state affect the reproduction of Aedes albopictus mosquitoes. J Am Mosq Contr Assoc 1993; 9:169-73.
108. Edman JD. Rate of digestion of three human blood fractions in Aedes aegypti (Diptera: Culicidae). Ann Entomol Soc Am 1970; 63:1778-9.
109. Downe AER. Internal regulation of rate of digestion of blood meals in the mosquito, Aedes aegypti. J Insect Physiol 1975; 21:1835-9.

110. Klowden MJ. The check is in the male: Male mosquitoes affect female physiology and behavior. J Am Mosq Contr Assoc 1999; 15(2):213-20.
111. Fuchs MS, Craig GB, Despommier DD. The protein nature of the substance inducing female monogamy in Aedes aegypti. J Insect Physiol 1969; 15:701-9.
112. Fuchs MS, Craig GB, Hiss EA. The biochemical basis of female monogamy in mosquitoes. I. Extraction of the active principles from Aedes aegypti. Life Sci 1968; 7:835-9.
113. Fuchs MS, Hiss EA. The partial purification and separation of the protein components of matrone from Aedes aegypti. J Insect Physiol 1970; 16:931-9.
114. Fowler K, Partridge L. A cost of mating in female fruitflies. Nature 1989; 338:760-1.
115. Partridge L, Ewing A, Chandler A. Male size and mating success in Drosophila melanogaster: The roles of male and female behaviour. Animal Behavior 1987; 35:555-62.
116. Partridge L, Fowler K. Non-mating costs of exposure to males in female Drosophila melanogaster. J Insect Physiol 1990; 36:419-25.
117. Chapman T, Liddle L, Kalb JM et al. Male seminal fluid components cause the cost of mating in Drosophila melanogaster females. Nature 1995; 373:241-4.
118. Marrelli M, Li C, Rasgon JL et al. Transgenic malaria-resistant mosquitoes have a fitness advantage when feeding on Plasmodium-infected blood. Proc Natl Acad Sci 2007; 104:5580-3.
119. James AA. Gene drive systems in mosquitoes: Rules of the road. Trends Parasitol 2005; 21:64-7.
120. Beatty J. Fitness: Theoretical contexts. In: Keller EF, Lloyd EA, eds. Keywords in Evolutionary Biology. Cambridge: Harvard University Press, 1992:115-9.
121. Paul D. Fitness: Historical perspective. In: Keller EF, Lloyd EA, eds. Keywords in Evolutionary Biology. Cambridge: Harvard University Press, 1992:112-4.
122. Munstermann LE. Unexpected genetic consequences of colonization and inbreeding: Allozyme tracking in Culicidae (Diptera). Ann Entomol Soc Am 1994; 87:157-64.
123. Mukhopadhyay J, Rangel EF, Ghosh K et al. Patterns of genetic variability in colonized strains of Lutzomyia longipalpis (Diptera: Psychodidae) and its consequences. Amer J Trop Med Hyg 1997; 57:216-21.
124. Tabachnick WJ. Reflections on the Anopheles gambiae genome sequence, transgenic mosquitoes and the prospect for controlling malaria and other vector-borne diseases. J Med Entomol 2003; 40:597-606.
125. Hartl DL, Clark AG. Principles of Population Genetics. Sunderland: Sinauer Associates, Inc, 1997.
126. Futuyma DJ. Evolutionary Biology. Sunderland: Sinauer Associates, Inc, 1998.
127. Grover KK, Curtis CF, Sharma VP et al. Competitiveness of chemosterilised males and cytoplasmically incompatible (IS31B) males of Culex pipiens fatigans in the field. Bull Ent Res 1976a; 66:469-80.
128. Grover KK, Suguna SG, Uppal DK et al. Field experiments on the competitiveness of males carrying genetic control systems for Aedes aegypti. Entomol Exp Appl 1976b; 20:8-18.
129. Curtis CF. Testing systems for the genetic control of mosquitoes. In: White D, ed. XV International Congress of Entomology. College Park: Entomological Society of America, 1977:106-16.
130. Rajagopalan PK, Curtis CF, Brooks GD et al. The density dependence of larval mortality of Culex pipiens fatigans in an urban situation and prediction of its effects on genetic control operations. Indian J Med Res 1977; 65:77-85.
131. Dame DA, Lowe RE, Williamson DL. Assessment of released sterile Anopheles albimanus and Glossina morsitans. In: Pal R, Kitzmiller B, Kanda T, eds. Cytogenetics and Genetics of Vectors Proceedings XVI International Congress of Entomology. Kyoto: Elsevier, 1981:231-41.
132. Lounibos LP. Genetic-control trials and the ecology of Aedes aegypti at the Kenya coast. In: Takken W, Scott TW, eds. Ecological Aspects for Application of Genetically Modified Mosquitoes. Springer Dordrecht: Frontis, 2003:33-43.
133. Catteruccia F, Godfray HCJ, Crisanti A. Impact of genetic manipulation on the fitness of Anopheles stephensi mosquitoes. Science 2003; 299:1225-7.
134. Moreira LA, Wang J, Collins FH et al. Fitness of anopheline mosquitoes expressing transgenes that inhibit plasmodium development. Genetics 2004; 166:1337-41.
135. Irvin N, Hoddle MS, O'Brochta DA et al. Assessing fitness costs for transgenic Aedes aegypti expressing the GFP marker and transposase genes. Proc Natl Acad Sci 2004; 101:891-6.
136. Carey JR. Applied Demography for Biologists with Special Emphasis on Insects. New York: Oxford University Press, 1993.
137. Scott TW, Naksathit A, Day JF et al. A fitness advantage for Aedes aegypti and the virus it transmits when females feed only on human blood. Am J Trop Med Hyg 1997; 57(2):235-9.
138. Harrington LC, Edman JD, Scott TW. Why do female Aedes aegypti (Diptera: Culicidae) feed preferentially and frequently on human blood? J Med Entomol 2001; 38:411-22.

139. Prout T. The relationship between fitness components and population prediction in Drosophila. I: The estimation of fitness components. Genetics 1971a; 68:127-49.
140. Prout T. The relationship between fitness components and population prediction in Drosophila. II. Population prediction. Genetics 1971b; 68:127-49.
141. Manly BFJ. The Statistics of Natural Selection. London: Chapman and Hall, 1985.
142. Endler JA. Natural Selection in the Wild. Princeton: Princeton University Press, 1986.
143. Lehmann T, Hawley WA, Kamau L et al. Genetic differentiation of Anopheles gambiae populations from East and West Africa: Comparison of microsatellite and allozyme loci. Heredity 1996; 77:192-208.
144. Lehmann TML, Gimnig JE, Hightower A et al. Spatial and temporal variation in kinship among Anopheles gambiae (Diptera: Culicidae) mosquitoes. J Med Entomol 2003; 40:421-9.
145. Coluzzi M, Sabatini A, Della Torre A et al. A polytene chromosome analysis of the Anopheles gambiae species complex. Science 2002; 298:1415-8.
146. Della Torre A, Tu Z, Petrarca V. On the distribution and genetic differentiation of Anopheles gambiae s.s. molecular forms. Insect Biochem Mol Biol 2005; 35:755-69.
147. Tripet F, Thiemann TC, Lanzaro GC. Effect of seminal fluids in mating between M and S forms of Anopheles gambiae. J Med Entomol 2005b; 42(596-603).
148. Ayala D, Goff G, Robert V et al. Population structure of the malaria vector Anopheles funestus (Diptera: Culicidae) in Madagascar and Comoros. Acta tropica 2006; 97:292-300.
149. Michel AP, Ingrasci MJ, Schemerhorn BJ et al. Rangewide population genetic structure of the African malaria vector Anopheles funestus. Mol Ecol 2005; 14:4235-48.
150. Edillo FE, Toure YT, Lanzaro GC et al. Spatial and habitat distribution of Anopheles gambiae and Anopheles arabiensis (Diptera: Culicidae) in Banambani village, Mali. J Med Entomol 2002; 39:70-7.
151. Tripet F, Dolo G, Lanzaro GC. Multilevel analyses of genetic differentiation in Anopheles gambiae s.s. reveal patterns of gene flow important for malaria-fighting mosquito projects. Genetics 2005a; 169:313-24.
152. Coluzzi M. Malaria vector analysis and control. Parasitol Today 1992; 8:113-8.
153. Yawson AE, Weetman D, Wilson MD et al. Ecological zones rather than molecular forms predict genetic differentiation in the malaria vector Anopheles gambaie s.s. in Ghana. Genetics 2007; 175:751-61.
154. Besansky NJ, Hill CA, Costantini C. No accounting for taste: Host preference in malaria vectors. Trends Parasitol 2004; 20:249-51.
155. Gillies MT, Coetzee M. A supplement to the Anophelinae of Africa south of the Sahara (Afrotropical region). Johannesburg: The South African Institute for Medical Research, 1987.
156. Diatta M, Spiegel A, Lochouarn L et al. Similar feeding preferences of Anopheles gambiae and A. arabiensis in Senegal. Trans Roy Soc Trop Med Hyg 1998; 92:270-2.
157. Tabachnick WJ. Evolutionary genetics and arthropod-borne disease: The yellow fever mosquito. Amer Entomol 1991; 37:14-24.
158. Gorrochotegui-Escalante N, Gomez-Machorro C, Lozano-Fuentes S et al. Breeding structure of Aedes aegypti populations in Mexico varies by region. Am J Trop Med Hyg 2002; 66:213-22.
159. Gorrochotegui-Escalante N, Munoz ML, Fernandez-Salas I et al. Genetic isolation by distance among Aedes aegypti populations along the northeastern coast of Mexico. Am J Trop Med Hyg 2000; 62:200-9.
160. Garcia-Franco F, Mde LM, Lozano-Fuentes S et al. Large genetic distances among Aedes aegypti populations along the South Pacific coast of Mexico. Am J Trop Med Hyg 2002; 66:594-8.
161. Bosio CF, Harrington LC, Jones J et al. Genetic structure of Aedes aegypti populations in Thailand using mtDNA. Am J Trop Med Hyg 2005; 72:434-42.
162. Nene V, Wortman JR, Lawson D et al. Genome sequence of Aedes aegypti, a major arbovirus vector. Science 2007, (in press).
163. Anonymous. Bugs in the system—issues in the science and regulation of genetically modified insects. Washington DC: PEW Initiative on Food and Biotechnology, 2004.
164. Lee JJ, Klowden MJ. A male accessory gland protein that modulates female mosquito (Diptera: Culicidae) host-seeking behavior. J Am Mosq Control Assoc 1999; 4-7.
165. Sirot LK, Poulson RL, McKenna MC, Girnary H, Wolfner MF, Harrington LC. Identity and transfer of male reproductive gland proteins of the dengue vector mosquito, Aedes aegypti: potential tools for control of female feeding and reproduction. Insect Biochem Mol Biol 2007 In press.
166. Dottorini T, Nicolaides L, Ranson H, Rogers DW, Crisanti A, Catteruccia F. A genome-wide analysis in Anopheles gambiae mosquitoes reveals 46 male accessory gland genes, possible modulators of female behavior. Proc Natl Acad Sci 2007; 16215-20.

Index

A

Acetic acid bacteria 51
Aedes aegypti 2, 6, 10, 16-19, 22-25, 27, 40, 67, 71, 72, 74, 75, 77, 79, 86, 87, 89, 96, 100, 106, 108, 127, 129, 134, 152, 155, 166-168
Age structure 104, 109, 118, 120, 126-129, 133, 135
Age-grading 126, 133, 134, 154
Alcaligenes xylosoxidans 143, 147
Alphavirus 19-24, 26, 28, 30
Anopheles 2, 19, 23, 25, 26, 45, 49, 50, 53, 56, 67, 71, 75-79, 85-89, 93, 95, 96, 106, 109, 132, 134, 152, 153, 159
Anopheles gambiae 19, 23, 25, 26, 71, 75-79, 86, 88, 89, 93, 106, 153, 159
Anopheles stephensi 95, 132
Arbovirus 26, 30
Asaia 49, 51-58

B

Beauveria 132
Beta-tubulin 87, 89
Biological control 1, 8, 9, 97, 109, 141

C

Ceratitis capitata 2, 10, 50, 63, 95, 96, 100, 107
Chemosterilant 85
Conditional lethality 1
Cre 6, 7
Culex pipiens 19, 23, 24, 26, 27, 104-106, 108, 109, 116, 124, 128, 134, 156
Cytoplasmic incompatibility (CI) 10, 43, 44, 56, 84, 85, 104-109, 115, 117, 118, 120, 122-124, 128-131

D

Dengue 20, 28, 71, 72, 79, 120, 126, 127, 129, 130, 133, 152, 158
Densovirus 10, 126, 127, 132-135
Developmental cycle 44, 85, 109
Drive mechanism 90, 108, 158
Drosophila melanogaster 2-7, 9-11, 27, 42, 43, 60, 61, 63, 66, 67, 72, 78, 84, 86, 87, 95, 96, 100, 106, 109, 115, 116, 120, 128, 142, 143, 156, 157
DsRed 95, 146-148

E

Enhanced green fluorescent protein (EGFP) 5, 8, 22, 41
Entomopathogenic fungi 126, 132, 134, 135
Extrinsic incubation period (EIP) 24, 119-121, 126, 127, 132, 135

F

FC31 6, 63
Feminization 104, 105, 109, 115, 128
FLP 6, 7, 66, 144
FRT 6, 7, 66

G

Gene drive 67, 68, 108, 115, 123
Gene therapy 60-63, 65, 68
Gene vector 2, 3, 61
Genetic control 62, 63, 65-67, 85, 87, 89, 101, 108, 141, 142, 151-155, 159, 163
Genetic drive 12, 44, 158
Genetically modified mosquitoes 151, 152
Genotype/phenotype association 77
Glassy-winged sharpshooter (GWSS) 143, 147, 148
Green fluorescent protein (GFP) 5, 21-26, 29, 30, 41, 49, 54-56, 142

H

Hermes 3, 4, 27, 63-65
Homing endonuclease 84, 94
Horizontal transmission 24, 57

I

Insect age 126
Insect pest 104, 108, 109, 134, 142
Insect vector species 109
Insecticide resistance 71, 72, 76, 78-80, 87, 93, 101, 134
Integrase 3, 6, 7, 10, 11, 61-64
Integrated pest management (IMP) 84, 90, 142
Irradiation 8, 44, 87, 96, 97, 99

L

Lethal gene 8, 9, 96, 97, 99, 141-143, 148
Linkage disequilibrium mapping 71, 77, 78

M

Malaria 11, 23, 24, 49-52, 55-58, 67, 71, 72, 79, 84, 90, 93, 109, 126, 127, 132, 135, 152, 153, 158-160
Male-killing 104, 115, 128
Mariner 3, 4, 11, 61, 63-65, 144
Marker 1-5, 7, 8, 20, 21, 29, 39, 54, 71-73, 76-79, 87-90, 95, 96, 100, 101, 131, 142, 144, 154, 155
Mathematical model 96, 114, 115, 119
Medfly 6, 9, 95, 96, 100, 107
Midgut 19, 22-27, 35-39, 42, 50, 51, 53-57, 79, 126
Minos 2-5, 7, 27, 63
Mos1 3, 27, 61, 63-65
Mosquito 1, 2, 6, 9-12, 19, 20, 22-30, 45, 49-57, 67, 68, 72, 77, 84, 86-90, 93-95, 99, 100, 104, 105, 107-109, 114-116, 118-121, 124, 126-135, 151-163
Mosquito fitness 50, 151
Mosquito mating behavior 151, 152
Mosquito population biology 152

O

Oocyte 37-39, 157

P

P element 2, 11, 12, 66, 115
P element paradigm 11
Parthenogenesis 104, 105, 109, 110, 115, 128
Pectinophora gossypiella 142
Pierce's disease 50, 141, 143, 148, 149
PiggyBac 2-8, 27, 60-63, 65, 67, 142
Pink bollworm 141-143, 146 149, 161
Plasmodium 27, 50, 51, 54-57, 67, 72, 78, 79, 126, 132
Popcorn (also see WMelPop) 109, 120, 128

Q

QTN mapping 71-73, 77-79
Quantitative trait nucleotide (QTN) 71

R

Recombinase-mediated cassette exchange (RMCE) 6, 7
Recombination 6, 7, 23, 41, 60, 73-76, 80, 86, 131, 132
Regulatory issue 94, 100, 146, 151, 152, 161, 162
Release-recapture 90, 133
Reproductive organ 53
RIDL 87, 93, 96-100
RNA interference 19, 27

S

Salivary gland 19, 22, 24, 26, 27, 35, 38, 42, 49, 51-55, 57, 126
Sexing 8, 85-87, 95-97, 101, 107, 108
Sindbis 10, 19, 20, 22, 30
Sleeping sickness 35, 88
Sporozoites 27, 55, 57
Sterile Insect Technique (SIT) 8, 12, 43, 44, 63, 65, 84-88, 90, 93-101, 104, 108, 109, 141, 142
Sterility 8, 9, 43, 44, 79, 85-87, 94, 105, 108, 142, 153
Symbiont 1, 2, 10, 35-45, 49-52, 54-57, 62, 104, 107-109, 114, 115, 127, 128, 143, 147
Symbiotic control 49-51, 54, 142, 143, 146-149

T

Temperature sensitive lethality 86
Tet-Off 9
Transformation 1, 2, 4-7, 10-12, 19, 27, 35, 38-41, 43, 45, 49-51, 54, 61-63, 67, 84, 88, 106, 109, 110, 122, 124, 131, 142, 143, 146, 151, 158, 159
Transformer 87

Transgene 1, 3, 5-9, 43, 45, 63, 66-68, 85-88, 115, 122, 123, 158, 159
Transgenesis 1, 2, 4, 5, 12, 27, 84, 85, 87, 88, 124, 144, 151, 158, 159
Transgenic mosquito 57
Transposable element 2-4, 11, 12, 27, 84, 88, 115, 144
Transposase 3, 5, 7, 8, 11, 60-68
Transposition 2, 3, 7, 12, 60-68
Transposon 1-5, 7-9, 11, 19, 41, 60-68, 108, 131, 144
Trypanosome 10, 27, 35, 38-40, 42, 88
Tsetse 35-39, 41-44, 87, 88, 94, 95, 129, 132

V

Vector control 11, 35, 36, 72, 86, 93-95, 101, 133, 134
Vector lifespan 120, 126, 132, 135
Vector-borne disease 90, 93, 101, 114, 117, 119, 124, 126, 127, 135, 151, 152, 155, 161-163
Vectorial capacity 36, 119-121, 126, 127, 132, 135, 160
Vertical transmission 13, 56, 133
Viral vector 10, 24, 62

W

WMelPop (also see Popcorn) 120, 128-131
Wolbachia 10, 36, 37, 39, 43, 44, 50, 56, 104-110, 114-124, 126-131, 134, 135

X

Xylella fastidiosa 50, 143, 144, 147